Essential Mathematics
for
Economists

Essential Mathematics
for
Economists
Second Edition

J. Black
Professor of Economic Theory,
University of Exeter

J. F. Bradley
Lecturer in Economics,
The Queen's University of Belfast

JOHN WILEY & SONS

Chichester · New York · Brisbane · Toronto

Library of Congress Cataloging in Publication Data:

Black, John, *b. 1931*
    Essential mathematics for economists.—
    2nd ed.
    1. Economics, Mathematical
    I. Title   II. Bradley, James F
    510′.2′433   HB135   79-40826

ISBN 0 471 27659 6 (cloth)
ISBN 0 471 27660 X (paper)

Photosetting by Thomson Press (India) Ltd., New Delhi.
Printed in Great Britain at the Pitman Press, Bath, Avon.

Preface

While writing this book the authors have been assisted by many people. We would like to thank for their help Mrs Alison Kershaw, who typed most of the first draft; many past and present colleagues in the Department of Economics of the University of Exeter, for their comments on successive drafts; Dr J. K. Ord of Bristol University for many helpful comments; and our pupils, on whom we have tried out the examples. We are not conscious of having made any direct use of other people's work; it is, however, only too likely that in preparing the text and examples we have unconsciously plagiarized the work of others. For this we offer our thanks and apologies.

We have expanded the second edition to include three new chapters on linear algebra and linear programming. We have also amended a few points in the first edition. In preparing this revised edition we have been greatly assisted by comments and suggestions from V. Daly, Dr J. K. Seade, and M. C. Timbrell; any remaining errors are our own.

Department of Economics,   J. BLACK
University of Exeter

Department of Economics,   J. F. BRADLEY
The Queen's University of Belfast

Contents

Addition—Matrix Subtraction—Multiplication of a Matrix by a
Scalar—Multiplication of a Matrix by a Matrix—The Logic
of Matrix Multiplication—The Existence of Matrix Products—
Multiplication of a Matrix by a Vector—Transposition of Matrices—Square Matrices—Symmetric Matrices—Diagonal Matrices—The Identity Matrix—Matrix Division—Economic Models
in Matrix Notation—A National Income Model—An Input–Output Model—Exercises—Solutions to Exercises

CHAPTER 1

Linear Functions

INTRODUCTION

Economics is concerned with the relations between various parts of the economy. Many of these relations are quantitative in nature; e.g. the relation between consumption and income. It is convenient to be able to state theories about the economy in mathematical terms, as an aid to rational thought. It is also essential that models of the economy be stated in mathematical terms if they are to be tested against reality, and if statistical methods are to be used to measure the various relations included in the models. At least some elementary mathematical methods are therefore to be found in an ever increasing proportion of even the most elementary economics texts.

This book is intended to equip students with the elementary mathematical skills they will need if they are to understand modern theories about how the economy works, and read intelligently modern quantitative studies based on these theories. The minimum of mathematical background is assumed. In the early chapters, especially, some very elementary points are spelled out in some detail. This is intended to help readers to whom they are new, and those for whom some reminder of rusty mathematical tools is useful.

To gain the maximum benefit from the book it is essential to work through the examples. To readers whose mathematical background is already satisfactory many parts of the early chapters may seem rather simple. What they should gain from reading through these parts is an ability to apply the mathematics they already know to the problems of economics. Some of the later chapters will probably present a real challenge even to the stronger mathematicians. They have been included because the theory of oscillations is essential for any serious study of economic dynamics.

The examples relate to economic problems which are very simple compared with those of the real economy. Many relations between the parts of the real economy are believed to be very complicated, so far as they are understood at all. To learn the methods necessary to analyse complex situations it is necessary to start with simpler ones. We have, however, tried to choose techniques which generalize to deal with more complex real-world problems.

It is assumed that the reader is interested in economics, and is concurrently following an economics course. The economic concepts used are therefore not discussed in detail. Mathematical concepts and notation are explained to the extent that is thought likely to help in following economics texts. Mathematical rigour and elegance for their own sake have not been pursued. Every effort has been made, however, to explain why the techniques taught are valid. This is not a 'cookbook'. It is worthwhile for economists to spend some effort on understanding the rationale of the mathematical techniques they need to use, because this makes them less likely to make silly mistakes, or to apply techniques blindly in situations where they are not valid. Stress is also placed on simple and unambiguous notation since this assists in understanding and the avoidance of trivial mistakes.

Every effort has been made to avoid mistakes of logic or typography, which are particularly baffling and irritating to people who are trying to master new techniques. If any slips remain this is the fault of the authors, who will be most grateful to anybody who will trouble to write in and point out any errors they may find.

Of all the chapters in this book, the first, while it covers the most elementary ground, has been the most difficult to write. This is because of the difficulty of guessing how much the reader already knows. As the book progresses, and can build on foundations laid or consolidated in earlier chapters, this problem diminishes. In an effort to make sure that everybody who wants to master economics is able to get started, Chapter 1 includes material which many readers will find familiar. Some may welcome a reminder of matters studied long ago and since neglected. Others who find Chapter 1 too elementary are urged to check that they can do the examples, and if they can then to skim on quickly through the earlier chapters and slow down when they get to something they find challenging.

THE DEMAND FUNCTION

Economics as a systematic study is only possible because different parts of the economy are related. These relations can be described in alternative ways, and the economist has frequently to translate between the various 'languages' in which his subject matter can be described. Different people will find different languages most familiar or congenial. Let us take a simple economic proposition and see how it can be stated in the various languages. We start with it in a verbal form.

Verbal Statements

'The amount of any given good which a consumer will buy will depend on its price, on his income, on the price of a rival good and on a variety of other factors, for example his age and marital status, which we sum up under the term "tastes".'

This type of statement has its merits, but precision is not one of them. It does not say how demand for a good depends on its price, or on the consumer's income. Which way, and how far, will demand move if price rises or income falls by any stated percentage?

Algebra

A second language is algebra. In this form, the proposition above appears as

$$Q = f(P, Y, \Pi, T)$$

This is a general equivalent of the verbal statement, where Q is the quantity demanded, P the price of the good, Y the consumer's income, Π the price of a rival good, and T a quantity describing some aspect of the consumer's 'tastes', e.g. his age. Q, P, Y, Π, and T are referred to as 'variables' because it is assumed that they can vary. The $f( )$ indicates that Q is dependent on the values of the variables listed inside the brackets; these are called the 'arguments' of the function, and are by convention separated by commas.

The general functional form does not allow us to answer the questions posed above; however, there are many particular forms the function can take which do allow us to answer them. Several forms of function are in common use in economics; for now we will concentrate on the simplest, in which

$$Q = a + bP + cY + d\Pi + eT$$

where a, b, c, d, and e are all fixed numbers, e.g.

$$Q = 100 - 3P + 0{\cdot}01\,Y + 0{\cdot}4\Pi + 0.05T$$

The numbers, a, b, etc. which describe the form of the function, are called 'parameters'.

With a function of any given specific form such as this we can discover what will happen to Q if P rises or Y falls. The point of having a general functional form available is that we may often want to state the hypothesis that variables are related without specifying the precise form of the relation, so that we can then consider alternatives and see which appears to fit the facts with least distortion.

Diagrams

The third language is that of diagrams. We cannot represent the function above in a diagram, but we can represent some interesting special cases. The simplest form of diagram represents the relation between one variable and another, assuming that all other relevant variables remain fixed. For example, in the case above, if we assume that Y, Π, and T do not change, then Q and P will be related. The general form of this relation can be written

$$Q = f(P)$$

Even though there is only one argument of the function $f(\ )$, the brackets do not mean multiplication of the contents of the bracket by the number f. In the particular case above,

$$Q = A + bP$$

where $A = a + cY + d\Pi + eT = 100 + 0\cdot01Y + 0\cdot4\Pi + 0\cdot05T$ and $b = -3$. If, say, $Y = 2000$, $\Pi = 2\cdot5$, and $T = 40$,

$$Q = 100 - 3P + 0\cdot01\,(2000) + 0\cdot4(2\cdot5) + 0\cdot05(40)$$
$$= 100 - 3P + 20 + 1 + 2$$
$$= 123 - 3P$$

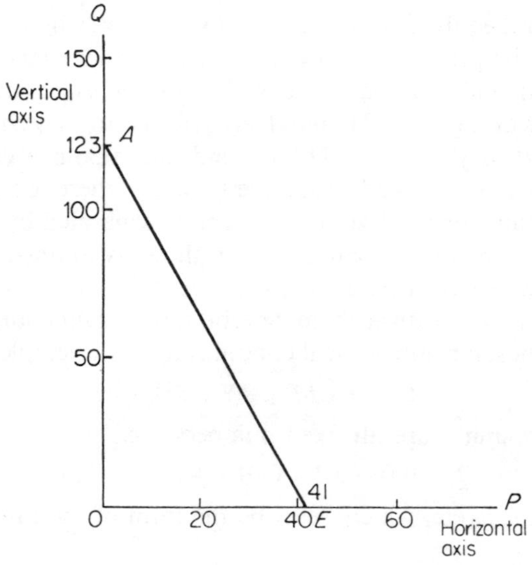

Figure 1   A linear demand curve

This can be represented in a diagram, as in Figure 1. In the diagram Q is measured vertically upwards from point 0, called the origin, and P is measured horizontally to the right from 0.

As price and quantity are measured in different units, there is no reason why the scales along the two axes should be the same. There is no one right choice of which variable to represent on which axis; the vertical axis usually represents the variable determined by the function, called the dependent variable, and the horizontal axis represents the variable forming the argument of the function, called the independent variable.

Plotting Graphs

To plot a graph we need to calculate values of Q which will correspond to various values of P, assuming that

$$Q = 123 - 3P$$

It is convenient to lay out the calculation in tabular form, i.e. to keep the figures in rows, across the page, and columns, reading down the page, so that we can see clearly what we are doing. For example,

if $P = 0$,　$Q = 123 - 3(0) = 123$
if $P = 1$,　$Q = 123 - 3(1) = 120$
if $P = 10$,　$Q = 123 - 3(10) = 93$
if $P = 20$,　$Q = 123 - 3(20) = 63$
etc.

Tables

This could be laid out in a formal table:

TABLE 1.1

P	$3P$	$Q = 123 - 3P$
0	0	123
1	3	120
10	30	93
20	60	63

To produce a graph we plot the points calculated, i.e. $P = 0$, $Q = 123$, etc. on a diagram, as in Figure 2. For convenience this point is written as (0, 123), i.e. the coordinates are listed, with that on the horizontal axis first.

We can calculate as many points as we like; if we join them up we will

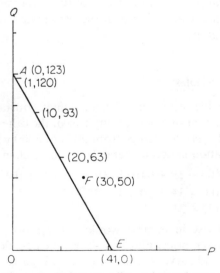

Figure 2　Points on a linear demand curve

find they all lie on a straight line running from point A, where $P = 0$ and $Q = 123$, to point E, where $P = 41$ and $Q = 0$. Because the points on a linear function will all lie on a straight line, it is only strictly necessary to calculate two of them in order to plot it; as a precaution against errors it is usually best to calculate a few more.

We can also check whether any given combination of P and Q satisfies the demand relation. To find whether the point $P = 30$, $Q = 50$ lies on the demand curve we can either substitute the value $P = 30$ into the expression $Q = 123 - 3P$ to get

$$Q = 123 - 3(30) = 123 - 90 = 33$$

and observe that $33 \neq 50$, where the symbol $\neq$ means 'is not equal to'; or we can plot the point $F = (30, 50)$ on Figure 2 and see that it lies above the demand curve AE.

LINEAR FUNCTIONS

Because of the fact that their graphs always turn out to be straight lines, functions like

$$Q = 123 - 3P$$

or more generally of the form

$$Q = A + bP$$

are called linear. Functions of the form

$$Q = a + bP + cY + d\Pi + eT$$

are also called linear, since if all but two of the variables Q, P, Y, Π, and T are held constant, the relation between the remaining two always produces a straight line as its graph.

Changes in Other Variables

If any of the variables which have been assumed constant were to change, this would alter the relation between any two of the others. For example, in the case above, if Y were to change from 2000 to 3000 while Π and T were unchanged, the relation between Q and P would be changed to

$$Q = 100 - 3P + 0{\cdot}01(3000) + 0{\cdot}4(2.5) + 0{\cdot}05(40)$$
$$= 100 - 3P + 30 + 1 + 2$$
$$= 133 - 3P$$

This shows that a rise in income would cause more of the good to be demanded at any given price. Figure 3 shows the new demand curve, $A'E'$, compared with the old curve AE, the same as in Figures 1 and 2.

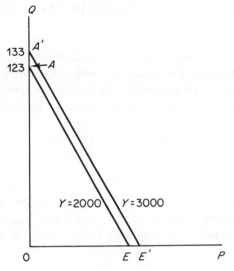

Figure 3   A shift in the demand curve

Non-linear Functions

To appreciate what is implied by a linear function, consider other possible functions. The demand curve showing the relation between Q and P, for given Y, Π, and T, could have been curved, as shown in Figure 4. Such a function is said to be non-linear or curvilinear; many forms of non-linear function are in common use in economics. However, it is convenient to start with simple functions, and linear functions are in fact very common.

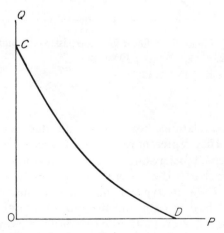

Figure 4   A non-linear demand curve

HOW LINEAR FUNCTIONS ARISE IN ECONOMICS

Tax Laws

Linear functions arise in economics in different ways. Some relations are linear by human design. For example, suppose a country has tax laws under which income tax takes a fixed proportion of any individual's pre-tax income Y. If D is disposable income, i.e. income left after paying income tax, then if t is the tax rate

$$D = Y - tY = (1 - t)Y$$

e.g. if $t = 0.3$,

$$D = (1 - 0.3)Y = 0.7Y$$

This is shown in Figure 5. Here we have an example of a linear function with a constant term equal to zero. D is a linear function of Y, given the tax rate, as the result of a conscious human decision, i.e. the income tax laws.

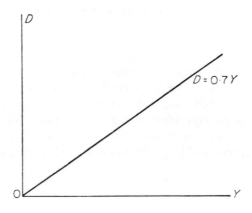

Figure 5   A linear function passing through
the origin

Statistical Relations

Many economic relations, however, are not due to any single conscious human decision. They represent relations inferred by statisticians from data on people's observed behaviour. One such relation is the consumption function. Suppose that D is disposable income and C is actual expenditure on consumption. These are typically not equal because people can save part of their incomes. The disposable income and consumption of various individuals are recorded, for example in expenditure surveys, and statisticians then fit a mathematical relation. They have to choose its form, and the linear function is often used because it is simple to fit and agrees with the data no worse than other, more complicated, functions. A linear consumption function takes the form

$$C = a + bD$$

where a and b are constants, e.g.

$$C = 100 + 0.8D$$

Derived Relations

Many economic relations are themselves derived from other relations. For example, suppose we know that $C = a + bD$ and that $D = (1 - t)Y$; we may wish to relate C to Y directly, without needing to refer to D, and can do so by arguing that

$$C = a + bD = a + b(1 - t)Y$$

e.g. if $C = 100 + 0.8D$ and $t = 0.3$, so that $D = (1 - 0.3)Y = 0.7Y$,

$$C = 100 + 0.8D$$
$$= 100 + 0.8(0.7Y)$$
$$= 100 + 0.56Y$$

Thus we have combined two linear functions to get a third linear function, relating C to Y.

Intercepts

We can see how various linear functions appear when we graph them. The graph of a linear function will cut the axes at points called intercepts. In Figure 1, point A where the demand curve cuts the Q-axis is called the Q-intercept, and point E where the demand curve cuts the P-axis is called the P-intercept. At the Q-intercept $P = 0$ and the intercept is given by the constant term A in the demand function $Q = A + bP$.

The constant term of a linear function may be zero, as shown in Figure 5 with portrays

$$D = f(Y)$$

in the case where

$$D = (1 - t)Y = (1 - 0.3)Y = 0.7Y$$

In this case the D and Y intercepts coincide at the origin.

THE SUPPLY FUNCTION

There is no requirement that the constant term in a linear function be positive. If we consider the supply function of a competitive industry, which shows the total amounts of a good the firms will choose to provide at various market prices, we may find that because there is some minimum price below which no production will cover its costs, the supply function takes the form

$$Q = \gamma + \delta P$$

where γ (Greek gamma) is a constant < 0. The symbol $<$ means that the quantity to the left of it is smaller than the quantity to the right: e.g.

$$Q = -100 + 4P$$

This is shown in Figure 6.

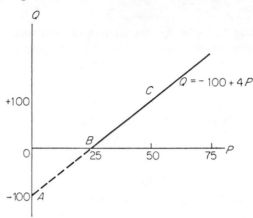

Figure 6   A linear supply curve

To discover what this minimum price is we need to find where the supply curve cuts the P-axis, i.e. to find its P-intercept. This means solving the equation

$$-100 + 4P = 0$$

The solution or root of the equation is the value of P, if there is one, for which this is true. In the present case we can find the solution, since

$$4P = 100$$

so

$$P = 25$$

This is shown by point B in Figure 6. Clearly, only the part of the line AC which lies above B will be economically relevant. The supply function should in fact read

$$Q = -100 + 4P \text{ provided } P \geqslant 25$$

(where the symbol $\geqslant$ means that the quantity to the left of it is larger than or equal to the quantity to the right), or

$$Q = 0 \text{ if } P < 25$$

This type of restriction is commonly assumed in economics texts, but not spelled out in detail, since it is simply assumed that everybody concerned realizes that prices and quantities cannot be negative, i.e. $P \geqslant 0$ and $Q \geqslant 0$.

EQUATIONS

In general, if

$$Q = \gamma + \delta P$$

$Q = 0$ when

$$\gamma + \delta P = 0$$

i.e. when

$$\delta P = -\gamma$$

so that

$$P = -\frac{\gamma}{\delta}$$

The solution to a linear equation can always be found by this method provided $\delta \neq 0$. The solution is clearly unique; no other value of P will satisfy the equation.

The equation is a mathematical form of a question; it allows us to ask the question, for what values of the variables will a given relation hold? We are not confined to the case where $Q = 0$. If we want to find what value of P will result in Q taking any given size $\bar{Q}$, where the bar over the Q is used to indicate that its value is being arbitrarily fixed, we need to solve the equation

$$\gamma + \delta P = \bar{Q}$$

i.e. to find P in terms of $\bar{Q}$. This is done in the linear case by separating out the terms involving P,

$$\delta P = \bar{Q} - \gamma$$

and dividing through by the coefficient of P, in this case δ, to get

$$P = \frac{\bar{Q} - \gamma}{\delta}$$

or

$$P = -\frac{\gamma}{\delta} + \frac{1}{\delta}\bar{Q}$$

Thus P is found to be a linear function of $\bar{Q}$. In our particular case

$$-100 + 4P = \bar{Q}$$

thus

$$4P = 100 + \bar{Q}$$

so

$$P = 25 + 0 \cdot 25\bar{Q}$$

If we let $\bar{Q} = 100$,

$$P = 25 + 0.25(100)$$
$$= 25 + 25 = 50$$

Inverse Functions

Turning round the function

$$Q = f(P)$$

to give a function in the form

$$P = g(Q)$$

where $g(\ )$ is used as an alternative notation for the function to point out that the functions $f(\ )$ and $g(\ )$ are not the same, is known as inverting or finding the inverse function of $f(\ )$. We have already seen that the inverse of a linear function is itself linear.

Implicit Functions

Often in economics we want to use alternative forms of stating the same relation for different purposes, or we may be uncertain of the direction of causation between different variables. For example, suppose there is a known relation between consumption and income; does this mean individuals are given their income and decide what they will spend, or that they assess their own wants and set out to earn enough income to satisfy them? In these cases it is found convenient to state functions in a form which is neutral as to which is the independent and which the dependent variable. This can be done by treating both variables symmetrically; thus if the consumption function is

$$C = \alpha + \beta Y$$

this can be stated in inverse form

$$Y = -\frac{\alpha}{\beta} + \frac{1}{\beta} C$$

or in 'implicit' form as

$$C - \beta Y = \alpha \quad \text{or} \quad Y - \frac{1}{\beta} C + \frac{\alpha}{\beta} = 0$$

with both variables on the same side of the $=$ sign.

BUDGET LINES

Another case where we may wish to use the implicit function form in order to avoid any imputation of causation either way is where a relation holds by definition. For example, if a consumer has a sum M to spend, on two goods X and Y which are available to him at fixed prices P_X and P_Y which he cannot affect, then the combinations of X and Y which he can get for M are described by

$$P_X.X + P_Y.Y = M$$

This is known as a budget line. The relation between X and Y is clearly linear since

$$P_X.X = M - P_Y.Y$$

and

$$X = \frac{M}{P_X} - \frac{P_Y}{P_X} \cdot Y$$

Subscripts

The use of subscripts to distinguish similar variables, in this case the prices of X and Y, and to relate the prices to the goods they refer to, is very common in economics. We could have said, let the price of good X be P and the price of good Y be S, and stated the budget line in the form

$$P.X + S.Y = M$$

In complex problems, however, the use of a separate letter for each variable would exhaust the English and Greek alphabets, and in any case there is an obvious attraction about the use of subscripts to relate variables to each other. P_X, where X is the subscript, is a form of notation extensively used in economics texts.

Multiplication

The process of multiplication can be denoted in various ways; multiplication signs, dots, brackets or merely placing the variables to be multiplied next to each other. Thus the same budget line could be written as

$$P_X \times X + P_Y \times Y = M$$
$$P_X.X + P_Y.Y = M$$
$$P_X(X) + P_Y(Y) = M$$

or

$$P_X X + P_Y Y = M$$

Which of these systems is used on any particular occasion is a matter of taste or convenience. Economics texts, and the present book, use various notations on different occasions.

GRAPHS OF NEGATIVE QUANTITIES

It should be noted that not all economic variables are required to be positive. For example, in considering the relation between disposable income and consumption, we can consider the savings function, where savings are defined as disposable income minus consumption. If

$$C = a + bD$$

then

$$S = D - C$$
$$= D - a - bD$$
$$= (1 - b)D - a$$

This savings function is shown in Figure 7.

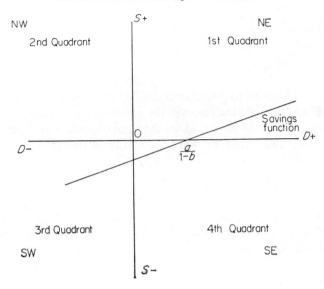

Figure 7   The four quadrants of a diagram

If disposable income falls to $D = a/(1 - b)$ then $S = 0$ and all disposable income is spent, while if D falls below $a/(1 - b)$ consumption will exceed disposable income, and savings will be negative. This can be achieved, at least for a limited time, by spending out of past savings or running into debt. For anybody whose income is earned, D cannot be negative; but for an individual who runs a business D itself may be negative in bad years. Other economic variables such as the profits of a company, the balance of trade of a country, or the excess of the demand for a good over the supply, may take either sign.

The Quadrants of a Diagram

It is possible to graph relations referring to these variables by making use of the negative parts of a diagram, as shown in Figure 7. To the left of the origin D is negative, and below the origin S is negative. The four parts of the diagram separated by the axes are referred to as 'quadrants'. By convention, the NE part where $D > 0$ and $S > 0$ is known as the first quadrant; the NW part where $D < 0$ and $S > 0$ is the second quadrant; the SW part where $D < 0$ and $S < 0$ is the third quadrant; and the SE part where $D > 0$ and $S < 0$ is the fourth quadrant. Which quadrants are used in any particular diagram depends on the economic restrictions on the particular functions represented.

EXERCISES

1. Given that quantity demanded per period of time is a function of price and that the relation takes the form

$$Q = 60 - \tfrac{1}{3}P$$

(i) Find quantity demanded when price is
   (a) 30,
   (b) 210,
   (c) zero.
Comment on (b).
(ii) Express the inverse function, i.e. give P as a function of Q and find price when quantity demanded is
   (a) zero,
   (b) 60,
   (c) 59.
Comment on (a).
(iii) Graph the relation.
(iv) The demand function now takes the form

$$Q = 40 - \tfrac{1}{3}P$$

Give possible economic explanations for this change, and graph the new relation on the same diagram as (iii).

2. Assume that the direct tax system takes 25 per cent of income Y, regardless of size, and that an individual's consumption is 500 plus 80 per cent of disposable income D. Disposable income D is income Y minus direct taxes.
   Express the relation $C = f(Y)$
(i) in algebra;
(ii) in a diagram.
(iii) Find C when Y is
   (a) zero,
   (b) 600,
   (c) 1200.
(iv) Express the inverse function, i.e. give Y as a function of C, and find Y when C is
   (a) 800,
   (b) 920.

3. The supply of tea is represented by the function

$$Q + 36 - \tfrac{2}{3}P = 0$$

where P is price and Q is quantity.
(i) Find Q when P is
   (a) 54,
   (b) zero,
   (c) 60.
Comment on (a) and (b).
(ii) Find P when Q is
   (a) zero,
   (b) 36.
(iii) Graph the supply function.

(iv) The supply function changes to take the form

$$Q = -40 + \tfrac{2}{3}P$$

Give possible economic explanations for this change and graph the new function on the same diagram as (iii).

4. (i) A firm's fixed costs are constant at 500 regardless of the number of units produced. Express this function
   (a) in algebra,
   (b) in a diagram.
   (ii) The firm's total costs TC are made up of fixed costs plus variable costs. If variable costs are 2 per unit produced, express the total cost function in algebra, and super-impose the graph of this function on graph (i).

   (iii) If variable costs equal 25 per cent of total revenue TR, express the TR function in algebra, and superimpose the graph of the TR function on graph (i).

   (iv) If profits S are equal to total revenue TR minus total costs TC, i.e. $S = TR - TC$, at what point on the graph are profits zero?

SOLUTIONS TO EXERCISES

It is particularly important when answering questions to make sure that the answer gives the information asked for, and that effort is not wasted on providing information which is not wanted. As an aid to checking that this has been done, the answers in the examples below are marked with an asterisk* in the right-hand margin. It is a good idea to make a habit of labelling the answers in worked examples.

1. (i) (a) When $P = 30$, $Q = 60 - 30/3 = 50$.   *
      (b) When $P = 210$, $Q = 60 - 210/3 = -10$.   *
      (c) When $P = 0$, $Q = 60 - 0/3 = 60$.   *
   Result (b) is not economically meaningful because Q cannot be negative.

   (ii) $Q = 60 - \tfrac{1}{3}P$
   therefore

   $$\tfrac{1}{3}P = 60 - Q \text{ and } P = 180 - 3Q$$

      (a) When $Q = 0$, $P = 180 - 3(0) = 180$.   *
      (b) When $Q = 60$, $P = 180 - 3(60) = 0$.   *
          See (i) (c) above.
      (c) When $Q = 59$, $P = 180 - 3(59) = 3$.   *
   From result (a) we see that there will be no demand for the good when P is equal to or greater than 180.

   (iii) See Figure 8.

   (iv) $Q = 60 - \tfrac{1}{3}P$ is a relation between two variables, i.e. price and quantity demanded. The linear relation has been drawn using the assumption that all other factors remain constant. However, factors other than price influence quantity demanded, e.g. income, tastes, the price of other goods, etc. Since the straight line demand function, $Q = 60 - \tfrac{1}{3}P$, assumes all of these factors constant then a change in any one of these factors will cause a shift in the curve, as distinct from a movement along it. A decrease in income could have caused the downward shift in this problem, as in Figure 8.

2. $D =$ disposable income so

$$D = Y - T$$

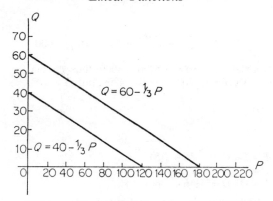

Figure 8 A downward shift in the demand function

where T = total direct taxes

$$C = 500 + 0.8D$$
$$= 500 + 0{\cdot}8(Y - T)$$
$$T = 0{\cdot}25\,Y$$

Thus

$$C = 500 + 0{\cdot}8(Y - 0{\cdot}25\,Y)$$
$$= 500 + 0{\cdot}8(0{\cdot}75\,Y) = 500 + 0{\cdot}6\,Y$$

(i) $C = 500 + 0{\cdot}6\,Y$ *

(ii) The function should take the form shown in Figure 9.   *

(iii) $C = 500 + 0{\cdot}6\,Y$

  (a) When $Y = 0$, $C = 500 + 0 = 500$.   *

  (b) When $Y = 600$, $C = 500 + 0{\cdot}6(600) = 860$.   *

  (c) When $Y = 1200$, $C = 500 + 0{\cdot}6(1200) = 1220$.   *

(iv) $C = 500 + 0{\cdot}6\,Y$
therefore

$$0{\cdot}6\,Y = C - 500$$

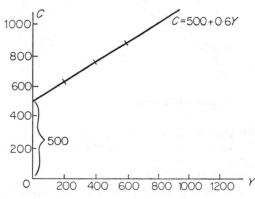

Figure 9 The consumption function

thus

$$Y = \frac{10}{6}C - \frac{5000}{6} \text{ or } 6Y = 10C - 5000$$

(a) When $C = 800$, $6Y = 8000 - 5000 = 3000$; thus $Y = 500$.   *
(b) When $C = 920$, $6Y = 9200 - 5000 = 4200$, thus $Y = 700$.   *

3. $Q + 36 - \frac{2}{3}P = 0$
    therefore

$$Q = -36 + \frac{2}{3}P$$

(i) (a) When $P = 54$, $Q = -36 + \frac{2}{3}(54) = 0$.   *
     (b) When $P = 0$, $Q = -36 + \frac{2}{3}(0) = -36$.   *
     (c) When $P = 60$, $Q = -36 + \frac{2}{3}(60) = 4$.   *

Result (b) is not economically meaningful as it stands, but (a) and (b) together mean that suppliers will withhold tea from the market whenever the price drops to, or below, 54.

(ii) $Q = -36 + \frac{2}{3}P$
therefore

$$\tfrac{2}{3}P = Q + 36$$

i.e.

$$P = 54 + 1{\cdot}5Q$$

(a) When $Q = 0$, $P = 54 + 1{\cdot}5(0) = 54$.   *
(b) When $Q = 36$, $P = 54 + 1{\cdot}5(36) = 108$.   *

(iii) and (iv) See Figure 10.

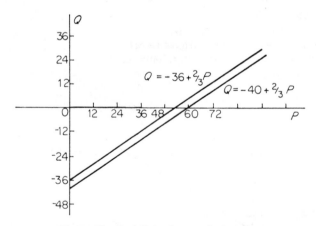

Figure 10   A shift in the supply function

If $Q = -36 + \frac{2}{3}P$ when $P = 0$, $Q = -36$ see (i)(b)
          when $Q = 0$, $P = 54$ see (ii)(a)
If $Q = -40 + \frac{2}{3}P$ when $P = 0$, $Q = -40$
          when $Q = 0$, $P = 60$

$Q = -36 + \frac{2}{3}P$ is a relation between two variables, assuming all other factors are held constant. Since factors other than price influence supply, e.g. the goals of the firm, the prices of other commodities, etc. a change in any one of these will cause the supply curve to shift. An increase in the prices of other commodities could have caused the shift in this problem; see Figure 10. If P is measured on the horizontal axis and Q on the vertical then $Q = -40 + \frac{2}{3}P$ lies to the right of $Q = -36 + \frac{2}{3}P$.

4. FC = fixed costs.
   VC = variable costs,
   TC = total costs,
   TR = total revenue and
   Q = quantity.

(i) (a) $FC = 500$.
   (b) See Figure 11.   *

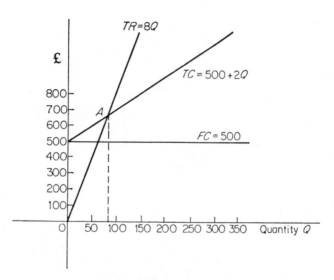

Figure 11   Total costs and revenue

(ii) $TC = FC + VC$.   *
Variable costs are 2 per unit, i.e.

$$\text{when } Q = 1 \quad \text{then } VC = 2(1)$$
$$\text{when } Q = 10 \text{ then } VC = 2(10)$$

thus

$$VC = 2Q$$

so

$$TC = 500 + 2Q$$ *

(iii) $VC = 25$ per cent of TR

thus

$$\text{revenue per unit} = 8$$

and

$$TR = 8Q$$

where Q can take any value.

(iv) Profit $= 0$ at point A in Figure 11

$$S = TR - TC$$
$$TR = 8Q \text{ and } TC = 500 + 2Q$$

therefore

$$S = 8Q - 500 - 2Q$$

When $S = 0$ then

$$8Q - 500 - 2Q = 0$$

thus

$$6Q = 500$$

i.e.

$$Q = 83\tfrac{1}{3}$$

                    *

CHAPTER 2

Slope and Elasticity

SLOPE

Straight lines have a slope, which may be either positive or negative. In Figure 1 the slope is negative, since a rise in price is associated with a fall in the quantity demanded. Quantity is said to be a decreasing function of price. If a function of the form

$$Q = A + bP$$

is to have a negative slope, this means that $b < 0$.

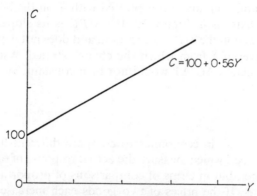

Figure 12   A linear consumption function

By convention, the slope of a line is defined as the amount by which its vertical coordinate changes for a unit increase along the horizontal axis. If we draw Q on the vertical and P on the horizontal axis, the slope of $Q = A + bP$ is equal to b, because an increase of one unit in P produces an increase of b units in Q. Figure 12 shows the consumption function

$$C = 100 + 0.56\,Y$$

This has a positive slope of 0·56 as consumption rises with income. Consumption is said to be an increasing function of income. If a function of the form

$$C = \alpha + \beta\,Y$$

is to have a positive slope this means that β must be positive, or $\beta > 0$.

THE MARGINAL PROPENSITY TO CONSUME

The slope of the graph of a function shows how the dependent variable will change if the independent variable alters. If the consumption function is

$$C = \alpha + \beta Y$$

if Y changes by an amount ΔY and C changes by an amount ΔC determined by the consumption function, then

$$C + \Delta C = \alpha + \beta(Y + \Delta Y)$$

so subtracting $C = \alpha + \beta Y$,

$$\Delta C = \beta \Delta Y$$

In our example where $C = 100 + 0\cdot56\,Y$

$$C + \Delta C = 100 + 0\cdot56(Y + \Delta Y)$$

so

$$\Delta C = 0\cdot56 \Delta Y$$

thus

$$\Delta C / \Delta Y = 0\cdot56$$

$\Delta C / \Delta Y$ gives the ratio of any change in consumption to the corresponding change in income, and is called the marginal propensity to consume (MPC). When the consumption function is plotted with Y on the horizontal and C on the vertical axis, as in Figure 12, the MPC is its slope. With a linear consumption function the MPC is constant, and does not depend on the size of ΔY or the level of Y from which the change started. With a curvilinear consumption function $\Delta C / \Delta Y$ would not be a constant, but would depend on both Y and ΔY.

ELASTICITY

We will often wish in economics to compare different functions not in terms of their slopes, which measure the actual response of one variable to a change in another, but in terms of comparisons of proportional changes in the two variables. If the prices of two goods each increase by 1 per cent, by what percentages will the amounts demanded or supplied be altered? Comparison of slopes alone cannot give us this information, since the slope of a supply or demand curve for any good depends on the units of measurement used.

PRICE ELASTICITY OF DEMAND

The price elasticity of demand is defined as

$$E_D = \frac{\text{percentage change in quantity demanded}}{\text{percentage change in price}}$$

$$= \frac{\text{proportional change in } Q_D}{\text{proportional change in } P}$$

If the demand function is given by

$$Q = A + bP$$

then

$$Q + \Delta Q = A + b(P + \Delta P)$$

so that

$$\Delta Q = b\Delta P$$

and $\Delta Q/\Delta P = b$. The proportional change in Q is $\Delta Q/Q$ and the proportional change in P is $\Delta P/P$, so

$$E_D = \frac{\Delta Q/Q}{\Delta P/P} = \frac{\Delta Q}{\Delta P} \cdot \frac{P}{Q} = b \cdot \frac{P}{Q}$$

This is clearly independent of the units in which we measure either prices or quantities, because if we make the units smaller, e.g. measure quantity in pounds instead of tons, this affects ΔQ in the same proportion as Q.

The Elasticity of Demand on the Linear Demand Curve $Q = A + bP$

We note that

$$\frac{P}{Q} = \frac{P}{A + bP}$$

which changes as P changes, so that with a linear demand function while $\Delta Q/\Delta P$ is constant regardless of the size of P and ΔP, although the elasticity of demand does not depend on the size of ΔP it does depend on P so that it changes as we move along the demand curve. As P approaches zero, written $P \to 0$,

$$E_D \to b\left(\frac{0}{A + b.0}\right) = 0$$

$Q = 0$ when $P = -A/b$, which is positive because $A > 0$ and $b < 0$, assuming a normal demand function. As $Q \to 0$, P/Q becomes indefinitely large. We call this 'P/Q tends to infinity', written

$$\frac{P}{Q} \to \infty$$

As $P \to -A/b$

$$E_D = b \cdot \frac{P}{Q} \to b \cdot \frac{P}{0} \to \infty$$

E_D can be written as

$$E_D = \frac{bP}{A + bP} = 1 \bigg/ \left[1 + \frac{A}{b}\left(\frac{1}{P}\right)\right]$$

As P increases, $(1/P)$ falls. A/b is negative, so as P increases

$$1 + \frac{A}{b}\left(\frac{1}{P}\right)$$

increases, and

$$1\Big/\left[1+\frac{A}{b}\left(\frac{1}{P}\right)\right]$$

decreases. E_D falls steadily from 0 towards $-\infty$ as P increases.

If any two demand curves actually intersect, the P/Q will be the same for both at the point of intersection, and if their slopes differ then their elasticities at this point will differ proportionally.

If the quantity demanded does not depend on price, i.e.

$$Q = A$$

then

$$\Delta Q / \Delta P = 0$$

and the elasticity of demand is zero.

If the quantity demanded is perfectly responsive to price, i.e.

$$P = K$$

where K is a constant, and for any $P > K$ none is demanded while for any $P < K$ demand becomes indefinitely large. As $\Delta P \to 0, \Delta Q / \Delta P \to -\infty$ and elasticity of demand $\to -\infty$.

ELASTICITY OF SUPPLY

This is defined as

$$E_S = \frac{\text{proportional change in quantity supplied}}{\text{proportional change in price}}$$

where the supply is decided by competitive firms, each of which treats the market price as given. If the supply function is given by

$$Q = \gamma + \delta P$$

then if P increases by ΔP,

$$Q + \Delta Q = \gamma + \delta(P + \Delta P)$$

so

$$\Delta Q = \delta \Delta P$$

The slope of the supply curve is thus $\Delta Q / \Delta P = \delta$, assuming that as in Figure 6 quantity is measured on the vertical axis and P on the horizontal axis. Thus

$$E_S = \frac{\Delta Q / Q}{\Delta P / P} = \frac{P}{Q} \cdot \frac{\Delta Q}{\Delta P} = \frac{P}{\gamma + \delta P} \cdot \delta$$

DEMAND AS A LINEAR FUNCTION OF SEVERAL VARIABLES

It is possible to apply the concept of elasticity to the reactions of the demand for a good to variables other than its own price. As in Chapter 1, let demand for a good be given by

$$Q = a + bP + cY + d\Pi + eT$$

e.g.

$$Q = 100 - 3P + 0{\cdot}01\,Y + 0{\cdot}4\Pi + 0{\cdot}05T$$

Given the values of the variables P, Y, Π and T at any point we can evaluate the proportional response of Q to a given proportional change in any one of them. Suppose $P = 10$, $Y = 2000$, $\Pi = 10$ and $T = 40$. Thus

$$Q = 100 - 3(10) + 0{\cdot}01(2000) + 0{\cdot}4(10) + 0{\cdot}05(40)$$
$$= 100 - 30 + 20 + 4 + 2 = 96$$

Suppose that P changes but Y, Π and T are constant; then

$$\frac{\Delta Q}{\Delta P} = -3 \quad \text{so} \quad E_D = \frac{P}{Q} \cdot \frac{\Delta Q}{\Delta P} = \frac{10}{96}(-3) = -\frac{5}{16}$$

This is the price elasticity of demand evaluated at the point we have chosen.

INCOME ELASTICITY OF DEMAND

Suppose that Y changes but P, Π and T remain constant; then $\Delta Q/\Delta Y = 0.01$ so the income elasticity of demand is

$$E_Y = \frac{Y}{Q} \cdot \frac{\Delta Q}{\Delta Y} = \frac{2000}{96}(0.01) = \frac{20}{96} = \frac{5}{24}$$

CROSS-ELASTICITY OF DEMAND

Suppose that Π changes but P, Y, and T remain constant; then $\Delta Q/\Delta \Pi = 0{\cdot}4$, thus the cross-elasticity of demand, defined as the ratio of the proportional change in demand for a good to the proportional change in the price of some other good, is given by

$$E_{XD} = \frac{\Pi}{Q} \cdot \frac{\Delta Q}{\Delta \Pi} = \frac{10}{96}(0{\cdot}4) = \frac{4}{96} = \frac{1}{24}$$

There will be a separate cross-elasticity of demand for every possible pair of goods.

AVERAGE AND MARGINAL PROPENSITY TO CONSUME

We may also use the elasticity concept to consider how total consumption expenditure will be affected by income changes. If

$$C = \alpha + \beta Y$$

e.g.

$$C = 100 + 0{\cdot}56\,Y$$

then the MPC, β or 0.56 may be compared with the average propensity to consume, APC. APC is given by

$$\frac{C}{Y} = \frac{\alpha + \beta Y}{Y} = \frac{\alpha}{Y} + \beta$$

or

$$\frac{C}{Y} = \frac{100 + 0 \cdot 56\,Y}{Y} = \frac{100}{Y} + 0 \cdot 56$$

APC is a decreasing function of Y because if $\alpha > 0$, α/Y falls as Y increases. The elasticity of response of C to Y is given by

$$\frac{Y}{C} \cdot \frac{\Delta C}{\Delta Y} = \frac{Y}{\alpha + \beta Y} \cdot \beta \quad \text{or} \quad \frac{Y}{100 + 0 \cdot 56\,Y}(0 \cdot 56)$$

These are clearly < 1 if $\alpha > 0$, but $\to 1$ as $Y \to \infty$.

EXERCISES

1. A consumer has an income of 100 per month and he wishes to spend all of it on two goods X and Y, whose prices are 4 and 5 respectively. All figures are in units of £1.

   (i) Express the budget line,
      (a) in a diagram,
      (b) in algebra.

   (ii) How will (i)(a) and (b) above be affected by

      (a) a doubling of money income, which leaves money prices constant,
      (b) a doubling of all absolute prices, leaving money income unchanged?

   (iii) Show that a doubling of all money prices has exactly the same effect on the budget line as halving money income.

   (iv) Show how a fall in the price of good X to 2, assuming the price of Y and money income remains constant, will affect the budget line

      (a) in algebra,
      (b) in a diagram.

   (v) If the consumer purchased 10 units of Y when the price of X was 4, give the two limiting points, between which more of both goods are available to him, if his money income remains unchanged when the price of X falls to 2.

2. An individual's consumption is 500 per annum when gross income Y is zero, and 1500 when $Y = 2000$, where all figures are in units of £1. Assuming the consumption function is linear,

   (i) express $C = f(Y)$

      (a) in algebra,
      (b) in a diagram.

   (ii) Show, using the algebraic form, that the average propensity to consume varies as Y changes.

   (iii) Assume that this individual's APC now depends upon the level of his income relative to the incomes of his neighbours rather than upon his absolute income level Y; i.e. his APC will change only if his income changes relative to other people's incomes. Show how a doubling of all individual incomes will affect (i)(a) and (b) above.

3. Calculate price elasticity of demand or supply for the following functions when
   (a) $P = 8$ and (b) $P = 6$

       (i) $P = 40 - 0.5Q$
       (ii) $Q = -4 + 0.75P$
       (iii) $Q - P + 2 = 0$
       (iv) $2P + 0.25Q = 40$
       (v) $P = 20 - 2Q$
       (vi) $4Q + 4P = 64$.

4. $Q_1 = 40 - 0.5P_1$ is the demand function of good 1. Assume

   (i) That the linear demand function for another commodity, good 2, intersects this function at the point $P_1 = 8$, and

   (ii) That the price elasticity of demand for good 1, E_{D1} is half that for good 2 at the point of intersection, and express the demand function for good 2 in algebraic form.

5. If the market demand schedule of a commodity can be represented by the function
   $$Q_1 = 20 - 2P_1 - 0.5P_2 + 0.01Y$$
   where Q_1 is the quantity demanded of commodity 1, P_1 is the price of commodity 1, P_2 is the price of another commodity, and Y is income.
   (i) Express
       (a) $Q_1 = f(P_1)$, when $P_2 = 10$ and $Y = 500$,
       (b) $Q_1 = f(P_2)$, when $P_1 = 10$ and $Y = 2000$,
       (c) $Q_1 = f(Y)$, when $P_1 = 5$ and $P_2 = 10$.
   (ii) Calculate
       (a) E_D for (i)(a) when $P_1 = 5$,
       (b) E_{XD} for (i)(b) when $P_2 = 10$,
       (c) E_Y for (i)(c) when $Y = 1000$.
   (iii) (a) Show how a change in P_2 to 12 and Y to 1000 will affect (i)(a).
       (b) Show how a change in P_1 to 5 and Y to 1500 will affect (i)(b).
       (c) Show how a change in P_1 to 3 and P_2 to 8 will affect (i)(c).

SOLUTIONS TO EXERCISES

1.   (i) (a) Given an income of 100, the consumer can spend all of his income on good X, purchasing 25 units, i.e. point A, Figure 13, or on good Y, purchasing 20 units, i.e. point B, Figure 13.
   The budget line is given by AB in Figure 13.   *

   (b) The budget line AB gives all the possible combinations of goods X and Y, open to the consumer, if he wishes to spend his entire income of 100. Thus
   $$P_X X + P_Y Y = 100$$
   When $P_X = 4$ and $P_Y = 5$, the budget line AB is defined by
   $$4X + 5Y = 100 \qquad *$$

   (ii) (a) Income is now 200, thus
   $$P_X X + P_Y Y = 200$$
   Consequently the algebraic form of the new budget line is $4X + 5Y = 200$ (see A_1B_1, Figure 13).   *
   Therefore a doubling of money income, assuming absolute prices are fixed, causes the budget line to shift away from the origin, parallel to itself.

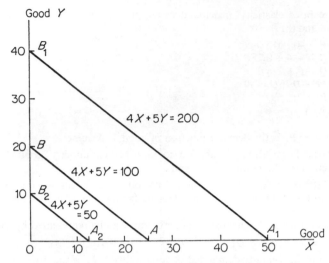

Figure 13   Budget lines with changes in real income

(b) $P_X = 8$ and $P_Y = 10$, but income remains at 100.
$$P_X X + P_Y Y = 100$$

thus

$$8X + 10Y = 100 \text{ or } 4X + 5Y = 50$$

is the algebraic form of the budget line (see $A_2 B_2$, Figure 13).   *

(iii) With an income of 50 the algebraic form of the budget line is

$$4X + 5Y = 50 \text{ (see (ii)(b) above)}$$ *

(iv) (a) When $P_X = 2, P_Y = 5$, and income is 100, the budget line is

$$2X + 5Y = 100$$ *

(b) See Figure 14, line $A_3 B$.
(c) The original budget line is given by AB in Figure 14. The consumer was originally at K, purchasing 10 units of Y. Since $4X + 5Y = 100$, when $Y = 10$ then $4X = 100 - 50$ so $X = 12.5$.

(v) The consumer's original combination was 10 units of Y and 12.5 units of X. $A_3 B$ gives all the new combinations open to the consumer when P_X changes to 2.

Given that the consumer was originally at K then more of both goods can be purchased along that part of the budget line between L and M (see Figure 14).

At L,
$X = 12.5$ and $2X + 5Y = 100$ thus $25 + 5Y = 100$, so $Y = 15$ *

At M,
$Y = 10$ and $2X + 5Y = 100$ thus $2X + 50 = 100$, so $X = 25$ *

2.   (i) (a) When $Y = 0, C = 500$.
When $Y = 2000, C = 1500$, thus
$$\Delta Y = 2000 \text{ and } \Delta C = 1000$$

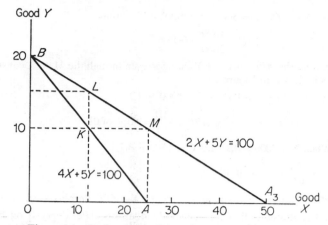

Figure 14   Budget lines with a change in relative prices

and

$$\frac{\Delta C}{\Delta Y} = \frac{1000}{2000} = 0.5 = \text{marginal propensity to consume}$$

Since the consumption function is linear then its slope, given by $\Delta C / \Delta Y$, is constant throughout and the consumption intercept is 500. Hence the consumption function is

$$C = 500 + 0.5\, Y \qquad\qquad *$$

See Figure 15.

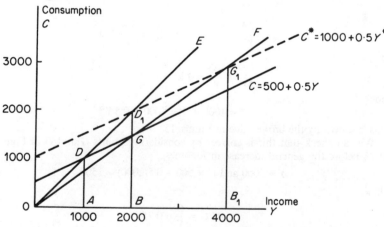

Figure 15   The consumption function with changing average incomes

(ii) The consumption function $C = 500 + 0.5\, Y$ gives

$$C = 500 + 0.5(1000) = 1000 \text{ when } Y = 1000.$$

$$APC = \frac{C}{Y} = \frac{1000}{1000} = 1 \text{ when } Y = 1000$$

When $Y = 2000$, $C = 500 + 0.5(2000) = 1500$, thus

$$APC = \frac{1500}{2000} = 0.75$$

Therefore the APC varies as Y changes, even though the MPC is constant at 0.5. This is shown in Figure 15.

When $Y = 1000 = 0A$, then $C = 1000 = AD$.

$$APC = \frac{AD}{0A} = \text{slope of } 0E$$

When $Y = 2000 = 0B$, then $C = 1500 = BG$,

$$APC = \frac{BG}{0B} = \text{slope of } 0F$$

$0F$ is flatter than $0E$, therefore as Y increases the APC decreases.

(iii) The hypothesis is that for each individual, his APC depends on his relative income, compared with other people. If all incomes were to double, then if his income doubled also his relative position would be the same as before, and so would his APC.

Before the change in incomes, $C = 500 + 0.5\,Y$ so

$$\frac{C}{Y} = \frac{500}{Y} + 0.5$$

Let the individual's income and consumption after the general change in incomes be Y^* and C^*. The hypothesis is that when all incomes double, i.e. $Y^* = 2Y$,

$$\frac{C^*}{Y^*} = \frac{C}{Y}$$

$$\frac{C^*}{Y^*} = \frac{C^*}{2Y} \quad \text{and} \quad \frac{C}{Y} = \frac{500}{Y} + 0.5$$

so

$$\frac{C^*}{2Y} = \frac{500}{Y} + 0.5$$

thus

$$C^* = 1000 + Y = 1000 + 0.5\,Y^*$$

This is shown by the broken line in Figure 15.

We can check that this is correct by considering points G and G_1 in Figure 15. At G, before the general increase in incomes,

$$Y = 2000 \text{ and } C = 500 + 0.5(2000) = 1500$$

Thus

$$APC = \frac{C}{Y} = 0.75$$

After all incomes double, we require that when $Y^* = 4000$, $C^*/Y^* = 0.75$, thus when $Y^* = 4000$, $C^* = 3000$; this is point G_1 which lies on $C^* = 1000 + 0.5\,Y^*$. A similar check can be made comparing points D and D_1 in Figure 15.

3. If $Q = K + aP$, where K and a are constants, then K is the quantity intercept and $\Delta Q/\Delta P = a$. Elasticity of demand or supply is defined as

$$E = \frac{\Delta Q}{\Delta P} \cdot \frac{P}{Q}$$

To calculate E_D or E_S all functions can be reduced to the form

$$Q = K + aP$$

This immediately gives $\Delta Q / \Delta P = a$.

(i) $P = 40 - 0{\cdot}5Q$ or $0{\cdot}5Q = 40 - P$

thus

$$Q = 80 - 2P$$

so

$$\frac{\Delta Q}{\Delta P} = -2$$

(a) When $P = 8$, $Q = 80 - 16 = 64$, thus

$$E_D = -2\left(\frac{8}{64}\right) = -0{\cdot}25 \qquad *$$

(b) When $P = 6$, $Q = 80 - 12 = 68$, thus

$$E_D = -2\left(\frac{6}{68}\right) = -\frac{3}{17} = -0{\cdot}176 \qquad *$$

(ii) $Q = -4 + 0{\cdot}75P$

thus

$$\frac{\Delta Q}{\Delta P} = 0{\cdot}75$$

(a) When $P = 8$, $Q = -4 + 6 = 2$, thus

$$E_S = 0{\cdot}75\left(\frac{8}{2}\right) = 3 \qquad *$$

(b) When $P = 6$, $Q = -4 + 4{\cdot}5 = 0{\cdot}5$, thus

$$E_S = 0{\cdot}75\left(\frac{6}{0{\cdot}5}\right) = 9 \qquad *$$

(iii) $Q - P + 2 = 0$ or $Q = -2 + P$

thus

$$\frac{\Delta Q}{\Delta P} = 1$$

(a) When $P = 8$, $Q = -2 + 8 = 6$, thus

$$E_S = 1\left(\frac{8}{6}\right) = 1.33 \qquad *$$

(b) When $P = 6$, $Q = -2 + 6 = 4$, thus

$$E_S = 1\left(\frac{6}{4}\right) = 1{\cdot}5 \qquad *$$

(iv) $2P + 0{\cdot}25Q = 40$ or $0{\cdot}25Q = 40 - 2P$

thus

$$Q = 160 - 8P$$

thus

$$\frac{\Delta Q}{\Delta P} = -8 \qquad *$$

(a) When $P = 8$, $Q = 160 - 64 = 96$, thus

$$E_D = -8\left(\frac{8}{96}\right) = -\frac{2}{3} = -0{\cdot}667 \qquad *$$

(b) When $P = 6$, $Q = 160 - 48 = 112$, thus

$$E_D = -8\left(\frac{6}{112}\right) = -\frac{3}{7} = -0.429 \qquad *$$

(v) $P = 20 - 2Q$ or $2Q = 20 - P$
thus

$$Q = 10 - 0.5P$$

so

$$\frac{\Delta Q}{\Delta P} = -0.5$$

(a) When $P = 8$, $Q = 10 - 4 = 6$, thus

$$E_D = -0.5\left(\frac{8}{6}\right) = -\frac{2}{3} = -0.667 \qquad *$$

(b) When $P = 6$, $Q = 10 - 3 = 7$, thus

$$E_D = -0.5\left(\frac{6}{7}\right) = -\frac{3}{7} = -0.429 \qquad *$$

(vi) $4Q + 4P = 64$ or $4Q = 64 - 4P$
thus

$$Q = 16 - P$$

so

$$\frac{\Delta Q}{\Delta P} = -1$$

(a) When $P = 8$, $Q = 16 - 8 = 8$, thus

$$E_D = -1\left(\frac{8}{8}\right) = -1 \qquad *$$

(b) When $P = 6$, $Q = 16 - 6 = 10$, thus

$$E_D = -1\left(\frac{6}{10}\right) = -0.6 \qquad *$$

4. $Q_1 = 40 - 0.5P_1$

$$\text{When } P_1 = 8 \text{ then } Q_1 = 40 - 4 = 36$$

$$E_D = \frac{P}{Q} \cdot \frac{\Delta Q}{\Delta P}$$

Where the demand curves for good 1 and good 2 intersect, at $P = 8$, $Q = 36$, P and Q are the same for both goods and thus to get $E_{D2} = 2E_{D1}$ requires that as

$$\frac{\Delta Q_{D1}}{\Delta P_1} = -0.5, \quad \frac{\Delta Q_{D2}}{\Delta P_2} = -1$$

The demand function for good 2 thus takes the form

$$Q_{D2} = K - P_2$$

and as this passes through $P = 8$, $Q = 36$, we know

$$36 = K - 8$$

so

$$K = 44$$

and the demand function for good 2 is given by

$$Q_{D2} = 44 - P_2 \qquad *$$

5. (i) (a) $P_2 = 10$ and $Y = 500$, thus
$$Q_1 = 20 - 2P_1 - 0{\cdot}5(10) + 0{\cdot}01(500)$$
i.e.
$$Q_1 = 20 - 2P_1 \qquad *$$

(b) When $P_1 = 10$ and $Y = 2000$, then
$$Q_1 = 20 - 2(10) - 0{\cdot}5P_2 + 0{\cdot}01(2000)$$
i.e.
$$Q_1 = 20 - 0{\cdot}5P_2 \qquad *$$

(c) When $P_1 = 5$ and $P_2 = 10$
$$Q_1 = 20 - 2(5) - 0{\cdot}5(10) + 0{\cdot}01 \, Y$$
i.e.
$$Q_1 = 5 + 0{\cdot}01 \, Y \qquad *$$

(ii) (a) From (i) (a) $\Delta Q_1 / \Delta P_1 = -2$.
When $P_1 = 5$, then $Q_1 = 20 - 10 = 10$, thus
$$E_D = -2\left(\frac{5}{10}\right) = -1 \qquad *$$

(b) From (i) (b) $\Delta Q_1 / \Delta P_2 = -0{\cdot}5$.
When $P_2 = 10$ then $Q_1 = 20 - 0{\cdot}5(10) = 15$, thus
$$E_{XD} = -0{\cdot}5\left(\frac{10}{15}\right) = -\frac{1}{3} = -0{\cdot}333 \qquad *$$

(c) From (i) (c) $\Delta Q_1 / \Delta Y = 0{\cdot}01$.
When $Y = 1000$, then $Q_1 = 5 + 10 = 15$, thus
$$E_Y = 0{\cdot}01\left(\frac{1000}{15}\right) = \frac{2}{3} = 0{\cdot}667 \qquad *$$

(iii) (a) When $P_2 = 12$ and $Y = 1000$
$$Q_1 = 20 - 2P_1 - 0{\cdot}5(12) + 0{\cdot}01(1000)$$
thus
$$Q_1 = 24 - 2P_1 \qquad *$$
The changes in P_2 and Y alter the constant term of the demand curve, and cause the curve to rise in this case.

(b) When $P_1 = 5$ and $Y = 1500$
$$Q_1 = 20 - 2(5) - 0{\cdot}5P_2 + 0{\cdot}01(1500)$$
i.e.
$$Q_1 = 25 - 0{\cdot}5P_2 \qquad *$$
The changes in P_1 and Y alter the constant term of this function, shifting it upwards.

(c) When $P_1 = 3$ and $P_2 = 8$
$$Q_1 = 20 - 2(3) - 0{\cdot}5(8) + 0{\cdot}01 \, Y$$
i.e.
$$Q_1 = 10 + 0{\cdot}01 \, Y \qquad *$$
The changes in P_1 and P_2 cause the income demand function to shift so that more is demanded at any given income.

CHAPTER 3

Simultaneous Linear Equations

MARKET EQUILIBRIUM

The demand function, $Q_D = K - aP$, gives quantity demanded Q_D at various prices, while the supply function $Q_S = -K_1 + bP$ gives information on the amount the firms in a competitive industry will supply as the price varies. The price at which supply and demand are equal and the market is in equilibrium is found by setting $Q_S = Q_D$ and solving for P. If the demand function is given by

$$Q_D = 60 - 2P$$

and the supply function is given by

$$Q_S = -20 + 3P$$

at equilibrium $Q_S = Q_D$ and thus

$$-20 + 3P = 60 - 2P$$

so

$$5P = 80$$

and

$$P = 16$$

When $P = 16$ the quantity supplied or demanded can be found by inserting the value of P into the supply and demand functions, thus

$$Q_S = -20 + 3(16) = 28$$

and

$$Q_D = 60 - 2(16) = 28$$

At equilibrium $Q_S = Q_D = 28$ and $P = 16$. The point with the coordinates (16, 28) must lie on both the supply and demand curves. As both these are linear the point of intersection is the only one which satisfies the conditions that $Q_S = Q_D$. To check that (16, 28) represents the point of intersection we show that both the supply and demand relations are satisfied. Taking the demand function first,

$$Q_D = 60 - 2P$$

At (16, 28) the left-hand side $LH = 28$ and the right-hand side $RH = 60 - 32 = 28$ also. Thus $LH = RH$ and (16, 28) lies on the demand curve. Taking the supply function

$$Q_S = -20 + 3P$$

at (16, 28), $LH = 28$ and $RH = -20 + 48 = 28$, so the point lies on the supply curve. The equilibrium point is thus at the intersection of the two curves.

DEMAND FOR TWO RELATED PRODUCTS

Suppose a firm has demand functions for two products given by

$$Q_M = K_1 - a_{11}P_M + a_{12}P_N$$

and

$$Q_N = K_2 + a_{21}P_M - a_{22}P_N$$

where Q_M and Q_N are the quantities demanded of goods M and N respectively. The subscripts of the parameters in the first function begin with 1, and those in the second begin with 2. Assuming all the parameters are positive, a demand surface of the type shown in Figure 16 can be constructed to represent the demand function

$$Q_M = K_1 - a_{11}P_M + a_{12}P_N$$

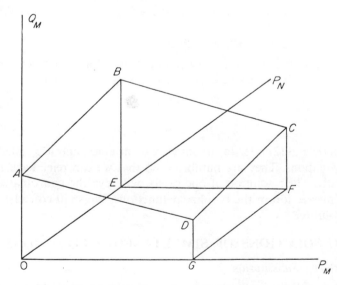

Figure 16   A three-dimensional diagram: demand as a
function of two prices

With Q_M measured along the vetical axis, the surface $ABCD$ gives the quantity demanded associated with various price combinations. When $P_M = P_N = 0$, $Q_M = K_1 = 0A$, i.e. Q_M is measured vertically from 0.

When $P_M = 0$ but $P_N = 0E$, $Q_M = EB$, which is greater than $0A$ since Q_M increases with P_N. When $P_N = 0$ but $P_M = 0G$, $Q_M = GD$, which is smaller than $0A$ since Q_M decreases as P_M rises. Other price combinations will yield different values for Q_M; e.g. when $P_M = 0G = EF$ and $P_N = 0E = GF$, $Q_M = FC$.

If we let $K_1 = 120$, $K_2 = 50$, $a_{11} = 6$, $a_{12} = 4$, $a_{21} = 2$ and $a_{22} = 5$ then

$$Q_M = 120 - 6P_M + 4P_N$$

and

$$Q_N = 50 + 2P_M - 5P_N$$

Suppose that we want to know what prices would have to be charged for the two goods so that Q_M would take a given value, say 48. There is no single pair of values of P_M and P_N which will make $Q_M = 48$. All combinations of P_M and P_N along the line defined by

$$48 = 120 - 6P_M + 4P_N$$

or

$$3P_M - 2P_N = 36$$

will give this result. If we want the combinations of values of P_M and P_N which will result in Q_N taking a certain value, say 30, the combinations of P_M and P_N consistent with this lie on the line defined by

$$30 = 50 + 2P_M - 5P_N$$

or

$$2P_M - 5P_N = -20$$

If we want to find the values of P_M and P_N which give both $Q_M = 48$ and $Q_N = 30$ then both the above conditions must be met at once, i.e.

$$3P_M - 2P_N = 36 \qquad (1)$$

and

$$2P_M - 5P_N = -20 \qquad (2)$$

Equations (1) and (2) refer to the same two variables, and are called simultaneous equations. They are numbered so that we can refer back to them conveniently. Their solution, if one exists, will give the coordinates of the point of intersection of the two linear functions. We shall consider how to find the solution.

FINDING SOLUTIONS OF SIMULTANEOUS EQUATIONS

First Method—Substitution

We begin with the method used above to calculate the equilibrium price and quantity in the market for a single good. This is to express one variable in terms of the other in each equation. Expressing P_N in terms of P_M, equation (1) becomes

$$P_N = -18 + 1{\cdot}5P_M$$

and equation (2) becomes

$$P_N = 4 + 0{\cdot}4P_M$$

As these expressions both equal P_N they must equal each other, so
$$-18 + 1.5P_M = 4 + 0.4P_M$$
and
$$1.1P_M = 22$$
so
$$P_M = 20$$
We can now find P_N by substituting $P_M = 20$ in $P_N = -18 + 1.5P_M$ to get
$$P_N = -18 + 1.5(20)$$
$$= -18 + 30 = 12$$
or by substituting $P_M = 20$ in $P_N = 4 + 0.4P_M$ to get
$$P_N = 4 + 0.4(20)$$
$$= 4 + 8 = 12$$
These two methods will give the same result. Thus we have found that when
$$Q_M = 48 \quad \text{and} \quad Q_N = 30$$
$$P_M = 20 \quad \text{and} \quad P_N = 12$$

Second Method

This is the same as the first method, except that we start by expressing P_M in terms of P_N in each equation. Equation (1) becomes
$$P_M = 12 + \tfrac{2}{3}P_N$$
and equation (2) becomes
$$P_M = -10 + 2.5P_N$$
so
$$12 + \tfrac{2}{3}P_N = -10 + 2.5P_N$$
thus
$$-\frac{11}{6}P_N = -22$$
and
$$P_N = 12$$
Substituting to get P_M
$$P_M = 12 + \tfrac{2}{3}P_N = 12 + \tfrac{2}{3}(12) = 12 + 8 = 20$$
or
$$P_M = -10 + 2.5P_N = -10 + 2.5(12) = -10 + 30 = 20$$
This gives the same results as the first method, i.e. it makes no difference which variable we choose to eliminate first.

Third Method

In this method equation (1) is multiplied by the coefficient of P_M in equation (2) and equation (2) is multiplied by the coefficient of P_M in equation (1). We have
$$3P_M - 2P_N = 36 \tag{1}$$

and

$$2P_M - 5P_N = -20 \tag{2}$$

To eliminate P_M multiply equation (1) by the coefficient of P_M in equation (2), i.e. by 2, to get

$$6P_M - 4P_N = 72 \tag{3}$$

and multiply equation (2) by the coefficient of P_M in equation (1), i.e. by 3, to get

$$6P_M - 15P_N = -60 \tag{4}$$

We have devised equations (3) and (4) so that they must have equal co-efficients for P_M, so subtraction of equation (4) from equation (3) will leave an equation in P_N only, i.e.

$$(6 - 6)P_M + (-4 + 15)P_N = 72 + 60$$

or

$$11P_N = 132$$

so

$$P_N = 12$$

This can be substituted in equation (1) or equation (2) to obtain $P_M = 20$.

Again, we could have eliminated P_N by multiplying equation (1) by the coefficient of P_N in equation (2), i.e. by -5, and multiplying equation (2) by the coefficient of P_N in equation (1), i.e. by -2. This gives two equations with equal coefficients for P_N,

$$-15P_M + 10P_N = -180$$

and

$$-4P_M + 10P_N = 40$$

so that on subtraction

$$-11P_M = -220$$

so

$$P_M = 20$$

and substitution in equation (1) or equation (2) gives $P_N = 12$. The results obtained by this method are exactly the same as those obtained earlier.

All three methods have given the same results. We can check that these are correct by inserting the values of P_M and P_N we have found in the original equations,

$$Q_M = 120 - 6P_M + 4P_N$$
$$= 120 - 120 + 48 = 48$$

as assumed, and

$$Q_N = 50 + 2P_M - 5P_N$$
$$= 50 + 40 - 60 = 30$$

as assumed.

The point (20, 12) gives the coordinates of the point of intersection of the lines represented by equations (1) and (2). Any of the three methods given above can be applied to any pair of linear equations in two variables.

CASES WHERE THE ABOVE PROCEDURES FAIL TO GIVE A UNIQUE SOLUTION

Inconsistency

Suppose equations (1) and (2) take the form

$$6P_M - 4P_N = 112 \tag{1a}$$

and

$$3P_M - 2P_N = 36 \tag{2a}$$

Method 1 gives

$$P_N = 1.5P_M - 28$$

from (1a) and

$$P_N = 1.5P_M - 18$$

from (2a). If there is to be a solution,

$$1.5P_M - 28 = 1.5P_M - 18$$

Clearly this is impossible since $-28 \neq -18$, consequently there is no solution. Methods 2 and 3 will give the same result. The reason for this is that the lines representing both equations have the same slope, namely 1.5, when P_N is measured along the vertical axis. Hence there is no position of intersection, see Figure 17. There is no solution because the equations are inconsistent.

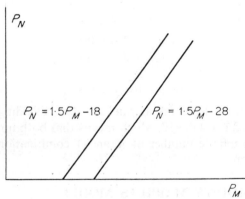

Figure 17   Parallel linear functions with no intersection

Linear Dependence

In Figure 17 the functions have the same slope but different intercepts. If the intercepts were the same both lines would coincide. The functions would then be 'linearly dependent', i.e. an infinite number of pairs of P_M and P_N would satisfy both equations.

Expressing our equations in general form will supply a general rule which can be used to discover whether the equations will have a unique solution or not. Suppose equations (1) and (2) now take the form

$$aX + bY = e$$

and

$$cX + gY = f$$

where a, b, c, g, e, and f are the parameters. Reducing each one to the form, $Y = mX + n$, gives

$$Y = -\frac{a}{b}X + \frac{e}{b}$$

and

$$Y = -\frac{c}{g}X + \frac{f}{g}$$

$-a/b$ and $-c/g$ give the slopes of both functions, where Y is plotted on the vertical axis, see Chapter 2. If $-a/b = -c/g$, i.e. $a/b = c/g$, then the equations will have no solution, as seen above. When $a/b = c/g$ cross-multiplying gives $ag = bc$. When $ag = bc$ then $a/c = b/g$. Consequently when $a/c = b/g$ there is no solution if $a/c \neq e/f$, see Figure 17.

When $a/c = b/g = e/f$ the equation for one line can be derived from that of the other, i.e. the are linearly dependent.

Let $a = 2, b = 4, e = 10, c = 4, g = 8$, and $f = 20$, i.e.

$$2X + 4Y = 10$$

and

$$4X + 8Y = 20$$
$$a/c = 2/4 = 0.5$$
$$b/g = 4/8 = 0.5$$

and

$$e/f = 10/20 = 0.5$$

The above equations have no unique solution. Dividing $4X + 8Y = 20$ across by 2 gives $2X + 4Y = 10$, which means that both functions coincide and there are an infinite number of X and Y combinations which satisfy this linear function.

A TWO COMMODITY MARKETS MODEL

At the beginning of this chapter market equilibrium was determined for a one commodity market model. Now let us examine the procedure which will enable us to find equilibrium prices and quantities for a market model of two interdependent commodities whose supply and demand functions take the following form:

$$Q_{DX} = \alpha_1 + \alpha_{11}P_X + \alpha_{12}P_Y$$
$$Q_{DY} = \alpha_2 + \alpha_{21}P_X + \alpha_{22}P_Y$$
$$Q_{SX} = \beta_1 + \beta_{11}P_X + \beta_{12}P_Y$$
$$Q_{SY} = \beta_2 + \beta_{21}P_X + \beta_{22}P_Y$$

X and Y are the two interdependent commodities, P_X and P_Y are the prices of both goods, and the αs and βs are the parameters of the model. The parameters are of a general form and signs have not been specified; certain restrictions would be necessary as a prerequisite to economically meaningful results.

The equilibrium condition for a one-commodity market model is that Q_S must equal Q_D. This gives one equation in one unknown from which it is possible to calculate the equilibrium price. Another way of expressing this condition is that excess demand or supply must equal zero, i.e.

$$Q_D - Q_S = 0 \text{ or } Q_S - Q_D = 0$$

The conditions for equilibrium in a two-commodity markets model, or indeed an n commodity markets model, are that the quantity demanded and supplied for each commodity must be equal, or that excess demand and supply must be zero for each commodity. In our example equilibrium will occur when

$$Q_{DX} = Q_{SX} \text{ and } Q_{DY} = Q_{SY}$$

When $Q_{DX} = Q_{SX}$, then

$$\alpha_1 + \alpha_{11} P_X + \alpha_{12} P_Y = \beta_1 + \beta_{11} P_X + \beta_{12} P_Y$$

or

$$(\alpha_1 - \beta_1) + (\alpha_{11} - \beta_{11})P_X + (\alpha_{12} - \beta_{12})P_Y = 0 \tag{5}$$

When $Q_{DY} = Q_{SY}$, then

$$\alpha_2 + \alpha_{21} P_X + \alpha_{22} P_Y = \beta_2 + \beta_{21} P_X + \beta_{22} P_Y$$

or

$$(\alpha_2 - \beta_2) + (\alpha_{21} - \beta_{21})P_X + (\alpha_{22} - \beta_{22})P_Y = 0 \tag{6}$$

Since the α's and β's are constants, these two conditions for equilibrium leave us with two equations, i.e. equations (5) and (6), in two unknowns, i.e. P_X and P_Y, which can be solved using any of the three methods discussed earlier. Once equilibrium prices are known it is merely a matter of substitution to find equilibrium quantities.

INPUT–OUTPUT MODELS

Suppose we take an economy with two sectors, say agriculture and industry, where the products of each sector are needed by final consumers, by the other sector as inputs and by the sector itself as inputs. Suppose each unit of gross output in agriculture Q_A requires b_{11} units of agricultural input and b_{21} units of industrial input; and each unit of gross output in industry Q_I requires b_{12} units of agricultural input and b_{22} units of industrial input. Assuming Y_A is final demand for agricultural products and Y_I is final demand for industrial products then

$$Q_A = b_{11} Q_A + b_{12} Q_I + Y_A$$

because $b_{11} Q_A =$ agricultural inputs used by the agricultural sector, and

$b_{12}Q_I$ = agricultural inputs used by the industrial sector, and

$$Q_I = b_{21}Q_A + b_{22}Q_I + Y_I$$

because $b_{21}Q_A$ = industrial inputs used by the agricultural sector, and $b_{22}Q_I$ = industrial inputs used by the industrial sector.

Thus we have

$$Q_A = b_{11}Q_A + b_{12}Q_I + Y_A$$

or

$$(1 - b_{11})Q_A - b_{12}Q_I = Y_A \tag{7}$$

and

$$Q_I = b_{21}Q_A + b_{22}Q_I + Y_I$$

or

$$-b_{21}Q_A + (1 - b_{22})Q_I = Y_I \tag{8}$$

If we know the final demands Y_A and Y_I and want to find the required gross output Q_A and Q_I this involves solving the two simultaneous equations, i.e. equations (7) and (8), in the ordinary way, since the bs take constant values. In the above model constant returns to scale and fixed technical coefficients i.e. constant values for the bs, have been assumed.

IS **AND** *LM* **SCHEDULES**

Taking a closed economic system with no external sector and no government activity gives us the identity

$$Y \equiv C + I$$

where $\equiv$ is the identically equals sign, Y is national income, C is consumption, and I is investment.

$$Y = C + I \quad \text{thus} \quad Y - C = I$$

but

$$Y - C = S$$

where S is savings, so

$$S = I$$

The Keynesian model assumes that S is a function of Y, and I is a function of the rate of interest r, i.e.

$$S = sY$$

where s is the marginal propensity to save, MPS, and

$$I = I^* - \alpha r$$

where I^* is a constant and α is the parameter of the investment function. Note that $\alpha > 0$ so that I decreases as r increases.

The *IS* schedule is a function along which $I = S$, or

$$I^* - \alpha r = sY$$

i.e.

$$sY + \alpha r = I^* \tag{9}$$

Hence the *IS* schedule is a linear function in Y and r. Along this schedule the goods market is in equilibrium. To determine the equilibrium Y and r for this economic system another function in Y and r is required. This is obtained from the money market.

The money market is in equilibrium when the supply of money M_S is equal to the demand for money M_D. In Keynesian terms M_D depends upon the transactions, precautionary and speculative motives for holding cash.

$$M_D = M_{D1} + M_{D2}$$

where M_{D1} = the transactions and precautionary demand for money, and M_{D2} = the speculative demand for money. Keynes assumes that M_{D1} is an increasing function of Y, and M_{D2} varies inversely with r. If $M_{D1} = \beta Y$ and $M_{D2} = K - gr$, where β, K and g are positive constants, and M_S is constant at M^*, then equilibrium in the money market will occur when

$$M_S = M_{D1} + M_{D2} \text{ i.e. } M_S = M_D$$

or

$$M^* = \beta Y + K - gr$$

i.e.

$$\beta Y - gr = M^* - K \tag{10}$$

This function gives all the possible combinations of Y and r which bring equilibrium to the money market and is called the *LM* schedule.

Equilibrium will occur in the money and goods market when equations (9) and (10) hold, i.e. when

$$sY + \alpha r = I^*$$

and

$$\beta Y - gr = M^* - K$$

We now have two equations in two unknowns, i.e. Y and r, whose solution, which can be found in the usual way, will give the equilibrium Y and r for this simple economic system. The model could be made more realistic by including the government sector, which would affect the *IS* schedule which in turn would affect the equilibrium values of Y and r.

EXERCISES

1. The market demand schedule for a certain good is given by the function

$$Q_D = 36 - \tfrac{1}{3}P$$

and the supply function is

$$Q_S + 9 - 0{\cdot}5P = 0$$

where Q_D and Q_S are quantities demanded and supplied and P is price.

(i) Calculate the equilibrium price and quantity in the market.

(ii) Calculate the price elasticity of demand and supply at the equilibrium price.

(iii) If the government imposes a tax of t per unit on quantity supplied and the producers adjust the supply function to include the tax:

(a) Calculate the price elasticity of demand and supply at the new equilibrium price, which includes the tax, when $t = 10$.

(b) Who pays the tax?

2. (i) Suppose that the quantity of a product demanded per period of time is Q_D, where

$$Q_D - E - 0 \cdot 5 U + P = 0$$

E is earned and U is unearned income per period, and P is the price of the product. The quantity supplied per period of time is given by

$$Q_S - 0 \cdot 5 P = 0$$

If total income per period is 300 units, how would this income have to be distributed between earned and unearned income in order to give an equilibrium price of:

(a) 100,

(b) 150,

(c) 200?

(ii) If the demand function changed to

$$Q_D - 0 \cdot 5 E - 0 \cdot 5 U + P = 0$$

what influence would the distribution of income between E and U have on the equilibrium price and quantity?

3. Suppose a firm has demand functions for two products given by

$$Q_{D1} - 16 = -2P_1 + 0 \cdot 5 P_2$$

and

$$Q_{D2} - 16 = 1 \cdot 5 P_1 - P_2$$

Q_{D1} and Q_{D2} are the quantities demanded of good 1 and 2 respectively, while P_1 and P_2 are the prices of both goods. If the demand for each good is 6, find the price which must be charged for each good.

4. Given the demand and supply functions for two interdependent commodities,

$$Q_{D1} = 8 - 2P_1 + P_2$$
$$Q_{D2} = 16 + P_1 - P_2$$
$$Q_{S1} = -5 + 3P_1$$

and

$$Q_{S2} = -1 + 2P_2$$

where $Q_{D1} = $ quantity demanded of good 1, $Q_{S1} = $ quantity supplied of good 1, etc., and P_1 and P_2 are the prices of good 1 and 2.

Calculate the equilibrium prices and quantities of the two-commodity market model.

5. Given the demand and supply functions for three interdependent commodities,

$$Q_{D1} = 45 - 2P_1 + 2P_2 - 2P_3$$
$$Q_{D2} = 16 + 2P_1 - P_2 + 2P_3$$
$$Q_{D3} = 30 - P_1 + 2P_2 - P_3$$
$$Q_{S1} = -5 + 2P_1$$
$$Q_{S2} = -4 + 2P_2$$

and

$$Q_{S3} = -5 + P_3$$

Calculate the equilibrium prices and quantities of this three-commodity market model.

6. Suppose an economy has two sectors, agriculture and industry, where each unit of gross output of agricultural products Q_A requires inputs of 0·2 unit of agricultural products and 0·3 unit of industrial products, while each unit of gross output of industrial products Q_I requires inputs of 0·1 unit of agricultural products and 0·4 unit of industrial products.

Find the required gross outputs when

(i) Final demands are given by $Y_A = 200$, $Y_I = 600$.
(ii) Final demands are given by $Y_A = 100$, $Y_I = 300$.
(iii) Final demands are given by $Y_A = 300$, $Y_I = 900$.

7. Given the following information about a closed economy:

$$\text{Consumption} \quad C = 100 + 0·8\,Y$$
$$\text{Investment} \quad I = 1200 - 30r$$

where r is the rate of interest.

Precautionary and transactions demand for money

$$M_{D1} = 0·25\,Y$$

Speculative demand for money

$$M_{D2} = 1375 - 25r$$

Money Supply

$$M_S = 2500$$

(i) Find the equilibrium values of Y and r.

(ii) Assume that a budget is now introduced into the system and that consumption is given by

$$C = 100 + 0·75\,Y_d$$

where Y_d is disposable income, i.e.

$$Y_d = Y - T$$

and taxes

$$T = 20 + 0·2\,Y$$

If government expenditure $G = 935$,

(a) Find the equilibrium values of Y and r.

(b) Show how a decrease of 43.75 in the autonomous part of the speculative demand for money function will affect Y and r.

(c) If starting from the position of (ii)(a) the autonomous part of the consumption function decreases by 35 show how this shift in the schedule will affect Y and r.

SOLUTIONS TO EXERCISES

1. $Q_D = 36 - \frac{1}{3}P$

$$Q_S + 9 - 0·5P = 0$$

thus

$$Q_S = -9 + 0·5P$$

(i) At equilibrium $Q_D = Q_S$, thus

$$36 - \tfrac{1}{3}P = -9 + 0·5P$$

i.e.
$$45 = \tfrac{5}{6}P$$
or
$$P = 54$$

When $P = 54$,
$$Q_D = 36 - 18 = 18$$
and
$$Q_S = -9 + 27 = 18 \qquad *$$

(ii) $E_D = \dfrac{\Delta Q_D}{\Delta P} \cdot \dfrac{P}{Q_D}$

$$Q_D = 36 - \tfrac{1}{3}P$$

thus
$$\frac{\Delta Q_D}{\Delta P} = -\frac{1}{3}$$

At equilibrium $P = 54$ and $Q_D = 18$, thus
$$E_D = -\frac{1}{3}\left(\frac{54}{18}\right) = -1 \qquad *$$

$$E_S = \frac{\Delta Q_S}{\Delta P} \cdot \frac{P}{Q_S}$$
$$Q_S = -9 + 0 \cdot 5P$$

thus
$$\frac{\Delta Q_S}{\Delta P} = 0 \cdot 5$$

and so
$$E_S = 0 \cdot 5\left(\frac{54}{18}\right) = 1.5 \qquad *$$

(iii) (a) If P is the market price to consumers then the effective price to producers when a tax t is imposed is $(P - t)$. The supply curve will thus take the form
$$Q_S = -9 + 0 \cdot 5(P - t)$$
When $t = 10$,
$$Q_S = -9 + 0 \cdot 5(P - 10) = -14 + 0 \cdot 5P$$
The demand function remains as before
$$Q_D = 36 - \tfrac{1}{3}P$$
At equilibrium $Q_D = Q_S$ thus
$$36 - \tfrac{1}{3}P = -14 + 0 \cdot 5P$$
so
$$50 = \tfrac{5}{6}P$$
and
$$P = 60$$
When $P = 60$,
$$Q_S = -14 + 30 = 16$$
and
$$Q_D = 36 - 20 = 16$$
$$E_D = \frac{\Delta Q_D}{\Delta P} \cdot \frac{P}{Q_D} = -\frac{1}{3}\left(\frac{60}{16}\right) = -1 \cdot 25 \qquad *$$

In calculating the elasticity of supply we consider the reaction of producers to the price they actually receive, $(P - t)$. Thus

$$E_S = \frac{\Delta Q_S}{\Delta P} \cdot \frac{(P - t)}{Q_S} = 0.5\left(\frac{50}{16}\right) = \frac{25}{16} = 1.562 \qquad *$$

(b) At the new equilibrium, $P = 60$, thus P has increased from 54 to 60 as the result of the tax of 10 per unit. Thus the consumers pay 60 per cent of the tax, and producers who received 54 before the tax and $60 - 10 = 50$ after, pay 40 per cent of the tax. This is shown in Figure 18.

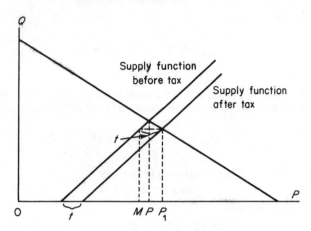

Figure 18   Division of the burden of taxation

P = original equilibrium price, i.e. 54 in our example.
P_1 = new equilibrium price, i.e. 60 in our example.
Price has risen by PP_1, which is less than the amount of tax MP_1.

2. (i) $E + U$ = total income = 300
At equilibrium $Q_S = Q_D$

$$Q_D - E - 0.5U + P = 0$$

so

$$Q_D = E + 0.5U - P$$

and

$$Q_S - 0.5P = 0$$

so

$$Q_S = 0.5P$$

At equilibrium,

$$E + 0.5U - P = 0.5P$$

but

$$U = 300 - E$$

thus

$$E + 0.5(300 - E) - P = 0.5P$$

so

$$0.5E + 150 = 1.5P$$

or

$$E = 3P - 300$$

(a) When $P = 100$

$$E = 300 - 300 = 0$$

When $E = 0, U = 300$. *

(b) When $P = 150$

$$E = 450 - 300 = 150$$

When $E = 150, U = 150$. *

(c) When $P = 200$

$$E = 600 - 300 = 300$$

When $E = 300, U = 0$. *

(ii) When

$$Q_D - 0.5E - 0.5U + P = 0$$

then

$$Q_D - 0.5(E + U) + P = 0$$

Since $E + U = 300$, i.e. the value of total income remains constant at 300, the value of $0.5(E + U)$ will remain the same, regardless of any redistribution between E and U.*

3.
$$Q_{D1} - 16 = -2P_1 + 0.5P_2 \qquad (11)$$
$$Q_{D2} - 16 = 1.5P_1 - P_2 \qquad (12)$$

When $Q_{D1} = Q_{D2} = 6$ then equation (11) becomes

$$-10 = -2P_1 + 0.5P_2 \qquad (13)$$

and equation (12) becomes

$$-10 = 1.5P_1 - P_2 \qquad (14)$$

Taking equation (13) and (14), we eliminate P_2 by multiplying equation (13) by 2 to give

$$-20 = -4P_1 + P_2 \qquad (15)$$

We now have

$$-20 = -4P_1 + P_2 \qquad (15)$$

and

$$-10 = 1.5P_1 - P_2 \qquad (14)$$

Adding equations (14) and (15) gives

$$-30 = -2.5P_1$$

thus

$$P_1 = 12 \qquad *$$

When $P_1 = 12$, from equation (14)

$$-10 = 1.5(12) - P_2$$

so

$$P_2 = 28 \qquad *$$

Thus when $Q_{D1} = Q_{D2} = 6$, $P_1 = 12$ and $P_2 = 28$. *

We can check that we have the correct answer and that Q_{D1} and Q_{D2} do in fact equal 6 by inserting $P_1 = 12$ and $P_2 = 28$ in the original demand functions. These give

$$Q_{D1} - 16 = -2(12) + 0.5(28) \quad \text{i.e.} \quad Q_{D1} = 6$$

and

$$Q_{D2} - 16 = 1.5(12) - 28 \quad \text{i.e.} \quad Q_{D2} = 6$$

4. In an isolated market model, the equilibrium condition consists of only one function, i.e. $Q_D = Q_S$ or $X = Q_D - Q_S = 0$, where X stands for excess demand. When several

interdependent commodities are considered equilibrium requires the absence of excess demand for each and every commodity.

Q_{D1} must equal Q_{S1} and Q_{D2} must equal Q_{S2}.

When $Q_{D1} = Q_{S1}$, then

$$8 - 2P_1 + P_2 = -5 + 3P_1$$

so

$$13 = 5P_1 - P_2 \qquad (16)$$

When $Q_{D2} = Q_{S2}$, then

$$16 + P_1 - P_2 = -1 + 2P_2$$

so

$$17 = -P_1 + 3P_2 \qquad (17)$$

Taking equations (16) and (17), let us eliminate P_2 by multiplying (16) by 3. This gives

$$39 = 15P_1 - 3P_2 \qquad (18)$$

We now have,

$$39 = 15P_1 - 3P_2 \qquad (18)$$
$$17 = -P_1 + 3P_2 \qquad (17)$$

Adding equations (17) and (18) gives $56 = 14P_1$, thus

$$P_1 = 4 \qquad *$$

When $P_1 = 4$, then from equation (16), $13 = 5(4) - P_2$, thus

$$-P_2 = -7$$

or

$$P_2 = 7 \qquad *$$

When $P_1 = 4$, $Q_{S1} = -5 + 12 = 7$. $\qquad *$

When $P_2 = 7$, $Q_{S2} = -1 + 14 = 13$. $\qquad *$

Let us check that $Q_{S1} = Q_{D1}$ and $Q_{S2} = Q_{D2}$.

$$Q_{D1} = 8 - 2(4) + 7 = 7 = Q_{S1}$$
$$Q_{D2} = 16 + 4 - 7 = 13 = Q_{S2}$$

5. At equilibrium, Q_{D1} must equal Q_{S1}, Q_{D2} must equal Q_{S2}, and Q_{D3} must equal Q_{S3}.

When $Q_{D1} = Q_{S1}$, then

$$45 - 2P_1 + 2P_2 - 2P_3 = -5 + 2P_1$$

i.e.

$$-50 = -4P_1 + 2P_2 - 2P_3 \qquad (19)$$

When $Q_{D2} = Q_{S2}$, then

$$16 + 2P_1 - P_2 + 2P_3 = -4 + 2P_2$$

i.e.

$$-20 = 2P_1 - 3P_2 + 2P_3 \qquad (20)$$

When $Q_{D3} = Q_{S3}$, then

$$30 - P_1 + 2P_2 - P_3 = -5 + P_3$$

i.e.

$$-35 = -P_1 + 2P_2 - 2P_3 \qquad (21)$$

Elimination of P_1 will give two equations in two unknowns, i.e. P_2 and P_3. Multiplying equation (20) by 2 gives

$$-40 = 4P_1 - 6P_2 + 4P_3 \qquad (22)$$

We have

$$-50 = -4P_1 + 2P_2 - 2P_3 \qquad (19)$$

Addition gives

$$-90 = -4P_2 + 2P_3 \qquad (23)$$

Multiplying equation (21) by 2 gives

$$-70 = -2P_1 + 4P_2 - 4P_3 \qquad (24)$$

We have

$$-20 = 2P_1 - 3P_2 + 2P_3 \qquad (20)$$

Addition gives

$$-90 = P_2 - 2P_3 \qquad (25)$$

Adding equations (23) and (25) will solve for P_2, i.e.

$$-180 = -3P_2 \text{ or } P_2 = 60 \qquad *$$

When $P_2 = 60$ then from equation (25)

$$-90 = 60 - 2P_3$$

or

$$P_3 = 75 \qquad *$$

When $P_2 = 60$ and $P_3 = 75$, then from equation (21)

$$-P_1 + 2(60) - 2(75) = -35$$

and

$$P_1 = 5 \qquad *$$

When $P_1 = 5, Q_{S1} = -5 + 2(5) = 5 = Q_{D1}$.   *

When $P_2 = 60, Q_{S2} = -4 + 2(60) = 116 = Q_{D2}$.   *

When $P_3 = 75, Q_{S3} = -5 + 75 = 70 = Q_{D3}$.   *

6. Q_A = gross agricultural output,
Q_I = gross industrial output,
$0.2Q_A$ gives agricultural inputs used in agricultural sector,
$0.1Q_I$ gives agricultural inputs used in industrial sector, and
Y_A gives final demand for agricultural products, thus

$$\text{gross agricultural output} = Q_A = 0.2Q_A + 0.1Q_I + Y_A$$

$0.3Q_A$ gives industrial inputs used in agricultural sector,
$0.4 Q_I$ gives industrial inputs used in industrial sector, and
Y_I gives final demand for industrial products, thus

$$\text{gross industrial output} = Q_I = 0.3Q_A + 0.4Q_I + Y_I$$

$$Q_A = 0.2Q_A + 0.1Q_I + Y_A$$

thus

$$0.8Q_A - 0.1Q_I = Y_A \qquad (26)$$
$$Q_I = 0.3Q_A + 0.4Q_I + Y_I$$

thus

$$-0.3Q_A + 0.6Q_I = Y_I \qquad (27)$$

To eliminate Q_I multiply equation (26) by 6

$$4.8Q_A - 0.6Q_I = 6Y_A \qquad (28)$$

We have

$$-0.3Q_A + 0.6Q_I = Y_I \qquad (27)$$

Addition gives

$$4.5Q_A = 6Y_A + Y_I$$

thus

$$Q_A = \frac{6Y_A + Y_I}{4 \cdot 5} = \frac{12Y_A + 2Y_I}{9}$$

To eliminate Q_A multiply equation (26) by 3 and equation (27) by 8.
From (26)

$$2 \cdot 4 Q_A - 0 \cdot 3 Q_I = 3 Y_A \qquad (29)$$

From (27)

$$- 2 \cdot 4 Q_A + 4 \cdot 8 Q_I = 8 Y_I \qquad (30)$$

Addition gives

$$4 \cdot 5 Q_I = 3 Y_A + 8 Y_I$$

thus

$$Q_I = \frac{3 Y_A + 8 Y_I}{4 \cdot 5} = \frac{6 Y_A + 16 Y_I}{9}$$

(i) When $Y_A = 200$ and $Y_I = 600$, then

$$Q_A = \frac{12(200) + 2(600)}{9} = 400 \qquad *$$

and

$$Q_I = \frac{6(200) + 16(600)}{9} = 1200 \qquad *$$

Check; when $Q_A = 400$ and $Q_I = 1200$
$$Y_A = 0 \cdot 8(400) - 0 \cdot 1(1200) = 200$$
from equation (26) and
$$Y_I = - 0 \cdot 3(400) + 0 \cdot 6(1200) = 600$$
from equation (27).

(ii) When $Y_A = 100$ and $Y_I = 300$, then

$$Q_A = \frac{12(100) + 2(300)}{9} = 200 \qquad *$$

and

$$Q_I = \frac{6(100) + 16(300)}{9} = 600 \qquad *$$

Check; when $Q_A = 200$ and $Q_I = 600$
$$Y_A = 0 \cdot 8(200) - 0 \cdot 1(600) = 100$$
from equation (26) and
$$Y_I = - 0 \cdot 3(200) + 0 \cdot 6(600) = 300$$
from equation (27).

(iii) When $Y_A = 300$ and $Y_I = 900$, then

$$Q_A = \frac{12(300) + 2(900)}{9} = 600 \qquad *$$

and

$$Q_I = \frac{6(300) + 16(900)}{9} = 1800 \qquad *$$

Check; when $Q_A = 600$ and $Q_I = 1800$
$$Y_A = 0 \cdot 8(600) - 0 \cdot 1(1800) = 300$$
and
$$Y_I = - 0 \cdot 3(600) + 0 \cdot 6(1800) = 900$$

7. (i) $Y = C + I$

$$C = 100 + 0.8\,Y$$

and

$$I = 1200 - 30r$$

thus

$$Y = 100 + 0.8\,Y + 1200 - 30r$$

i.e.

$$(1 - 0.8)\,Y = 1300 - 30r$$

so

$$0.2\,Y = 1300 - 30r$$

i.e.

$$Y = 6500 - 150r \qquad (31)$$

The money market is in equilibrium when $M_S = M_D$ but $M_D = M_{D1} + M_{D2}$, thus $M_S = M_{D1} + M_{D2}$ in equilibrium so

$$2500 = 0.25\,Y + 1375 - 25r$$

i.e.

$$0.25\,Y = 1125 + 25r$$

thus

$$Y = 4500 + 100r \qquad (32)$$

Thus

$$6500 - 150r = 4500 + 100r$$

from equations (31) and (32), i.e.

$$250r = 2000$$

so

$$r = 8$$

When $r = 8$, $Y = 5300$.   *

(ii) (a) $Y = C + I + G$.

But

$$C = 100 + 0.75\,Y_d = 100 + 0.75(Y - 20 - 0.2\,Y)$$
$$= 100 + 0.75(1 - 0.2)\,Y - 15 = 85 + 0.6\,Y$$

thus

$$Y = 85 + 0.6\,Y + 1200 - 30r + 935$$

i.e.

$$(1 - 0.6)Y = 2220 - 30r$$

so

$$Y = 5550 - 75r \qquad (33)$$

But

$$Y = 4500 + 100r \qquad (32)$$

thus

$$5550 - 75r = 4500 + 100r$$

from equation (32) and (33), i.e.

$$175r = 1050$$

and

$$r = 6$$

When $r = 6$, then $Y = 5100$.   *

(b) $M_{D2} = 1375 - 43.75 - 25r$
$$= 1331.25 - 25r$$

At equilibrium $M_S = M_{D1} + M_{D2}$, i.e.
$$2500 = 0.25\,Y + 1331.25 - 25r$$
thus
$$0.25\,Y = 1168.75 + 25r$$
i.e.
$$Y = 4675 + 100r \tag{34}$$
$$Y = 5550 - 75r \tag{33}$$
thus
$$4675 + 100r = 5550 - 75r$$
from equations (33) and (34), i.e.
$$175r = 875$$
and
$$r = 5$$
When $r = 5$, then $Y = 5175$. *

(c) From (ii) (a) we saw that $C = 85 + 0.6\,Y$, thus
$$C = 85 + 0.6\,Y - 35 = 50 + 0.6\,Y$$
and
$$Y = 50 + 0.6\,Y + 1200 - 30r + 935$$
since $Y = C + I + G$, or
$$0.4\,Y = 2185 - 30r$$
so
$$Y = 5462.5 - 75r \tag{35}$$
But
$$Y = 4500 + 100r \tag{32}$$
thus
$$4500 + 100r = 5462.5 - 75r$$
from equations (32) and (35), i.e.
$$175r = 962.5$$
so
$$r = 5.5$$
When $r = 5.5$, then $Y = 5050$. *

CHAPTER 4

Curvilinear Functions

Linear functions at one stage of an economic model often give rise to curvilinear functions at another stage. In our discussion of linear demand functions the demand relation took the form

$$P = K - \alpha Q \text{ where } \alpha > 0$$

If it is assumed that $P = K - \alpha Q$ is the demand function for a particular firm then the total revenue function TR of that firm is obtained by multiplying quantity by price, i.e.

$$TR = P.Q$$
$$P = K - \alpha Q$$

thus

$$TR = P.Q = (K - \alpha Q)Q$$
$$= KQ - \alpha Q^2$$

Q^2 is used to denote $Q.Q$, or 'Q squared'. Q is said to be raised to the power 2.

If the TR function is plotted on a graph it is clearly seen to be curvilinear. Thus linear demand functions give rise to curvilinear TR functions.

If the firm's total cost function takes the form $TC = F + \beta Q$, where F stands for fixed costs and β gives variable costs per unit, then $TR - TC$ will provide us with a profit function in Q.

PROFITS

If we denote profit by S then

$$S = TR - TC = KQ - \alpha Q^2 - F - \beta Q$$

or

$$S = -\alpha Q^2 + (K - \beta)Q - F$$

This profit function is also curvilinear.

THE QUADRATIC EQUATION

To find the Q which makes $S = 0$, it is necessary to solve the equation

$$-\alpha Q^2 + (K - \beta)Q - F = 0$$

This is a quadratic equation in Q or an equation of the second degree, i.e. an equation in Q^2, since the degree of a variable is the value of the non-negative integer power to which it is raised. A quadratic equation is also known as a second-degree polynomial. The degree of an equation is the degree of its term of highest degree.

Assume that the parameters K, F, α, and β take the values 18, 50, 1, and 3 respectively, i.e. the demand function is $P = 18 - Q$, and the TC function is $TC = 50 + 3Q$. The profit function becomes

$$S = -1Q^2 + (18 - 3)Q - 50$$

i.e.

$$S = -Q^2 + 15Q - 50$$

This is a quadratic function which can be plotted if values for S are calculated for values of $Q > 0$, see Table 4.1. Figure 19 gives a graph of this profit function. This graph of a quadratic is called a parabola.

TABLE 4.1

Q	$-Q^2$	$15Q$	-50	$-Q^2 + 15Q - 50 = S$
1	-1	15	-50	-36
2	-4	30	-50	-24
3	-9	45	-50	-14
4	-16	60	-50	-6
5	-25	75	-50	0
6	-36	90	-50	$+4$
7	-49	105	-50	$+6$
8	-64	120	-50	$+6$
9	-81	135	-50	$+4$
10	-100	150	-50	0
11	-121	165	-50	-6
12	-144	180	-50	-14

To find the Q which makes $S = 0$ involves finding the roots of the equation

$$-Q^2 + 15Q - 50 = 0$$

The roots of this equation are the quantity ordinates of the points of intersection of the line $S = 0$ and the quadratic function $S = -Q^2 + 15Q - 50$, i.e. points R and T in Figure 19, since $S = 0$ along the quantity axis. From the graph it is obvious that the roots are 5 and 10. Consequently, profits are zero when $Q = 5$ or 10.

Factors

An alternative and simpler method for finding the roots of a quadratic involves factorization. When $S = 0$, then

$$-Q^2 + 15Q - 50 = 0$$

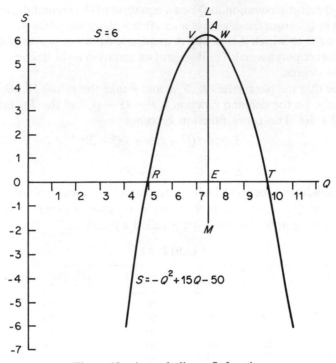

Figure 19   A parabolic profit function

or

$$Q^2 - 15Q + 50 = 0$$

To find factors we need to find numbers to fill the blanks in the expression

$$(Q - \_)(Q - \_) = 0$$

so that when we multiply out we obtain the original equation. Trial and error, using different pairs of integers which multiply to 50 such as 1 and 50, 2 and 25, 5 and 10, etc. shows that in the present case 5 and 10 are the numbers which fit. Thus

$$(Q - 5)(Q - 10) = 0$$

so that

$$Q - 5 = 0 \quad \text{or} \quad Q - 10 = 0$$

i.e.

$$Q = 5 \quad \text{or} \quad 10$$

as above. $Q = 5$ and $Q = 10$ are the two roots so that if Q takes any other value then $-Q^2 + 15Q - 50 \neq 0$.

To find the value for Q which will make $S = 6$, it is necessary to solve the equation

$$-Q^2 + 15Q - 50 = 6$$

The solution of this equation will provide the quantity ordinates of the points of intersection of the line $S = 6$ and the quadratic

$$S = -Q^2 + 15Q - 50$$

i.e. points V and W on the graph in Figure 19. When

$$-Q^2 + 15Q - 50 = 6$$

then

$$Q^2 - 15Q + 50 = -6$$

or

$$Q^2 - 15Q + 56 = 0$$

Trial and error gives

$$(Q - 7)(Q - 8) = 0$$

thus

$$Q = 7 \text{ or } 8$$

Figure 19 provides the same result.

Solution by Means of the Quadratic Formula

Factorization is fast and simple but it does not always work. In most economic problems the coefficients of Q^2 and Q are unlikely to be convenient whole numbers, and a lot of time can be wasted on the factorization method when this is the case. To solve equations of this nature we shall use an algebraic formula which can be derived from the general quadratic equation

$$aX^2 + bX + c = 0$$

where a, b, and c are constants.

When $aX^2 + bX + c = 0$, $a \neq 0$, so we can divide both sides by it, giving

$$\frac{a}{a}X^2 + \frac{b}{a}X + \frac{c}{a} = 0$$

thus

$$X^2 + \frac{b}{a}X = -\frac{c}{a}$$

Adding $b^2/4a^2$ to both sides gives

$$X^2 + \frac{b}{a}X + \frac{b^2}{4a^2} = -\frac{c}{a} + \frac{b^2}{4a^2}$$

thus

$$\left(X + \frac{b}{2a}\right)^2 = \frac{-4ac + b^2}{4a^2}$$

Taking square roots on both sides,

$$X + \frac{b}{2a} = \pm \sqrt{\left(\frac{-4ac + b^2}{4a^2}\right)}$$

so

$$X = -\frac{b}{2a} \pm \frac{\sqrt{(b^2 - 4ac)}}{2a} = \frac{-b \pm \sqrt{(b^2 - 4ac)}}{2a}$$

When $b^2 > 4ac$,

$$X = \frac{-b \pm \sqrt{\text{positive number}}}{2a}$$

Therefore X will take two values, one using the positive and the other the negative root.

When $b^2 < 4ac$, X will have no real roots since the solution will involve the square root of a minus quantity. We return to this problem in Chapter 16.

When $b^2 = 4ac$,

$$X = \frac{-b \pm \sqrt{0}}{2a} = -\frac{b}{2a}$$

Consequently X will take only one value.

Since all quadratic equations can be reduced to the form $aX^2 + bX + c = 0$, this formula should be used when factorization fails. We can now find the Q which gives zero profits when

$$S = -\alpha Q^2 + (K - \beta)Q - F$$

The roots of the equation

$$-\alpha Q^2 + (K - \beta)Q - F = 0$$

can be found by substituting, $-\alpha$ for a, $(K - \beta)$ for b, and $-F$ for c. This gives

$$Q = \frac{-(K - \beta) \pm \sqrt{[(K - \beta)^2 - 4(-\alpha)(-F)]}}{2(-\alpha)}$$

$$= \frac{K - \beta \pm \sqrt{[(K - \beta)^2 - 4\alpha F]}}{2\alpha}$$

To give economically meaningful results Q must be $\geqslant 0$. If $4\alpha F > (K - \beta)^2$ then there is no output at which $S = 0$, i.e. no real roots.

An Alternative Approach: Solution by Graphical Methods

A systematic trial and error method can be used to find the roots of a quadratic or an equation of higher degree. By way of illustration let us use this method to find the Q which gives $S = 5$, assuming,

$$S = -Q^2 + 15Q - 50$$

When $S = 5$, then

$$Q^2 - 15Q + 55 = 0$$

The value of $Q^2 - 15Q + 55$ varies as Q changes.

The roots of the equation are those values of Q which make $Q^2 - 15Q + 55 = 0$. In Table 4.2 the value of $Q^2 - 15Q + 55$ has been calculated for values of

TABLE 4.2

$$f(Q) = Q^2 - 15Q + 55$$

When $Q = 1$, $f(1) = 1 - 15 + 55 = +41$
When $Q = 2$, $f(2) = 4 - 30 + 55 = +29$
                  $f(3) = 9 - 45 + 55 = +19$
                  $f(4) = 16 - 60 + 55 = +11$
                  $f(5) = 25 - 75 + 55 = +5$
                  $f(6) = 36 - 90 + 55 = +1$
                  $f(7) = 49 - 105 + 55 = -1$
                  $f(8) = 64 - 120 + 55 = -1$
                  $f(9) = 81 - 135 + 55 = +1$
                $f(10) = 100 - 150 + 55 = +5$

Q varying from 1 to 10. When $Q = 6$, the value of the function is $+1$, whereas when Q increases to 7 the value of the function becomes -1. The function will therefore equal zero when $7 > Q > 6$, i.e. when Q takes a value between 6 and 7. By calculating the value of the function for some Q within this interval, the range, within which the solution lies, can be reduced. Table 4.2 also shows that the value of the function goes from negative to positive as Q changes from 8 to 9. Therefore the second root lies between $Q = 8$ and $Q = 9$. The range within which the exact root lies can be reduced if values of Q between 8 and 9 are taken. This procedure is essential for equations of degree three or higher. However, since we are dealing with a quadratic equation, the general formula gives

$$Q = \frac{15 \pm \sqrt{[(-15)^2 - 4(1)(55)]}}{2} = \frac{15 \pm \sqrt{5}}{2}$$

Therefore the roots are $Q = 8 \cdot 62$ and $Q = 6 \cdot 38$ approximately.

In applying the trial and error method to the above problem, values of Q from 1 to 10 were taken and the roots were found to lie between these values. However, there are an infinite number of values of Q we could try when seeking the roots of an equation, and to find an initial value near to the root is rather a hit and miss affair. In addition, this method cannot assure us that there are no more roots besides the ones it has found.

Checking Solutions

If we wish to check that the roots we have found are correct, consider the factors of the original quadratic. To check that the roots of

$$Q^2 - 15Q + 50 = 0$$

are $Q = 5$ and $Q = 10$, take

$$(Q - 5)(Q - 10)$$

This will be zero if $Q = 5$ or $Q = 10$, and when multiplied out gives $Q^2 - 15Q + 50$, our original expression.

THE PARABOLA APPROACH TO PROFIT MAXIMIZATION

Given the profit function $S = -Q^2 + 15Q - 50$, let us now calculate the output which gives maximum profit and the level of profit associated with it.

The coordinates of point A, i.e. the vertex of the parabola in Figure 19, will provide the necessary information. Consequently, if the graph is fairly exact, we can read off these coordinates.

LM is the axis of the parabola in Figure 19 and the axis bisects RT perpendicularly. Since the coordinates of R and T are 5 and 10 respectively it is possible to calculate the coordinates of their mid point. In this case E is a distance 7·5 from the origin. Consequently the quantity ordinate of the vertex is 7·5. To obtain the maximum profit level at this output, 7·5 is inserted into the profit function, i.e.

$$S = -(7·5)^2 + 15(7·5) - 50 = 6·25$$

This method is fine provided the coordinates of the points of intersection of the profits function, and the axis $S = 0$, are known. If the coordinates of these points are not readily available then an alternative method can be used.

Completing the Square

$$S = -Q^2 + 15Q - 50$$
$$= -(Q^2 - 15Q) - 50$$
$$= -\left(Q^2 - 15Q + \frac{225}{4}\right) - 50 + \frac{225}{4}$$

Adding $-225/4$ and $+225/4$ leaves the right-hand side unchanged. $225/4$ is found by dividing the coefficient of Q by 2 and squaring it, i.e.

$$\left(\frac{15}{2}\right)^2 = \frac{225}{4}$$

thus

$$S = -(Q - 7·5)^2 + 6·25$$

$(Q - 7·5)^2 = 0$ when $Q = 7·5$ and is positive when $Q \neq 7·5$. Profits equal 6·25 minus the squared term, so they have a maximum when the squared term is zero, i.e. when $Q = 7·5$. Thus S has a maximum of 6·25 when $Q = 7·5$. In general, as we saw earlier, when $S = aX^2 + bX + c$, this can be rewritten as

$$S = a\left(X^2 + \frac{b}{a}X + \frac{b^2}{4a^2}\right) + \left(c - \frac{b^2}{4a}\right)$$
$$= a\left(X + \frac{b}{2a}\right)^2 + \left(c - \frac{b^2}{4a}\right)$$

This can be checked by inserting $a = -1, b = 15$ and $c = -50$. Whenever $a < 0$, S has a maximum at $X + b/2a = 0$, and whenever $a > 0, S$ has a minimum at $X + b/2a = 0$. This standard method can be applied to any quadratic function. In Chapter 6 calculus will be used to solve maximization problems of this nature.

CURVILINEAR DEMAND AND SUPPLY FUNCTIONS

Approximations to observed demand and supply functions are frequently curvilinear and these functions can be represented by the first quadrant part of a parabola. Suppose the market supply function is

$$P = Q^2 + 4Q + 1$$

and the market demand function takes the form

$$P = -Q^2 - Q + 4$$

Both of these functions are plotted in Figure 20. Since quantity demanded or quantity supplied cannot be negative, the portion AB of the parabola $P = Q^2 + 4Q + 1$ represents this supply function and the portion LM of the parabola $P = -Q^2 - Q + 4$ represents the demand function. Point X gives equilibrium price and quantity in the market. Since

$$P = -Q^2 - Q + 4 \quad \text{demand function}$$

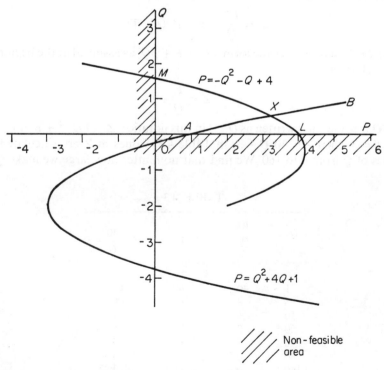

Figure 20   Parabolic supply and demand functions

and

$$P = Q^2 + 4Q + 1 \qquad \text{supply function}$$
$$-Q^2 - Q + 4 = Q^2 + 4Q + 1$$

at equilibrium. Thus

$$-2Q^2 - 5Q + 3 = 0$$

i.e.

$$2Q^2 + 5Q - 3 = 0$$

so

$$(2Q - 1)(Q + 3) = 0$$

i.e.

$$Q = 0.5 \quad \text{or} \quad -3$$

Since Q cannot be negative, equilibrium quantity is 0·5. When $Q = 0.5$, $P = -0.25 - 0.5 + 4 = 3.25$.

AVERAGE COST FUNCTIONS

Average costs AC are costs per unit of output. To obtain a firm's average costs, total costs TC are divided by quantity produced.

Let $AC = Z$ then

$$Z = \frac{TC}{Q} \quad \text{or} \quad Z.Q = TC$$

If the TC function takes the form $TC = F + \beta Q$, as assumed at the beginning of the chapter

$$Z = \frac{F}{Q} + \beta$$

$1/Q$ is sometimes written as Q^{-1}; in this notation $Z = FQ^{-1} + \beta$. If $F = 40$ and $\beta = 3$, then $Z = 40/Q + 3$. In Table 4.3, values of Z are calculated for values of Q from 0 to 160. We find that no matter how large we make Q, Z

TABLE 4.3

Q	$\dfrac{40}{Q}$	3	$Z = \dfrac{40}{Q} + 3$
0	∞	3	∞
1	40	3	43
5	8	3	11
10	4	3	7
20	2	3	5
40	1	3	4
80	0·5	3	3·5
160	0·25	3	3·25

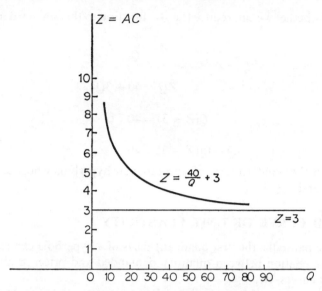

Figure 21   A hyperbolic cost function

will never reach 3. When Q is zero, Z is undefined. The graph of this function, shown in Figure 21, is a hyperbola and $Z = 40/Q + 3$ is a hyperbolic function. $Q = 0$ and $Z = 3$ are the asymptotes of this hyperbola, i.e. when $Q = 0$, Z is undefined and when $Z = 3$, Q is undefined.

$$Z = \frac{40}{Q} + 3$$

When $Q = 0, 40/0 \rightarrow \infty$, therefore Z is undefined.

The inverse function shows that Q is undefined when $Z = 3$.

$$ZQ = 40 + 3Q$$

thus

$$Q(Z - 3) = 40 \quad \text{or} \quad Q = \frac{40}{Z - 3}$$

When $Z = 3$, $Q = 40/0 \rightarrow \infty$, therefore Q is undefined.

A hyperbola always has two branches. In this case only one branch is economically meaningful since the second would lie in the second and third quadrants. This is irrelevant since Q cannot be < 0.

The above function represents a type of hyperbola called an equilateral or rectangular hyperbola. A rectangular hyperbola is one whose asymptotes are perpendicular to each other and parallel to the coordinate axes as above. A hyperbola of this type will take the form,

$$(X - u)(Y - m) = r$$

where u, m, and r are constants. $X = u$ and $Y = m$ are the asymptotes.

see whether we can reduce the AC function to this standard form,

$$Z = \frac{40}{Q} + 3$$

thus

$$ZQ = 40 + 3Q$$

so

$$Q(Z - 3) = 40$$

i.e.

$$(Q - 0)(Z - 3) = 40$$

This then is the standard form of a rectangular hyperbola whose asymptotes are $Q = 0$ and $Z = 3$.

DEMAND CURVE OF UNIT ELASTICITY

Like the parabola, the first quadrant parts of a hyperbola can represent a curvilinear relation between quantity demanded and price, product transformation curves, etc.

A demand curve of unit elasticity throughout its range means that total revenue remains constant while price and quantity vary.

$$TR = P.Q$$

so the demand curve of unit elasticity will take the form

$$P.Q = K$$

where K is a positive constant. This relation will trace out a rectangular hyperbola when plotted. Reducing it to the standard form will give

$$(P - 0)(Q - 0) = K$$

Therefore $Q = 0$ and $P = 0$, i.e. the price and quantity axes, are the asymptotes of a demand curve of unit elasticity.

EXERCISES

1. Suppose the demand function of a firm is given by
$$Q + P - 20 = 0$$
and its costs by $TC - 48 - 4Q = 0$. Find the largest Q it can produce consistent with:
   (i)   breaking even, i.e. profits = 0,
   (ii) making profits of 12,
   (iii) making a loss of 20.

2. Given the market demand function
$$P + Q^2 + Q = 11$$
and the market supply function
$$2P - 2Q^2 + Q - 4 = 0$$
calculate the equilibrium price and quantity in the market.

3. A firm produces amounts X and Y of two different grades of tin, using the same production process. The production transformation curve, showing the maximum output of either good attainable for any given level of output of the other, is

$$X^2 + 2X = 10 - Y$$

(i) What are the largest amounts
   (a) of X,
   (b) of Y

that can be produced?

(ii) What amounts of X and Y should be produced to have
   (a) $0.25\,Y = X$,
   (b) $Y = 2X$?

4. The total costs of a firm are 550 when output is 100 and fixed costs are 50. Assuming that the total cost function is linear;

(i) Graph the average cost function.

(ii) What are the asymptotes of the AC function?

(iii) Find the inverse form of the AC function and give the values for which it is valid.

(iv) If total costs decrease to 450, assuming fixed costs and output remain unchanged, calculate the AC function and say for what values its inverse function is valid.

5. The demand function of a profit maximizing monopolist is

$$Q + 2P = 40$$

and his average cost function is

$$AC = 20Q^{-1} + 4$$

(i) At what output will the monopolist maximize his profits?

(ii) Calculate price elasticity of demand when profits are maximized.

(iii) Suppose the government imposes a lump sum tax, T, on the monopolist, show how this will affect the profit maximizing output when $T = 48$.

6. The demand function of a profit maximizing monopolist is

$$3Q = 15 - P$$

and his average cost function is

$$AC - Q - 6 = 0$$

(i) Calculate the output at which profits will be maximized.

(ii) If a tax of t per unit of quantity produced is imposed on the monopolist, calculate the output at which profits will be maximized when $t = 5$.

SOLUTIONS TO EXERCISES

1. $Q + P - 20 = 0$
   thus

$$P = 20 - Q$$

$$TR = P.Q$$

   thus

$$TR = (20 - Q)Q = 20Q - Q^2$$
$$TC - 48 - 4Q = 0$$

thus
$$TC = 48 + 4Q$$
Profits $= S = TR - TC = 20Q - Q^2 - 48 - 4Q = -Q^2 + 16Q - 48$

(i) When $S = 0$,
$$-Q^2 + 16Q - 48 = 0$$
or
$$Q^2 - 16Q + 48 = 0$$
thus
$$(Q - 4)(Q - 12) = 0$$
$$Q = 4 \text{ or } 12 \qquad\qquad *$$

When $S = 0$, $Q = 12$ gives the largest Q.

Check; when $Q = 12$,
$$TR = 20(12) - (12)^2$$
$$= 96$$
$$TC = 48 + 4(12) = 96$$
thus
$$S = TR - TC = 96 - 96 = 0$$

(ii) When $S = 12$, then
$$-Q^2 + 16Q - 48 = 12$$
or
$$Q^2 - 16Q + 48 = -12$$
so
$$Q^2 - 16Q + 60 = 0$$
thus
$$(Q - 6)(Q - 10) = 0$$
i.e.
$$Q = 6 \text{ or } 10$$

When $S = 12$, then $Q = 10$ gives the largest Q. *

Check; when $Q = 10$,
$$TR = 20(10) - (10)^2$$
$$= 100$$
$$TC = 48 + 4(10) = 88$$
thus
$$S = TR - TC = 100 - 88 = 12$$

(iii) When $S = -20$, then
$$-Q^2 + 16Q - 48 = -20$$
or
$$Q^2 - 16Q + 48 = 20$$
so
$$Q^2 - 16Q + 28 = 0$$
thus
$$(Q - 2)(Q - 14) = 0$$
i.e.
$$Q = 2 \quad \text{or} \quad 14 \qquad\qquad *$$

When $S = -20$, then $Q = 14$ gives the largest Q.

Check; when $Q = 14$,
$$TR = 20(14) - (14)^2$$
$$= 84$$
$$TC = 48 + 4(14) = 104$$

thus
$$S = TR - TC = 84 - 104 = -20$$

2. We want to find positive values for P and Q which will make
$$P + Q^2 + Q = 11$$
and
$$2P - 2Q^2 + Q - 4 = 0$$

This provides us with two simultaneous equations, whose solution will give the coordinates of the points of intersection of the two curves, i.e. equilibrium price and quantity, when the point of intersection is within the first quadrant.
$$P + Q^2 + Q = 11$$
thus
$$P = -Q^2 - Q + 11$$
$$2P - 2Q^2 + Q - 4 = 0$$
thus
$$2P = 2Q^2 - Q + 4$$
or
$$P = Q^2 - 0.5Q + 2$$
so
$$-Q^2 - Q + 11 = Q^2 - 0.5Q + 2$$
i.e.
$$2Q^2 + 0.5Q - 9 = 0$$
or
$$4Q^2 + Q - 18 = 0$$

We have seen that when $aX^2 + bX + c = 0$
$$X = \frac{-b \pm \sqrt{(b^2 - 4ac)}}{2a}$$

In this example $a = 4$, $b = 1$ and $c = -18$.
Using the formula gives
$$Q = \frac{-1 \pm \sqrt{[1 - (4)(4)(-18)]}}{2(4)} = \frac{-1 \pm \sqrt{289}}{8}$$
$$= \frac{-1 \pm 17}{8} = -2.25 \text{ or } 2$$

Q cannot take a negative value so
$$\text{equilibrium } Q = 2 \qquad *$$

When $Q = 2$, then $P + 4 + 2 = 11$, from the demand function, or
$$P = 5 \qquad *$$

Check; when $Q = 2$, $P = 5$, then
$$P + Q^2 + Q = 5 + 4 + 2 = 11$$
as given, and
$$2P - 2Q^2 + Q - 4 = 10 - 8 + 2 - 4 = 0$$
as given.

3. (i) (a) X is as large as possible when $Y = 0$, i.e. when
$$X^2 + 2X = 10 - 0$$
or
$$X^2 + 2X - 10 = 0$$

Using the formula gives

$$X = \frac{-2 \pm \sqrt{(4+40)}}{2} = \frac{-2 \pm \sqrt{44}}{2} = \frac{-2 \pm 2\sqrt{11}}{2}$$

$$= -1 \pm \sqrt{11}$$

X must be positive, thus $-1 + \sqrt{11}$ gives the largest possible amount of X.   *

Check; when $X = -1 + \sqrt{11}$, then

$$(-1 + \sqrt{11})^2 + 2(-1 + \sqrt{11}) = 10 - Y$$

or

$$1 - 2(\sqrt{11}) + 11 - 2 + 2\sqrt{11} = 10 - Y$$

i.e.

$$Y = 0$$

(b) Y is as large as possible when $X = 0$, *i.e.* when

$$0 + 0 = 10 - Y$$

or

$$Y = 10$$ *

Check; when $Y = 10$, then

$$X^2 + 2X = 10 - 10$$

or

$$X^2 + 2X = 0$$
$$X(X + 2) = 0$$

i.e.

$$X = 0 \quad \text{or} \quad X = -2$$

X cannot take a negative value, thus $X = 0$ when $Y = 10$.

(ii) (a) When $0.25\,Y = X$ or $Y = 4X$, then

$$X^2 + 2X = 10 - 4X$$

or

$$X^2 + 6X - 10 = 0$$

Using the quadratic formula gives

$$X = \frac{-6 \pm \sqrt{(36+40)}}{2} = \frac{-6 \pm \sqrt{76}}{2} = \frac{-6 \pm 2\sqrt{19}}{2}$$

$$= -3 \pm \sqrt{19}$$

X must be positive, thus

$$X = -3 + \sqrt{19} \quad \text{when} \quad 0.25\,Y = X$$ *

and

$$Y = 4X = -12 + 4\sqrt{19}$$

Check; when $X = -3 + \sqrt{19}$, then

$$(-3 + \sqrt{19})^2 + 2(-3 + \sqrt{19}) = 10 - Y$$

or

$$9 - 6(\sqrt{19}) + 19 - 6 + 2\sqrt{19} = 10 - Y$$

i.e.

$$12 - 4\sqrt{19} = -Y$$

or

$$Y = -12 + 4\sqrt{19} = 4(-3 + \sqrt{19}) = 4X$$

thus

$$X = 0.25\,Y$$

(b) When $Y = 2X$, then

$$X^2 + 2X = 10 - 2X$$

or

$$X^2 + 4X - 10 = 0$$

Using the quadratic formula gives

$$X = \frac{-4 \pm \sqrt{(16 + 40)}}{2} = \frac{-4 \pm \sqrt{56}}{2} = \frac{-4 \pm 2\sqrt{14}}{2}$$

$$= -2 \pm \sqrt{14}$$

X must be positive, thus $X = -2 + \sqrt{14}$ when $Y = 2X$ *

$$Y = 2X = -4 + 2\sqrt{14}$$ *

Check; when $X = -2 + \sqrt{14}$, then

$$(-2 + \sqrt{14})^2 + 2(-2 + \sqrt{14}) = 10 - Y$$

or

$$4 - 4(\sqrt{14}) + 14 - 4 + 2\sqrt{14} = 10 - Y$$

i.e.

$$4 - 2\sqrt{14} = -Y$$

or

$$Y = -4 + 2\sqrt{14} = 2(-2 + \sqrt{14}) = 2X$$

4. Total costs = total variable costs + fixed costs, i.e.

$$TC = TVC + FC$$

When output = 100, $550 = TVC + 50$, thus

$$TVC = 500$$

Since the TC function is linear then variable costs are 5 per unit, i.e.

$$TVC = 5Q$$

where Q is quantity produced, thus

$$TC = 5Q + 50$$

$$AC = \frac{TC}{Q}$$

thus

$$AC = \frac{5Q + 50}{Q} \text{ or } 5 + \frac{50}{Q}$$

(i) See Figure 22.

(ii) Let $AC = Z$

$$Z = 5 + \frac{50}{Q}$$

thus

$$ZQ = 5Q + 50 \text{ or } Q(Z - 5) = 50$$

We have seen that when a rectangular hyperbola is reduced to the form

$$(X - u)(Y - m) = r$$

this implies that $X = u$ and $Y = m$ are the asymptotes.

$$Q(Z - 5) = 50 \text{ or } (Q - 0)(Z - 5) = 50$$

thus $Q = 0$ and $Z = 5$ are the asymptotes of this hyperbola, see Figure 22. *

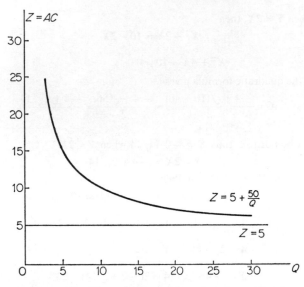

Figure 22   The average cost function of question 4

(iii) $Q(Z - 5) = 50$, thus

$$Q = \frac{50}{Z - 5}$$

This inverse function is valid for $Z > 5$.

(iv) $450 = TVC + 50$

when output $= 100$, thus

$$TVC = 4Q$$

where Q is output, so

$$Z = 4 + \frac{50}{Q}$$

i.e. $ZQ = 4Q + 50$ or $Q(Z - 4) = 50$

$$Q = \frac{50}{Z - 4}$$

The inverse function is valid for $Z > 4$.

5. (i) $Q + 2P = 40$, thus

$$P = 20 - 0\cdot5Q$$
$$TR = P.Q = (20 - 0\cdot5Q)Q = 20Q - 0\cdot5Q^2$$
$$AC = \frac{TC}{Q}$$
$$TC = Q.AC$$
$$= Q(20Q^{-1} + 4) = Q\left(\frac{20}{Q} + 4\right) = 20 + 4Q$$

$$\text{Profit} = S = TR - TC$$
$$S = 20Q - 0{\cdot}5Q^2 - 20 - 4Q$$
$$= 16Q - 0{\cdot}5Q^2 - 20$$
$$= -0{\cdot}5(Q^2 - 32Q) - 20$$
$$= -0{\cdot}5(Q^2 - 32Q + 256) - 20 + 0{\cdot}5(256)$$

Adding -128 and $+128$ leaves the right-hand side unchanged. 256 is obtained by dividing the coefficient of Q by 2 and squaring it.

$$S = -0{\cdot}5(Q - 16)^2 + 108$$

The first term on the right-hand side is a square, i.e. $(Q - 16)^2$, and must always be positive or zero.

$S = 108$ minus the squared term.

S will be at a maximum when the squared term is zero, i.e. when

$$Q - 16 = 0$$

or

$$Q = 16$$

Profits are maximized at a level of 108 when $Q = 16$.   *

(ii) When $Q = 16$,

$$16 + 2P = 40 \text{ or } P = 12$$

from the demand function.

$$E_D = \frac{\Delta Q}{\Delta P} \cdot \frac{P}{Q}$$
$$Q = 40 - 2P$$
$$\frac{\Delta Q}{\Delta P} = -2$$
$$E_D = -2\left(\frac{12}{16}\right) = -1{\cdot}5$$ *

(iii) A lump sum tax will cause total costs to increase by the amount of the tax. This results in a decrease in the constant term of the profit function, i.e.

$$S = -0{\cdot}5(Q - 16)^2 + 108 - 48$$
$$= -0{\cdot}5(Q - 16)^2 + 60$$

Hence the profit maximizing output will remain at 16 but the level of profit will decrease by the amount of the tax.   *

6. (i) $3Q = 15 - P$, thus

$$P = 15 - 3Q$$
$$TR = (15 - 3Q)Q = 15Q - 3Q^2$$
$$AC - Q - 6 = 0$$

thus

$$AC = Q + 6$$

$AC = TC/Q$ so $TC = (Q + 6)Q = Q^2 + 6Q$,
Profits $= S = TR - TC = 15Q - 3Q^2 - Q^2 - 6Q$
$$= -4Q^2 + 9Q$$
$$= -4(Q^2 - \tfrac{9}{4}Q)$$
$$= -4\left(Q^2 - \frac{9}{4}Q + \frac{81}{64}\right) + 4 \cdot \frac{81}{64}$$

Adding $-81/16$ and $+81/16$ leaves the right-hand side unchanged. $81/64$ is obtained by dividing the coefficient of Q by 2 and squaring it. Thus

$$S = -4\left(Q - \frac{9}{8}\right)^2 + \frac{81}{16}$$

The first term on the right is a square and must take a positive or zero value. Thus

$$S = \frac{81}{16} \text{ minus a squared term}$$

S is at a maximum when the squared term is zero, i.e. when

$$Q - \frac{9}{8} = 0 \text{ or } Q = \frac{9}{8} = 1{\cdot}125 \qquad *$$

(ii) If a tax of t per unit of quantity produced is levied, then this affects total costs, i.e.

$$TC = Q^2 + 6Q + tQ$$

When $t = 5$, $TC = Q^2 + 6Q + 5Q = Q^2 + 11Q$, thus

$$
\begin{aligned}
S &= 15Q - 3Q^2 - Q^2 - 11Q \\
&= -4Q^2 + 4Q \\
&= -4(Q^2 - Q) \\
&= -4(Q^2 - Q + \tfrac{1}{4}) + 4.\tfrac{1}{4} \\
&= -4(Q - \tfrac{1}{2})^2 + 1
\end{aligned}
$$

Thus S is at a maximum when

$$Q - \tfrac{1}{2} = 0 \text{ or } Q = 0{\cdot}5$$

$S = 1$, when $Q = 0{\cdot}5$. $\qquad *$

CHAPTER 5

Differentiation

CURVED FUNCTIONS

In Chapter 2 we saw that a change in price ΔP will cause quantity to change by ΔQ and $\Delta Q/\Delta P$ will take a constant value when the demand or supply function is linear. When the functions are curvilinear $\Delta Q/\Delta P$ will vary with the size of the price change ΔP, see Figure 23. A is a point on the supply

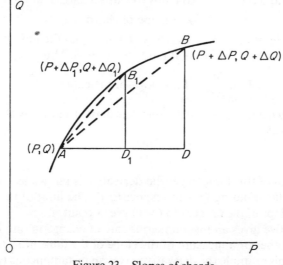

Figure 23   Slopes of chords

function whose coordinates are (P, Q). When price increases by AD, quantity supplied increases by DB and the coordinates of point B are $(P + \Delta P, Q + \Delta Q)$, where $AD = \Delta P$ and $DB = \Delta Q$. AB is a chord, i.e. a straight line drawn between two points on a curve.

$$\Delta Q/\Delta P = \text{slope of the chord } AB$$

However, when price increases by AD_1, quantity supplied increases by $B_1 D_1$ and

$$\Delta Q_1/\Delta P_1 = \text{slope of the chord } AB_1$$

73

where $\Delta P_1 = AD_1$ and $\Delta Q_1 = B_1 D_1$. The chord AB is flatter than AB_1 thus

$$\frac{\Delta Q}{\Delta P} \neq \frac{\Delta Q_1}{\Delta P_1}$$

i.e. $\Delta Q/\Delta P$ does not take a constant value when the function is curvilinear. Consequently our price elasticity formula

$$E_S = \frac{\Delta Q}{\Delta P} \cdot \frac{P}{Q}$$

cannot determine elasticity at a particular price when the supply function is curvilinear. It can only provide a measure of E_S over a range. From Figure 23 we can see that $(\Delta Q/\Delta P) \cdot (P/Q)$ will give a measure of E_S over the price range AD. Economists, however, are usually interested in a much more precise measure of elasticity, namely, elasticity at a particular price.

To calculate E_S at a point A in Figure 23 it is necessary to bring ΔP towards zero as a result of which ΔQ will also tend towards zero, i.e. $\Delta Q \to 0$. As $P + \Delta P$ approaches A, i.e. as $\Delta P \to 0$, in Figure 23, the point B also approaches A and the chord AB tends towards a tangent at point A.

$$\Delta Q/\Delta P = \text{slope of chord } AB$$

Therefore the limit of $\Delta Q/\Delta P$ as $\Delta P \to 0$ must equal the limit of the slope of the chord AB as $\Delta P \to 0$, i.e. as B tends towards A. Thus

$$\operatorname*{Lim}_{\Delta P \to 0} \frac{\Delta Q}{\Delta P} = \operatorname*{Lim}_{\Delta P \to 0} \text{slope of the chord } AB$$

The limit of $\Delta Q/\Delta P$ as $\Delta P \to 0$ is called the derivative of Q with respect to P, and is written as

$$\frac{dQ}{dP}$$

The evaluation of this limit to get the derivative is known as differentiation. We talk of differentiating Q with respect to P. The limit of the slope of the chord is the slope of the tangent to the curve at point A.

The derivative gives an instantaneous rate of change in one variable with respect to another or a measure of the slope of the tangent to the particular function. In this example the derivative gives an instantaneous rate of change in quantity with respect to price or the slope of the tangent to the supply function.

$$\frac{dQ}{dP}$$

is a single symbol and is not a ratio of dQ to dP. It may be written in the form dQ/dP or $f'(P)$.

To find E_S at point A, i.e. at price P, it is necessary to use the formula

$$E_S = \frac{dQ}{dP} \cdot \frac{P}{Q}$$

i.e. P/Q multiplied by the slope of the tangent to the curve at A. At B, dQ/dP will take a different value since the slope of the tangent to the curve changes at every point. The supply function slopes upwards from left to right so that the slope of the tangent is always positive, i.e. dQ/dP is always positive. With a normal demand function dQ/dP will take a negative sign.

When we talk of E_s or E_D at a point we do not assume that ΔP ever reaches zero. If it did then this would give

$$\frac{\Delta Q}{\Delta P} = \frac{0}{0}$$

which is undefined. However, although ΔP never actually reaches zero it is possible to make ΔP so small that for all practical purposes we are operating at a point, and can use the derivative dQ/dP.

MARGINAL REVENUE AND MARGINAL COST

Marginal revenue MR is often defined as the change in total revenue TR per unit change in sales and marginal costs MC as the change in total costs TC per unit change in quantity produced. Figure 24 gives a TR function.

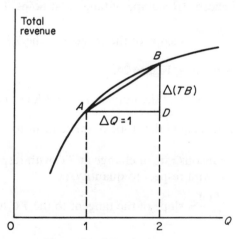

Figure 24   Marginal revenue

Figure 24 shows that as a result of a unit change in quantity sold, TR increases by DB

$$\Delta Q = AD = 1 \quad \text{and} \quad \Delta(TR) = DB$$

As $\Delta Q = 1$,

$$MR = \frac{\Delta(TR)}{\Delta Q} = \Delta(TR) = \frac{DB}{AD} = \text{slope of chord } AB$$

Clearly this measure of MR is unsatisfactory since the value of $\Delta(TR)/\Delta Q$ or the slope of the chord depends upon the size of the interval ΔQ. MR is treated as an average in that it is the change in TR divided by the change in Q. There is no obvious correct unit on which the slope will provide an exact measure of MR. The correct measure is found by taking the limit of $\Delta(TR)/\Delta Q$ as $\Delta Q \to 0$, i.e. by differentiating TR with respect to quantity. MR is thus the instantaneous rate of change in TR with respect to quantity sold or the slope of the tangent to the TR function at a particular point, i.e.

$$MR \neq \frac{\Delta(TR)}{\Delta Q}$$

but

$$MR = \lim_{\Delta Q \to 0} \frac{\Delta(TR)}{\Delta Q}$$

Figure 24 shows that $\Delta(TR)/\Delta Q$ = slope of chord AB

$$\lim_{\Delta Q \to 0} \frac{\Delta(TR)}{\Delta Q} = \lim_{\Delta Q \to 0} \text{slope of chord } AB$$

$$\lim_{\Delta Q \to 0} \frac{\Delta(TR)}{\Delta Q} = \frac{d(TR)}{dQ}$$

and $\lim_{\Delta Q \to 0}$ slope of chord AB = slope of tangent at point A. Thus

$$MR = \frac{d(TR)}{dQ} = \text{slope of the tangent to the } TR \text{ curve}$$

When $Q = 1$, i.e. at point A, Figure 24,

$$MR = \frac{d(TR)}{dQ} = \text{slope of the tangent to the } TR \text{ curve at point } A$$

When $Q = 20$, MR will equal the slope of the tangent to the TR function at this point.

MC is the instantaneous rate of change in TC with respect to quantity or the derivative of TC with respect to quantity, i.e.

$$MC = \frac{d(TC)}{dQ} = \text{slope of the tangent to the } TC \text{ function}$$

DIFFERENTIATION RULES

Throughout the previous pages the derivative has been treated from a geometrical point of view and no attempt has been made to obtain a value for the derivative.

Let us now take the curvilinear function $Q = P^2$ and calculate dQ/dP or the slope of the tangent to this function. Let ΔQ be the small change in Q which results from a small change ΔP in P. Thus

$$Q + \Delta Q = (P + \Delta P)^2 = P^2 + 2P(\Delta P) + (\Delta P)^2$$
$$Q = P^2$$

thus

$$\Delta Q = 2P(\Delta P) + (\Delta P)^2$$

Dividing across by ΔP gives

$$\frac{\Delta Q}{\Delta P} = \frac{2P(\Delta P)}{\Delta P} + \frac{(\Delta P)^2}{\Delta P}$$
$$= 2P + \Delta P$$

$$\operatorname*{Lim}_{\Delta P \to 0} \frac{\Delta Q}{\Delta P} = \operatorname*{Lim}_{\Delta P \to 0} 2P + \Delta P$$

thus

$$dQ/dP = 2P$$

since $2P + \Delta P \to 2P$ as $\Delta P \to 0$. dQ/dP gives the slope of the tangent to the function $Q = P^2$.

When $P = 2$

$dQ/dP = 2P = 2(2) = 4 =$ slope of tangent to the function at point $P = 2$

When $P = 6$

$dQ/dP = 2P = 2(6) = 12 =$ slope of tangent to the function at point $P = 6$

The above method for finding the derivative can be extended. Applying it to the function $Q = P^n$ will give

$$dQ/dP = n . P^{n-1}$$

(see Appendix 1). This general formula holds provided the exponent n is a real number, i.e. n can be a positive or negative integer, or a positive or negative fraction. If $Y = X^{-1}$, the formula gives

$$\frac{dY}{dX} = (-1)X^{-2} = \frac{-1}{X^2}$$

The same result can be obtained directly. If ΔY is the small change in Y which results from a small change ΔX in X

$$Y + \Delta Y = (X + \Delta X)^{-1} = \frac{1}{X + \Delta X}$$

$$Y = X^{-1} = \frac{1}{X}$$

so

thus

$$\Delta Y = \frac{1}{X + \Delta X} - \frac{1}{X} = \frac{X - X - \Delta X}{X(X + \Delta X)} = \frac{-\Delta X}{X(X + \Delta X)}$$

and

$$\frac{\Delta Y}{\Delta X} = \frac{-1}{X(X + \Delta X)}$$

$$\frac{dY}{dX} = \operatorname*{Lim}_{\Delta X \to 0} \frac{\Delta Y}{\Delta X} = \frac{-1}{X^2}$$

as above.

Constants Rule

The derivative of any constant multiple of a function, e.g. $Q = KP^2$, where K is a constant, instead of $Q = P^2$, will be the constant times the derivative of the original function.

Taking $Q = KP^2$ and assuming ΔQ is the small change in Q which results from a small change ΔP in P gives

$$Q + \Delta Q = K(P + \Delta P)^2 = KP^2 + K.2P(\Delta P) + K(\Delta P)^2$$
$$Q = KP^2$$

so

$$\Delta Q = K.2P(\Delta P) + K(\Delta P)^2$$

Dividing across by ΔP gives

$$\frac{\Delta Q}{\Delta P} = \frac{K.2P(\Delta P)}{\Delta P} + \frac{K(\Delta P)^2}{\Delta P}$$
$$= K.2P + K(\Delta P)$$
$$\underset{\Delta P \to 0}{\text{Lim}} \frac{\Delta Q}{\Delta P} = \underset{\Delta P \to 0}{\text{Lim}} K.2P + K(\Delta P)$$

Thus $dQ/dP = K.2P$, i.e. K times the derivative of the original function. If $K = 3$, i.e. $Q = 3P^2$,

$$dQ/dP = 3(2P) = 6P$$

Addition Rule

Suppose a demand function takes the form

$$Q = K - aP - bP^2$$

where K, a, and b are positive constants, and we want to find dQ/dP in order to calculate E_D at a particular price. To find dQ/dP let ΔQ be the small change in Q which results from a small change ΔP in P.

$$Q + \Delta Q = K - a(P + \Delta P) - b(P + \Delta P)^2$$
$$= K - aP - a.\Delta P - bP^2 - b.2P(\Delta P) - b(\Delta P)^2$$
$$Q = K - aP - bP^2$$

thus

$$\Delta Q = -a.\Delta P - b.2P(\Delta P) - b(\Delta P)^2$$

so

$$\Delta Q/\Delta P = -a - b.2P - b.\Delta P$$

and

$$dQ/dP = -a - b.2P$$

since $b.\Delta P \to 0$ as $\Delta P \to 0$. We can see that the derivative of the function is the sum of the derivatives of the parts calculated separately, i.e.

$$dK/dP = 0$$

since the derivative of any constant is zero. In this case K is fixed whatever the size or change in P, i.e. $\Delta K = 0$, therefore

$$\Delta K/\Delta P \quad \text{or} \quad dK/dP = 0$$

$$\frac{d(-aP)}{dP} = -a$$

$$\frac{d(-bP^2)}{dP} = -b.2P$$

SOME FURTHER APPLICATIONS OF THE DERIVATIVE

Isocost Functions

Suppose a producer has total costs of 100 and purchases labour, L, and capital, K, at prices $P_L = 4$ and $P_K = 5$. The isocost function gives all the combinations of factor inputs available at any given total cost, and is defined by

$$4L + 5K = 100$$

thus

$$L = 25 - 1 \cdot 25K$$

and

$$dL/dK = -1 \cdot 25$$

The derivative of a linear function is a constant. Note that $dL/dK = \Delta L/\Delta K$ when the function is linear, i.e. the size of the increments which are compared does not matter.

Average Costs

Since the derivative is a measure of the slope of the tangent to a curve, then its sign will tell us which way the relation slopes. $d(TR)/dQ$, $d(TC)/dQ$ and $d(MC)/dQ$ will provide information on whether the TR, TC, or MC functions are decreasing or increasing functions of quantity. If an AC function is taken the derivative of AC with respect to Q will give the slope of the tangent to the AC function. If

$$AC = \frac{K}{Q} + b$$

where K and b are positive constants, then

$$\frac{d(AC)}{dQ} = \frac{d(K/Q)}{dQ} + \frac{d(b)}{dQ}$$

$$\frac{K}{Q} = KQ^{-1}$$

so

$$\frac{d(KQ^{-1})}{dQ} = -KQ^{-2} = \frac{-K}{Q^2}$$

and

$$\frac{d(AC)}{dQ} = \frac{-K}{Q^2}$$

since the derivative of the constant b is zero.

Consequently the slope of the tangent to the function will vary as Q changes but it will take a negative sign for all $Q > 0$. Therefore this AC function must slope downwards from left to right for all values of $Q > 0$.

U-shaped Average Cost Curves

If

$$AC = \frac{K}{Q} + a + bQ$$

where K, a, and b are positive constants, then

$$\frac{d(AC)}{dQ} = \frac{-K}{Q^2} + b$$

The slope of the tangent to this AC function is negative when $K/Q^2 > b$ and positive when $K/Q^2 < b$. This means that for certain values of Q, i.e.

$$Q < \sqrt{(K/b)}$$

the AC function slopes downwards and for values of Q for which

$$Q > \sqrt{(K/b)}$$

the AC function is upward sloping. This AC function is therefore U-shaped.

Indifference Curves and Isoquants

Suppose a consumer purchases two goods X and Y and that his utility function takes the form

$$U = X^\alpha Y^\beta$$

where U is utility and α and β are positive constants.

The meaning of X^α where α is not an integer is as follows. Suppose that $\alpha = \gamma/\delta$ where γ and δ are integers. If $Z = X^{1/3}$ then $Z^3 = X$, i.e. Z is the cube root of X. If $Z = X^{1/\delta}$ then $Z^\delta = X$, and Z is the δth root of X, i.e. the number which gives X when multiplied by itself δ times.

If $W = X^{3/4}$ then $W^4 = X^3$, i.e. W is the fourth root of X^3. If $W = X^{\gamma/\delta}$ then $W^\delta = X^\gamma$, i.e. W is the δth root of X^γ.

It is commonly assumed that utility will rise less than in proportion to an all-round rise in consumption. Thus we require that $\alpha + \beta < 1$, and as $\alpha > 0$ and $\beta > 0$, we have $\alpha < 1$ and $\beta < 1$, i.e. α and β are proper fractions.

Indifference curves will give all the combinations of X and Y which will yield various given levels of utility, e.g. Z. Any one indifference curve will take the form

$$X^\alpha Y^\beta = Z$$

To determine which way this relation slopes we need to find dY/dX, or dX/dY.

$$Y^\beta = \frac{Z}{X^\alpha} = ZX^{-\alpha}$$

so

$$Y = Z^{1/\beta} X^{-\alpha/\beta}$$

thus

$$dY/dX = -\frac{\alpha}{\beta} Z^{1/\beta} X^{-(\alpha/\beta)-1} = -\frac{\alpha}{\beta} \frac{Z^{1/\beta}}{X^{(\alpha+\beta)/\beta}} < 0$$

Consequently this indifference curve slopes downwards throughout its length. The reader can check by a similar procedure that dX/dY is also negative.

If we have output instead of utility and factor inputs of L and K instead of goods, then the relation

$$L^\alpha K^\beta = Z$$

will give us an isoquant, which is the set of combinations of L and K which can be used to produce any given level of output. Again dL/dK or dK/dL can be found, to give the negative slope of the isoquant.

Inverse Function Rule

Demand and supply functions are often expressed in the form $P = f(Q)$. To find a value for dQ/dP one can express $Q = f(P)$ and differentiate. This is relatively simple when the function is linear, but it can be cumbersome to express the inverse form of a function of degree 2 or higher. To cope with this the inverse function rule states that since

$$\frac{\Delta P}{\Delta Q} = \frac{1}{\Delta Q/\Delta P}$$

then the limit

$$\frac{dP}{dQ} = \frac{1}{dQ/dP} \quad \text{or} \quad \frac{dQ}{dP} = \frac{1}{dP/dQ}$$

Suppose the demand function takes the form $P = K - aQ - bQ^2$, where K, a, and b are positive constants and we wish to calculate dQ/dP then this rule states that

$$\frac{dQ}{dP} = \frac{1}{dP/dQ}$$

and since

$$\frac{dP}{dQ} = -a - 2bQ$$

then
$$\frac{dQ}{dP} = \frac{1}{-a - 2bQ}$$

Function of a Function Rule

If Y is a function of a variable U and U itself is a function of some other variable, say X, then a small change ΔX in X will cause U and Y to change by ΔU and ΔY respectively.

$$\frac{\Delta Y}{\Delta X} = \frac{\Delta Y}{\Delta U} \cdot \frac{\Delta U}{\Delta X} \tag{1}$$

The ΔU's cancel out and we are left with

$$\frac{\Delta Y}{\Delta X} = \frac{\Delta Y}{\Delta X}$$

Therefore the limit of the left-hand side of equation (1) is equal to the limit of the right-hand side as $\Delta X \rightarrow 0$, i.e.

$$\frac{dY}{dX} = \frac{dY}{dU} \cdot \frac{dU}{dX}$$

Suppose a firm's TR function is $TR = 9 - (Q - 3)^2$. If we let $Q - 3 = U$ then $TR = 9 - U^2$. TR and U are both functions of Q. Using the function of a function rule gives

$$\frac{d(TR)}{dQ} = \frac{d(TR)}{dU} \cdot \frac{dU}{dQ} = -2U(1) = -2U$$

$$= -2(Q - 3)$$

because

$$d(TR)/dU = -2U \quad \text{and} \quad dU/dQ = 1$$

If the firm's TR function takes the form $TR = 9 - (Q - 3)^4$ and we let $Q - 3 = U$, then

$$TR = 9 - U^4$$

and

$$d(TR)/dQ = -4U^3(1) = -4U^3 = -4(Q - 3)^3$$

because

$$d(TR)/dU = -4U^3 \quad \text{and} \quad dU/dQ = 1$$

If either derivative is zero and the other is finite or if both are zero, then their product is zero. If either tends to infinity and the other is finite, or if both tend to infinity, then their product is infinite. However, if one tends to infinity and the other equals zero then the product must be investigated further.

Product Rule

If Y is a function of two variables U and V so that $Y = U.V$, where U and V are functions of another variable X, then a small change ΔX in X will cause

U, V, and Y to change by ΔU, ΔV, and ΔY respectively. We have

$$Y + \Delta Y = (U + \Delta U)(V + \Delta V) = U.V + U.\Delta V + V.\Delta U + \Delta U.\Delta V$$
$$Y = U.V$$

thus

$$\Delta Y = U.\Delta V + V.\Delta U + \Delta U.\Delta V$$

Dividing across by ΔX gives

$$\frac{\Delta Y}{\Delta X} = U.\frac{\Delta V}{\Delta X} + V.\frac{\Delta U}{\Delta X} + \Delta U.\frac{\Delta V}{\Delta X}$$

The limit of the left-hand side is equal to the limit of the right-hand side as $\Delta X \to 0$, i.e.

$$\frac{dY}{dX} = U.\frac{dV}{dX} + V.\frac{dU}{dX}$$

since $\Delta U.\Delta V/\Delta X \to 0$ as ΔX and $\Delta U \to 0$.

Suppose a firm's demand function is $P = 12 - 2Q$, then $TR = (12 - 2Q)Q$. If we let $12 - 2Q = U$ and $Q = V$ then $TR = U.V$, where U and V are functions of Q. Using the product rule gives

$$\frac{d(TR)}{dQ} = U.\frac{dV}{dQ} + V.\frac{dU}{dQ}$$
$$= (12 - 2Q)(1) + Q(-2) = 12 - 4Q$$

As a check on the product rule one can reduce the TR function to $TR = 12Q - 2Q^2$ and differentiate in the ordinary way, i.e.

$$d(TR)/dQ = 12 - 4Q$$

THE RELATION BETWEEN AC AND MC

$$TC = Q.AC$$
$$MC = d(TC)/dQ = d(Q.AC)/dQ$$

If we assume $Q = U$ and $AC = V$, since each is a function of Q the product rule gives

$$\frac{d(Q.AC)}{dQ} = Q.\frac{d(AC)}{dQ} + AC.\frac{dQ}{dQ}$$

thus

$$MC = Q.\frac{d(AC)}{dQ} + AC$$

since $dQ/dQ = 1$. Therefore the relation between MC and AC depends upon the sign of $d(AC)/dQ$, assuming Q is positive.

When $d(AC)/dQ > 0$, $MC > AC$, see interval XB, Figure 25.
When $d(AC)/dQ < 0$, $MC < AC$, see interval AX, Figure 25.
When $d(AC)/dQ = 0$, $MC = AC$, point X, Figure 25.

In Chapter 6 we will see that AC is at a minimum when $d(AC)/dQ = 0$. Therefore $MC = AC$ at the minimum point of the AC function, see Figure 25.

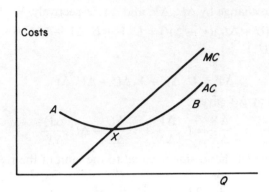

Figure 25   Average and marginal cost

Rule for Finding the Derivative of a Ratio of Two Functions

If Y is a function of two variables U and V so that $Y = U/V$, where U and V are functions of another variable X, then a small change ΔX in X will cause U, V, and Y to change by ΔU, ΔV and ΔY respectively. We have

$$Y + \Delta Y = \frac{U + \Delta U}{V + \Delta V}$$

and

$$Y = \frac{U}{V}$$

thus

$$\Delta Y = \frac{U + \Delta U}{V + \Delta V} - \frac{U}{V} = \frac{U.V + V.\Delta U - U.V - U.\Delta V}{V(V + \Delta V)}$$

$$= \frac{V.\Delta U - U.\Delta V}{V(V + \Delta V)}$$

Dividing across by ΔX gives

$$\frac{\Delta Y}{\Delta X} = \frac{V.\Delta U/\Delta X - U.\Delta V/\Delta X}{V(V + \Delta V)}$$

In the limit

$$\frac{dY}{dX} = \frac{V.dU/dX - U.dV/dX}{V^2}$$

since $V.\Delta V \to 0$ as ΔX and $\Delta V \to 0$.

If a firm's AC function takes the form

$$AC = \frac{40}{Q} + 3 \quad \text{or} \quad AC = \frac{40 + 3Q}{Q}$$

and we let $40 + 3Q = U$ and $Q = V$, then use of the above rule gives

$$\frac{d(AC)}{dQ} = \frac{Q(3) - (40 + 3Q)(1)}{Q^2} = -\frac{40}{Q^2}$$

because

$$dU/dQ = 3 \quad \text{and} \quad dV/dQ = 1$$

Clearly this AC function is downward sloping for all values of $Q > 0$.

Continuity Assumption

We have assumed that all functions dealt with in this chapter are continuous. This implies that variables such as price vary continuously and can take any value, e.g. 10·032 etc. Even though most economic variables are likely to be discontinuous functions, this assumption is essential if one wishes to use the derivative. Since dY/dX is the limit of $\Delta Y/\Delta X$ as $\Delta X \to 0$, this implies that for every possible value of X there is a corresponding value for Y. If the function is continuous and smooth the derivative is a reliable measure of the rate of change. However, if at some points a function is discontinuous, or has sharp bends, the derivative is not defined at such points.

EXERCISES

1. (i) When the demand function is $2Q - 24 + 3P = 0$, find the marginal revenue when $Q = 3$.

(ii) Given the demand function $0·1Q - 10 + 0·2P + 0·02P^2 = 0$, calculate the price elasticity of demand when $P = 10$.

(iii) If supply is related to price by the function $P = 0·25Q + 10$, find the price elasticity of supply when $P = 20$.

(iv) Given the demand function $aQ + bP - k = 0$, where a, b, and k are positive constants, show that price elasticity of demand is minus one when $MR = 0$.

(v) When the demand function is $P = 20/(4 + Q)$, calculate price elasticity of demand when $P = 4$.

2. (i) A firm's total costs are 500 when output is 100. If the TC function is linear and fixed costs (FC) are 200, find the marginal cost when $Q = 4$, 5, and 6.

(ii) The following are estimates of TC and AC functions for various firms. Calculate the MC function in each case and say whether, or under what conditions, the MC function is economically meaningful.

(a) $AC = \dfrac{20}{Q} + 3 + 0·5Q$

(b) $AC - 2 = \dfrac{100}{Q} + 0·2Q$

(c) $TC - 100 - 2Q + 2Q^2 = Q^3$

(d) $Q \cdot AC = a + bQ - cQ^2 + gQ^3$

where a, b, c, and g are positive constants.

Give the MC when $Q = 4$ for (a), (b), and (c).

3. The following are AC and TC functions for various firms

(i) $AC = \dfrac{140}{Q} + 20$

(ii) $AC - \dfrac{a}{Q} = k$

(iii) $TC - 10 = 2Q + 0 \cdot 1Q^2$

(iv) $TC - k - \beta Q = cQ^2$

where a, k, β, and c are positive constants.

   (a) Find the expression for the gradients of the AC functions.

   (b) For what values of Q will AC be decreasing?

   (c) Which functions will give U-shaped AC curves?

4. The indifference curve of a consumer takes the form

$$K = X^{0 \cdot 4} Y^{0 \cdot 6}$$

where K represents a constant level of utility and X and Y are two goods. If $K = 9$, $P_X = 2$, and $P_Y = 3$:

(i) Show that this indifference curve slopes downwards from left to right.

(ii) Assuming the consumer wishes to achieve the level of satisfaction given by this indifference curve, calculate the minimum level of income which is necessary to achieve this and express the budget line in algebraic form.

Hint: what is the position in relation to the indifference curve of the budget line which represents the minimum income requirement?

5. A firm's isoquant curve is given by

$$10 = L^{3/8} K^{5/8}$$

where 10 is the level of output along the isoquant. L and K stand for the inputs labour and capital respectively. Assuming that $P_L = 3$ and $P_K = 5$, calculate the minimum level of costs which is necessary to produce this level of output and express the isocost function in algebraic form.

Hint: what is the position relative to the isoquant curve of the lowest attainable isocost function?

SOLUTIONS TO EXERCISES

1. (i) $2Q - 24 + 3P = 0$

thus

$$P = 8 - \tfrac{2}{3}Q$$
$$TR = P.Q = (8 - \tfrac{2}{3}Q)Q = 8Q - \tfrac{2}{3}Q^2$$
$$MR = d(TR)/dQ = 8 - 2(\tfrac{2}{3})Q = 8 - \tfrac{4}{3}Q \qquad *$$

When $Q = 3$, then $MR = 8 - (\tfrac{4}{3})3 = 4$.

(ii) $0 \cdot 1Q - 10 + 0 \cdot 2P + 0 \cdot 02P^2 = 0$

thus

$$Q = 100 - 2P - 0 \cdot 2P^2$$
$$E_D = \frac{dQ}{dP} \cdot \frac{P}{Q}$$
$$dQ/dP = -2 - 0 \cdot 4P$$

When $P = 10$, then $dQ/dP = -2 - 4 = -6$.

When $P = 10$, $Q = 100 - 2(10) - 0.2(100) = 60$, thus

$$E_D = -6\left(\frac{10}{60}\right) = -1$$

*

(iii) $P = 0.25Q + 10$, thus

$$Q = 4P - 40$$

$$E_S = \frac{dQ}{dP} \cdot \frac{P}{Q}$$

$$dQ/dP = 4$$

When $P = 20$, $Q = 80 - 40 = 40$, thus

$$E_S = 4\left(\frac{20}{40}\right) = 2$$

*

(iv) $aQ + bP - k = 0$
thus

$$P = \frac{k}{b} - \frac{a}{b}Q$$

so

$$TR = \left(\frac{k}{b} - \frac{a}{b}Q\right)Q = \frac{k}{b}Q - \frac{a}{b}Q^2$$

$$MR = d(TR)/dQ = \frac{k}{b} - \frac{2a}{b}Q$$

When $MR = 0$, then

$$\frac{k}{b} - \frac{2a}{b}Q = 0$$

i.e.

$$\frac{2a}{b}Q = \frac{k}{b} \text{ or } Q = \frac{k}{2a}$$

When $Q = k/2a$, then

$$P = \frac{k}{b} - \left(\frac{a}{b}\right)\left(\frac{k}{2a}\right)$$

$$= \frac{k}{b}\left(1 - \frac{a}{2a}\right) = \frac{k}{2b}$$

$$E_D = \frac{dQ}{dP} \cdot \frac{P}{Q}$$

$$\frac{dQ}{dP} = \frac{1}{dP/dQ} = \frac{1}{-a/b} = -b/a$$

thus

$$E_D = -\frac{b}{a} \cdot \frac{k}{2b} \cdot \frac{2a}{k} = -1$$

*

(v) $P = \dfrac{20}{4 + Q}$

When $Y = U/V$, where U and V are functions of X, then

$$\frac{dY}{dX} = \frac{V.dU/dX - U.dV/dX}{V^2}$$

$$P = \frac{20}{4 + Q}$$

thus

$$\frac{dP}{dQ} = \frac{(4+Q)(0) - 20(1)}{(4+Q)^2} = -\frac{20}{(4+Q)^2}$$

$$\frac{dQ}{dP} = \frac{1}{dP/dQ} = -\frac{(4+Q)^2}{20}$$

When $P = 4$, then $4 = 20/(4+Q)$, i.e. $Q = 1$.

$$E_D = \frac{dQ}{dP} \cdot \frac{P}{Q}$$

$$dQ/dP = -\frac{(5)^2}{20} = -1.25$$

thus

$$E_D = -1.25\left(\frac{4}{1}\right) = -5 \qquad *$$

2. (i) $TC = FC + TVC$

$$FC = 200$$

When output is 100, $TVC = 300$, thus $TVC = 3Q$, where Q is quantity produced and

$$TC = 200 + 3Q$$
$$MC = d(TC)/dQ = 3$$

This cost function gives constant MC *

(ii) (a) $AC = \dfrac{20}{Q} + 3 + 0.5Q$

thus

$$TC = 20 + 3Q + 0.5Q^2$$
$$MC = d(TC)/dQ = 3 + Q \qquad *$$

This MC function is economically plausible since MC is always positive and increases as Q increases.

When $Q = 4$, $MC = 3 + 4 = 7$. *

(b) $AC - 2 = \dfrac{100}{Q} + 0.2Q$

thus

$$AC = 2 + \frac{100}{Q} + 0.2Q$$

or

$$TC = 2Q + 100 + 0.2Q^2$$
$$MC = d(TC)/dQ = 2 + 0.4Q$$

This MC function is economically plausible, giving an upward sloping function.

When $Q = 4$, $MC = 2 + 0.4(4) = 3.6$. *

(c) $TC - 100 - 2Q + 2Q^2 = Q^3$

thus

$$TC = 100 + 2Q - 2Q^2 + Q^3$$
$$MC = d(TC)/dQ = 2 - 4Q + 3Q^2$$

To give economically meaningful results MC must be greater than zero. Does $2 - 4Q + 3Q^2$ provide us with a positive MC?

$$2 - 4Q + 3Q^2 = 2 - 4Q + 2Q^2 + Q^2$$
$$= 2(1 - 2Q + Q^2) + Q^2$$
$$= 2(1 - Q)^2 + Q^2$$

Q will either be positive or zero. Therefore $(1 - Q)^2$ is always positive or zero. Consequently MC is always positive.

When $Q = 4$, $MC = 2 - 4(4) + 3(16) = 34.$ *

(d) $Q.AC = a + bQ - cQ^2 + gQ^3$

Since $TC = Q.AC$, so $TC = a + bQ - cQ^2 + gQ^3$

$$MC = d(TC)/dQ = b - 2cQ + 3gQ^2 \qquad *$$

To be economically plausible MC must be greater than zero for all values of Q. For $MC > 0$ the quadratic $b - 2cQ + 3gQ^2 = 0$ must have no real roots,

i.e.

$$(-2c)^2 \text{ must be } < 4(3g)(b)$$

or

$$4c^2 < 12bg$$

i.e.

$$c^2 < 3bg \qquad *$$

3. (i) (a) $AC = \dfrac{140}{Q} + 20 = 140Q^{-1} + 20$

thus

$$d(AC)/dQ = -140Q^{-2} = -\dfrac{140}{Q^2} \qquad *$$

(b) $\dfrac{d(AC)}{dQ} = -\dfrac{140}{Q^2} < 0$ for all values of Q

Therefore the AC is downward sloping throughout its length.   *

(ii) (a) $AC - \dfrac{a}{Q} = k$

or

$$AC = k + \dfrac{a}{Q} = k + aQ^{-1}$$

thus

$$d(AC)/dQ = -aQ^{-2} = -\dfrac{a}{Q^2} \qquad *$$

(b) $d(AC)/dQ = -\dfrac{a}{Q^2} < 0$ for all values of Q

Therefore the AC function is decreasing for all values of Q.

(iii) (a) $TC - 10 = 2Q + 0 \cdot 1Q^2$

i.e.

$$TC = 10 + 2Q + 0 \cdot 1Q^2$$

thus

$$AC = \dfrac{10}{Q} + 2 + 0 \cdot 1Q$$

$$d(AC)/dQ = -\dfrac{10}{Q^2} + 0 \cdot 1 \qquad *$$

(b) $d(AC)/dQ < 0$ when

$$\frac{10}{Q^2} > 0.1$$

or

$$10 > 0.1Q^2$$

i.e.

$$Q < 10$$

Thus the AC function decreases until quantity produced rises to 10.   *

(c) $d(AC)/dQ > 0$ when

$$\frac{10}{Q^2} < 0.1$$

i.e., when

$$Q > 10$$

When $Q < 10$ average costs will fall, i.e. $d(AC)/dQ < 0$.

When $Q > 10$ average costs will rise, i.e. $d(AC)/dQ > 0$.

Therefore $TC - 10 = 2Q + 0.1Q^2$ will give a U-shaped average cost curve.

(iv) (a) $TC - k - \beta Q = cQ^2$

or

$$TC = k + \beta Q + cQ^2$$

thus

$$AC = \frac{k}{Q} + \beta + cQ$$

$$d(AC)/dQ = -\frac{k}{Q^2} + c$$ *

(b) $d(AC)/dQ < 0$ when

$$\frac{k}{Q^2} > c$$

or

$$k > cQ^2$$

i.e. when $\sqrt{(k/c)} > Q$.

Thus the AC function decreases until quantity produced rises to $\sqrt{(k/c)}$.   *

(c) $d(AC)/dQ > 0$ when

$$\frac{k}{Q^2} < c$$

or

$$\sqrt{(k/c)} < Q$$

When $Q < \sqrt{(k/c)}$ average costs will fall, i.e. $d(AC)/dQ < 0$.

When $Q > \sqrt{(k/c)}$ average costs will rise, i.e. $d(AC)/dQ > 0$.

Therefore $TC = k + \beta Q + cQ^2$ gives a U-shaped AC function.

4. (i) $9 = X^{0.4} Y^{0.6}$

so

$$X^{0.4} = \frac{9}{Y^{0.6}} = 9 Y^{-0.6}$$

and

$$X = 9^{1/0.4} Y^{-0.6/0.4} = 9^{2.5} Y^{-1.5}$$

thus

$$dX/dY = 9^{2.5}(-1.5) Y^{-2.5}$$

dX/dY gives the slope of the indifference curve, and is always negative since $Y^{-2\cdot5}$ is always positive; thus the indifference curve slopes downwards throughout its length.

(ii) Income will be at a minimum when the consumer's budget line is tangential to this indifference curve. Since $P_X = 2$ and $P_Y = 3$ the budget line will take the form $2X + 3Y = M$, where M stands for income.

When this budget line is tangential to the indifference curve, the slope of the tangent to the indifference curve at this point, i.e. the marginal rate of commodity substitution, $MRCS$, will equal the slope of the budget line.

Part (i) gives the slope of the tangent to the indifference curve or the $MRCS$, i.e.

$$dX/dY = 9^{2\cdot5}(-1\cdot5)Y^{-2\cdot5}$$

Since $2X + 3Y = M$, $X = -1\cdot5Y + 0\cdot5M$, and along the budget line $dX/dY = -1\cdot5$. For maximum consumer satisfaction the indifference curve and budget line have the same slope, hence

$$-1\cdot5 = 9^{2\cdot5}(-1\cdot5)Y^{-2\cdot5}$$

i.e.

$$1 = 9^{2\cdot5}Y^{-2\cdot5}$$

so

$$Y^{2\cdot5} = 9^{2\cdot5} \quad \text{and} \quad Y = 9$$

When $Y = 9$ substitution gives

$$9 = X^{0\cdot4}9^{0\cdot6}$$

so

$$X^{0\cdot4} = 9^{0\cdot4} \quad \text{and} \quad X = 9$$

When $Y = X = 9$ then $2X + 3Y = 2(9) + 3(9) = 45$. Therefore the minimum level of income required is 45   *
and the algebraic form of the budget line is

$$2X + 3Y = 45 \qquad\qquad *$$

5. $10 = L^{3/8}K^{5/8}$

thus

$$L^{3/8} = \frac{10}{K^{5/8}} = 10K^{-5/8}$$

or

$$L = 10^{8/3}K^{(-5/8)(8/3)} = 10^{8/3}K^{-5/3}$$

thus

$$dL/dK = 10^{8/3}\left(-\frac{5}{3}\right)K^{-8/3}$$

dL/dK gives the slope of the isoquant or a measure of the marginal rate of technical substitution.

Costs will be at a minimum when the firm's isocost function is tangential to this isoquant. Since $P_L = 3$ and $P_K = 5$, the isocost function will take the form $3L + 5K = N$, where N stands for costs. Along the isocost line dL/dK is thus $-5/3$.

Costs are at a minimum when the slope of the tangent to the isoquant is equal to the slope of the isocost function, i.e. when

$$10^{8/3}\left(-\frac{5}{3}\right)K^{-8/3} = -\frac{5}{3}$$

or
$$K = 10$$

When $K = 10$, $10 = L^{3/8}10^{5/8}$, or $L = 10$.

When $K = L = 10$, then $3L + 5K = 3(10) + 5(10) = 80$.

Therefore the minimum level of costs is 80 and the algebraic form of the isocost function is $3L + 5K = 80$.   *

CHAPTER 6

Maxima and Minima

THE SECOND DERIVATIVE

The derivative of TR with respect to quantity gives MR. When $P = K - aQ$, where K and a are positive constants, then

$$TR = KQ - aQ^2$$

and

$$MR = d(TR)/dQ = K - 2aQ$$

MR is itself a linear function. Since MR varies with Q it is possible to talk of a rate of change in MR or the derivative of MR with respect to Q, i.e. $d(MR)/dQ$. This will give the slope of the tangent to the MR function for all values of Q. Using the above example, $d(MR)/dQ = -2a$, i.e. the slope of the tangent to the MR function is constant since MR is linear.

Here, $d(MR)/dQ$ is the derivative of $d(TR)/dQ$, i.e.

$$\frac{d(MR)}{dQ} = \frac{d(d(TR)/dQ)}{dQ} = \frac{d^2(TR)}{dQ^2}$$

This is the second derivative of TR with respect to Q. The second derivative can be written

$$d^2(TR)/dQ^2 \quad \text{or} \quad f''(Q)$$

Like $d(TR)/dQ$ it is a single symbol and not a ratio of two distinct quantities. The exact same rules apply to the second derivative. Higher derivatives are defined in the same way, e.g.

$$d^3(TR)/dQ^3 = \frac{d}{dQ}\left(\frac{d^2(TR)}{dQ^2}\right)$$

Taking a TC function of the form $TC = K + aQ - bQ^2 + cQ^3$, where $K, a, b,$ and c are positive constants gives

$$MC = d(TC)/dQ = a - 2bQ + 3cQ^2$$

The MC function is thus curvilinear. Consequently the slope of the tangent to this function varies with Q, i.e.

$$d(MC)/dQ = d^2(TC)/dQ^2 = -2b + 6cQ$$

In this case the third derivative of TC is a constant, i.e.

$$d^3(TC)/dQ^3 = 6c$$

RELATIVE MAXIMA AND MINIMA

Consider the function $Y = f(X)$ given in Figure 26. This function has relative maxima at the points P and R, and relative minima at Q and S. P is a maximum relative to the other points in the vicinity of P, and S is a minimum relative to the other points around S. It is clear from Figure 26 that the absolute maximum is at point F and the absolute minimum at point K.

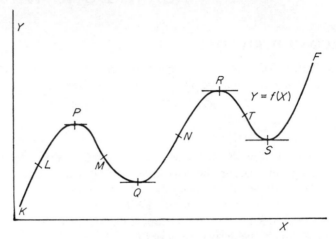

Figure 26   Local maxima and minima

CONDITIONS FOR A RELATIVE MAXIMUM

First-Order Condition

As Y increases from K to P in Figure 26 the slope of the tangent to this interval of the curve is positive for all values of X. Therefore from K to P, the derivative dY/dX is positive. As Y decreases from P to Q its derivative dY/dX is negative. At point P the tangent is parallel to the X-axis and consequently its slope is zero.

$$dY/dX = 0 \quad \text{at} \quad P$$

Consequently the first-order condition for a maximum is that the first derivative is zero.

Second-Order Condition

At L, dY/dX is positive.
At P, dY/dX is zero.
At M, dY/dX is negative.
Consequently dY/dX is decreasing form L to M, i.e. positive at $L \rightarrow$ zero

at $P \to$ negative at M. We have seen that when a function is decreasing its derivative is negative. Thus

$$\frac{d(dY/dX)}{dX} \quad \text{or} \quad \frac{d^2Y}{dX^2} < 0$$

is the second-order condition for a maximum.

When the second derivative of a function is negative the curve is said to be concave downwards or convex upwards. The function $Y = f(X)$ in Figure 26 is concave downwards from L to M and from N to T. A curve will be concave downwards in the neighbourhood of a maximum.

CONDITIONS FOR A RELATIVE MINIMUM

First-Order Condition

As Y decreases from P to Q, in Figure 26, its derivative dY/dX is negative. As Y increases from Q to R its derivative dY/dX is positive. At Q, the slope of the tangent to the function is zero.

Thus $dY/dX = 0$ at Q. This is the first-order condition for a minimum.

Second-Order Condition

At M, dY/dX is negative.
At Q, dY/dX is zero.
At N, dY/dX is positive.
Therefore dY/dX is increasing from M to N, i.e. negative at $M \to 0$ at $Q \to$ positive at N. We have seen that when a function is increasing its derivative is positive. Thus

$$\frac{d(dY/dX)}{dX} \quad \text{or} \quad \frac{d^2Y}{dX^2} > 0$$

is the second-order condition for a minimum.

When the second derivative of a function is positive the curve is said to be concave upwards or convex downwards. The function $Y = f(X)$ in Figure 26 is concave upwards from M to N. A curve will be concave upwards in the neighbourhood of a minimum.

NECESSARY AND SUFFICIENT CONDITIONS FOR A MAXIMUM OR MINIMUM

The first-order condition for a relative maximum or minimum, i.e. $dY/dX = 0$ is a necessary condition. However, the second-order condition, i.e. $d^2Y/dX^2 < 0$ for a maximum or > 0 for a minimum, is not necessary, but is sufficient, assuming the first-order condition holds, to guarantee a relative maximum or minimum. This can be seen by considering the function $Y = X^4$.

For a maximum or minimum $dY/dX = 0$

$$dY/dX = 4X^3$$

When $dY/dX = 0$ $4X^3 = 0$ or $X = 0$

When $X = 0$ $d^2Y/dX^2 = 12X^2 = 0$

This still does not tell us whether $X = 0$ is a relative maximum or minimum. The simplest procedure is to calculate values of Y for values of X above and below $X = 0$.

When $X = 0, Y = 0$

When $X > 0, Y_1 > 0$ since $Y = X^4$

When $X < 0, Y_2 > 0$ since $Y = X^4$

Thus in this case there is a minimum at $X = 0$.

The general rule is that if $Y = f(X)$ has a stationary value Y_0 at $X = X_0$, then we calculate $Y_1 = f(X_1)$ for $X_1 < X_0$ and $Y_2 = f(X_2)$ for $X_2 > X_0$. If Y_1 and Y_2 are both smaller than Y_0, then Y has a maximum at X_0. If Y_1 and Y_2 are both larger than Y_0, then Y has a minimum at X_0. If Y_1 is larger and Y_2 is smaller than Y_0, or vice versa, then $Y = f(X)$ has neither a maximum nor a minimum at X_0.

In the case of $Y = X^4$, $X_0 = 0$, Y_1 and Y_2 are both positive and Y thus has a minimum at $X = 0$ even though $d^2Y/dX^2 = 0$ at $X = 0$.

MAXIMUM REVENUE

When the demand function takes the form $P = K - aQ$, then

$$TR = KQ - aQ^2$$

This is a relation between TR and Q and TR will be at a maximum when the first- and second-order conditions, hold, i.e. when

$$d(TR)/dQ = 0 \quad \text{and} \quad d^2(TR)/dQ^2 < 0$$

When $d(TR)/dQ = 0$,

$$K - 2aQ = 0 \quad \text{or} \quad Q = \frac{K}{2a}$$

$$d^2(TR)/dQ^2 = -2a < 0$$

Thus TR is a maximum when $Q = K/2a$.

Maximum Revenue and Elasticity

The first-order condition for a maximum is $d(TR)/dQ = 0$, or $MR = 0$ since $d(TR)/dQ = MR$. When $MR = 0$, TR remains unchanged as Q changes and $E_D = -1$.

In the above example TR is at a maximum when $Q = K/2a$

$$E_D = \frac{P}{Q} \cdot \frac{dQ}{dP} = \frac{K - aQ}{Q}\left(-\frac{1}{a}\right)$$

since

$$P = K - aQ \quad \text{and} \quad dQ/dP = -1/a$$

When $Q = K/2a$,

$$E_D = \frac{K - a.K/2a}{K/2a}\left(-\frac{1}{a}\right) = -1$$

Thus when TR is at a maximum $E_D = -1$.

MINIMUM AVERAGE COSTS

If a firm's TC function is $TC = K + aQ + bQ^2$, where K, a, and b are positive constants then

$$AC = \frac{K}{Q} + a + bQ$$

Average costs are at a minimum when $d(AC)/dQ = 0$ and $d^2(AC)/dQ^2 > 0$. When $d(AC)/dQ = 0$,

$$-\frac{K}{Q^2} + b = 0$$

i.e.

$$Q^2 = \frac{K}{b} \quad \text{or} \quad Q = \sqrt{(K/b)}$$

Before we can say that average costs are at a minimum at this point it is necessary to check the second-order condition.

$$d^2(AC)/dQ^2 = \frac{2K}{Q^3} > 0 \quad \text{for all } Q > 0$$

Thus AC is at a minimum when $Q = \sqrt{(K/b)}$.

MINIMUM MARGINAL COSTS

If a firm's TC function is $TC = K + aQ - bQ^2 + cQ^3$, where K, a, b, and c are positive constants, then

$$d(TC)/dQ \quad \text{or} \quad MC = a - 2bQ + 3cQ^2$$

MC is at a minimum when $d(MC)/dQ = 0$ and $d^2(MC)/dQ^2 > 0$. When $d(MC)/dQ = 0$, $-2b + 6cQ = 0$, i.e. $Q = b/3c$

$$d^2(MC)/dQ^2 = 6c > 0$$

Thus MC is at a minimum when $Q = b/3c$.

PROFIT MAXIMIZATION

In Chapter 4 the properties of the parabola were used to determine the output of a profit maximizing firm. The conditions for a maximum derived in this chapter can now be used.

$$\text{Profits } S = TR - TC$$

If $TC = K + aQ + bQ^2$ and the demand function takes the form $P = L - nQ$, where K, a, b, L, and n are positive constants,

$$TR = LQ - nQ^2$$

thus
$$S = LQ - nQ^2 - K - aQ - bQ^2$$
$$= -(b + n)Q^2 + (L - a)Q - K$$

S is now a function of Q and takes a maximum when $dS/dQ = 0$ and $d^2S/dQ^2 < 0$.

When $dS/dQ = 0$, $-2(b + n)Q + (L - a) = 0$, i.e.

$$Q = \frac{L - a}{2(b + n)}$$

$d^2S/dQ^2 = -2(b + n) < 0$ since b and n are positive.

Therefore profits are maximized when $Q = (L - a)/2(b + n)$.

Applying these conditions to the specific profits function

$$S = -Q^2 + 15Q - 50$$

used in Chapter 4, gives

$$dS/dQ = -2Q + 15$$

When $dS/dQ = 0$, $-2Q + 15 = 0$ or $Q = 7{\cdot}5$
$$d^2S/dQ^2 = -2 < 0$$

Thus profits are maximized when $Q = 7{\cdot}5$; see the parabola approach to profit maximization, Chapter 4.

The Profit Maximization Conditions

$$S = TR - TC$$

and

$$dS/dQ = \frac{d(TR)}{dQ} - \frac{d(TC)}{dQ}$$

since S, TR, and TC are all functions of Q. When $dS/dQ = 0$,

$$\frac{d(TR)}{dQ} - \frac{d(TC)}{dQ} = 0 \quad \text{or} \quad MR - MC = 0$$

Consequently the first-order condition for a maximum is $MR = MC$.

We have shown that revenue is maximized when $MR = 0$ or $E_D = -1$. A profit maximizing firm will be in equilibrium when $MR = MC$ and since MC must be positive MR will take a positive value in equilibrium. Consequently a profit maximizing firm will be in equilibrium when $-E_D > 1$. When $d^2S/dQ^2 < 0$,

$$\frac{d^2(TR)}{dQ^2} - \frac{d^2(TC)}{dQ^2} \quad \text{or} \quad \frac{d(MR)}{dQ} - \frac{d(MC)}{dQ} \quad \text{must be } < 0$$

Hence profits are maximized when $MR = MC$ and $d(MR)/dQ - d(MC)/dQ$ is negative. Figure 27 shows that profits are maximized at M. At M,

$MC = MR$. MR is downwards sloping, i.e. the slope of the tangent to the MR curve is negative at every point, thus

$$d(MR)/dQ < 0 \quad \text{at } M$$

MC is upward sloping at M, thus

$$d(MC)/dQ > 0$$

and

$$\frac{d(MR)}{dQ} - \frac{d(MC)}{dQ} < 0$$

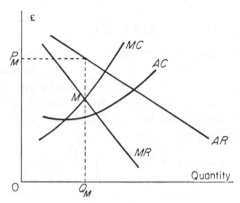

Figure 27 Marginal costs and revenue

In general, however, we require only that $d(MR)/dQ - d(MC)/dQ < 0$, so that $d(MC)/dQ$ need not be positive.

EFFECTS OF TAXATION ON THE OUTPUT OF A PROFIT MAXIMIZING FIRM

Earlier in this chapter we saw that profits are maximized when

$$Q = \frac{L - a}{2(b + n)}$$

given that $TC = K + aQ + bQ^2$ and $P = L - nQ$. The levying of a lump sum tax, i.e. a tax which does not depend on output, on this firm will have no effect on the equilibrium output. This tax will increase the value of the constant K in the TC function and consequently decrease profits by the amount of the tax. However, the equilibrium output will remain unchanged since the constant term disappears with differentiation.

A tax which varies with output will affect the equilibrium output and profits. If the government imposes a tax of t per unit of quantity produced then TC will increase by tQ or S will decrease by tQ, i.e.

$$S = -(b + n)Q^2 + (L - a)Q - K - tQ$$
$$= -(b + n)Q^2 + (L - a - t)Q - K$$

When $dS/dQ = 0$, $-2(b + n)Q + L - a - t = 0$, thus

$$Q = \frac{L - a - t}{2(b + n)}$$

$$d^2S/dQ^2 < 0$$

Thus profits are maximized when $Q = (L - a - t)/2(b + n)$. The optimum output falls as result of the per unit tax; t is subtracted from the numerator and has the effect of decreasing Q. A per unit subsidy would have the opposite effect.

MAXIMIZATION OF TAX REVENUE

A per unit tax on quantity produced causes a profit maximizing firm to cut back production. Since total tax revenue T depends upon the tax rate t and the output level Q, i.e. $T = tQ$, it is possible to find the tax rate which maximizes total tax revenue from the point of view of the exchequer.

In the previous example, equilibrium quantity depended on the tax rate t, i.e. $Q = (L - a - t)/2(b + n)$

$$T = tQ = t\left(\frac{L - a - t}{2(b + n)}\right) = \frac{Lt - at - t^2}{2(b + n)}$$

We now have T as a function of t, alternatively T could be expressed as a function of Q.

Tax revenue T is maximized when $dT/dt = 0$ and $d^2T/dt^2 < 0$. When $dT/dt = 0$,

$$\frac{L - a - 2t}{2(b + n)} = 0 \quad \text{i.e.} \quad t = \frac{L - a}{2}$$

$$d^2T/dt^2 = \frac{-2}{2(b + n)} = \frac{-1}{b + n} < 0$$

Thus T is maximized when $t = (L - a)/2$.

When $t = (L - a)/2$,

$$Q = \frac{L - a - (L - a)/2}{2(b + n)} = \frac{L - a}{4(b + n)}$$

$T = tQ$, thus

$$T = \frac{L - a}{2} \cdot \frac{L - a}{4(b + n)} = \frac{(L - a)^2}{8(b + n)}$$

THE SUPPLY FUNCTION AND A PER UNIT TAX

Consider an industry with the following demand and supply functions

$$P = K - \alpha Q_D$$

and

$$P = -K_1 + \beta Q_S$$

where K, K_1, α, and β are positive constants. If the government imposes a tax of t per unit of quantity produced and P is the market price to consumers then the effective price for producers is $(P - t)$. Therefore the supply function which includes the tax is $(P - t) = -K_1 + \beta Q_S$ or $P = -K_1 + \beta Q_S + t$.

At equilibrium $Q_S = Q_D$, so

$$K - \alpha Q = -K_1 + \beta Q + t \quad \text{i.e.} \quad Q = \frac{K + K_1 - t}{\alpha + \beta}$$

Equilibrium Q is reduced as a result of the tax. To find the tax rate which maximizes total tax revenue T one can express T as a function of t. Alternatively, T can be expressed as a function of Q. At equilibrium

$$t = K + K_1 - (\alpha + \beta)Q$$

$T = tQ$, thus
$$T = KQ + K_1 Q - (\alpha + \beta)Q^2$$

T is at a maximum when $dT/dQ = 0$ and $d^2T/dQ^2 < 0$. When $dT/dQ = 0$,

$$K + K_1 - 2(\alpha + \beta)Q = 0 \quad \text{i.e.} \quad Q = \frac{K + K_1}{2(\alpha + \beta)}$$

$$d^2T/dQ^2 = -2(\alpha + \beta) < 0$$

since α and β are positive constants. Thus T is maximized when

$$Q = (K + K_1)/2(\alpha + \beta)$$

Once Q is known, t and T can be found by substitution.

The conditions for a maximum or minimum discussed above apply only to continuous functions whose derivatives are also continuous and smooth, i.e. without sharp bends. Most of the functions dealt with in this chapter are quadratics so that the relative maximum or minimum of any of these functions is also the absolute maximum or minimum. With functions of degree 3 the relative maximum or minimum will not be the absolute maximum or minimum.

Cubic Functions

$$Y = aX + bX^2 + cX^3$$

where $c > 0$; thus
$$dY/dX = a + 2bX + 3cX^2$$

This function will have a maximum and minimum when

$$a + 2bX + 3cX^2 = 0$$

Assuming real roots, the values for X which satisfy this equation will only provide a relative maximum or minimum. There is no limit to the value of Y in the upward or downward direction, i.e. no absolute maximum or minimum, since larger positive values for X will keep increasing Y and larger negative values will keep decreasing Y.

EXERCISES

1. The market demand function of a firm is given by
$$4P + Q - 16 = 0$$
and the AC function takes the form
$$AC = \frac{4}{Q} + 2 - 0{\cdot}3Q + 0{\cdot}05Q^2$$

Find the Q which gives
(i) maximum revenue,
(ii) minimum marginal costs,
(iii) maximum profits.

Use the second derivative test in each case.

2. A monopolist's demand function is
$$P = 30 - 0{\cdot}75Q$$
and his AC function takes the form
$$AC - \frac{30}{Q} = 9 + 0{\cdot}3Q$$

(i) Find the Q which gives
(a) maximum revenue,
(b) minimum average costs,
(c) maximum profits.
In each case check the second-order conditions.

(ii) Find the Q which gives maximum profits
(a) when the government imposes a lump sum tax of 10,
(b) when a tax of 8·4 per unit of quantity produced is imposed on the monopolist,
(c) when a subsidy of 4·2 per unit of quantity produced is paid to the monopolist.

3. Given the demand function
$$Q_D = 40 - 2P$$
and the supply function
$$2P - Q_S = 20$$
assume that the government imposes a tax of t per unit on quantity supplied and the producers adjust the supply function to include the tax, and calculate
(i) the tax rate which will maximize tax revenue,
(ii) the maximum revenue which can be obtained from taxation.

4. The demand function of a profit maximizing monopolist is
$$P + 3Q - 30 = 0$$
and his TC function is
$$TC = 2Q^2 + 10Q$$
If a tax of t per unit of quantity produced is imposed on the monopolist, calculate the maximum tax revenue obtainable by the government.

5. A discriminating monopolist can separate his consumers into two distinct markets with the following demand functions:

Market I $\quad Q_1 = 16 - 0.2P_1$

Market II $\quad P_2 = 180 - 20Q_2$

Assume the monopolist's TC function takes the form

$$TC - 20Q - 20 = 0$$

where $Q = Q_1 + Q_2$, and that he wishes to maximize profits, and calculate

(i) the prices the monopolist will charge in each market,

(ii) the monopolist's profit *with* and *without* price discrimination.

SOLUTIONS TO EXERCISES

1. (i) $4P + Q - 16 = 0$

thus

$$P = 4 - 0.25Q$$
$$TR = P.Q = (4 - 0.25Q)Q = 4Q - 0.25Q^2$$

Revenue is at a maximum when $d(TR)/dQ = 0$ and $d^2(TR)/dQ^2 < 0$.

$$d(TR)/dQ = 4 - 0.5Q$$

When $d(TR)/dQ = 0$, then

$$4 - 0.5Q = 0$$

i.e.

$$Q = 8$$
$$d^2(TR)/dQ^2 = -0.5 < 0$$

Thus revenue is at a maximum when $Q = 8$. *

(ii) $AC = \dfrac{4}{Q} + 2 - 0.3Q + 0.05Q^2$

thus

$$TC = 4 + 2Q - 0.3Q^2 + 0.05Q^3$$
$$MC = d(TC)/dQ = 2 - 0.6Q + 0.15Q^2$$

Marginal costs are at a minimum when $d(MC)/dQ = 0$ and $d^2(MC)/dQ^2 > 0$.

$$d(MC)/dQ = -0.6 + 0.3Q$$

When $d(MC)/dQ = 0$, then

$$-0.6 + 0.3Q = 0$$

i.e.

$$Q = 2$$
$$d^2(MC)/dQ^2 = 0.3 > 0$$

Thus marginal costs are at a minimum when $Q = 2$. *

(iii) Profits $S = TR - TC$, thus

$$S = 4Q - 0.25Q^2 - 4 - 2Q + 0.3Q^2 - 0.05Q^3$$
$$= -0.05Q^3 + 0.05Q^2 + 2Q - 4$$

Profits are maximized when $dS/dQ = 0$ and $d^2S/dQ^2 < 0$.

$$dS/dQ = -0.15Q^2 + 0.1Q + 2$$

When $dS/dQ = 0$, then

$$-0.15Q^2 + 0.1Q + 2 = 0$$

i.e.

$$Q = \frac{-0{\cdot}1 \pm \sqrt{[(0{\cdot}1)^2 - 4(-0{\cdot}15)(2)]}}{2(-0{\cdot}15)}$$

$$= \frac{-0{\cdot}1 \pm 1{\cdot}1}{-0{\cdot}3} = \frac{-1{\cdot}2}{-0{\cdot}3} \quad \text{or} \quad \frac{1{\cdot}0}{-0{\cdot}3}$$

Q cannot be negative, so $Q = 4$.

$$d^2 S/dQ^2 = -0{\cdot}3Q + 0{\cdot}1$$

When $Q = 4$,

$$d^2 S/dQ^2 = -1{\cdot}2 + 0{\cdot}1 = -1{\cdot}1 < 0$$

Thus profits are maximized when $Q = 4$.   *

2. (i) (a) $P = 30 - 0{\cdot}75Q$

$$TR = P{\cdot}Q = (30 - 0{\cdot}75Q)Q = 30Q - 0{\cdot}75Q^2$$

Revenue is at a maximum when $d(TR)/dQ = 0$ and $d^2(TR)/dQ^2 < 0$.

$$d(TR)/dQ = 30 - 1{\cdot}5Q$$

thus

$$Q = 20 \quad \text{when} \quad d(TR)/dQ = 0.$$
$$d^2(TR)/dQ^2 = -1{\cdot}5 < 0$$

Thus revenue is at a maximum when $Q = 20$.   *

(b) $AC - \dfrac{30}{Q} = 9 + 0{\cdot}3Q$

i.e.

$$AC = \frac{30}{Q} + 9 + 0{\cdot}3Q$$

AC is at a minimum when $d(AC)/dQ = 0$ and $d^2(AC)/dQ^2 > 0$.

$$d(AC)/dQ = -\frac{30}{Q^2} + 0{\cdot}3$$

When $d(AC)/dQ = 0$,

$$-\frac{30}{Q^2} + 0{\cdot}3 = 0$$

i.e.

$$Q^2 = \frac{30}{0{\cdot}3} = 100 \quad \text{or} \quad Q = 10$$

$$d^2(AC)/dQ^2 = \frac{60}{Q^3} > 0 \quad \text{for all} \quad Q > 0$$

Thus average costs are at a minimum when $Q = 10$.   *

(c) Profits $S = TR - TC$.

$$AC = \frac{30}{Q} + 9 + 0{\cdot}3Q$$

thus

$$TC = 30 + 9Q + 0{\cdot}3Q^2$$

and

$$S = 30Q - 0{\cdot}75Q^2 - 30 - 9Q - 0{\cdot}3Q^2$$

i.e.

$$S = -1{\cdot}05Q^2 + 21Q - 30$$

Profits are at a maximum when $dS/dQ = 0$ and $d^2S/dQ^2 < 0$.

$$dS/dQ = -2\cdot1Q + 21$$

When $dS/dQ = 0$, then

$$-2\cdot1Q + 21 = 0$$

i.e.

$$Q = 10$$
$$d^2S/dQ^2 = -2\cdot1 < 0$$

Thus profits are at a maximum when $Q = 10$.   *

(ii) (a) Total costs will increase by the amount of the lump sum tax. Therefore the constant term, -30, in the profits function will become -40. However, this will not alter the value of dS/dQ and Q will remain at 10.

(b) With a per unit tax, total costs will increase by tQ, where t is the tax rate and Q is the quantity produced. The profit function becomes

$$S = -1\cdot05Q^2 + 21Q - 30 - 8\cdot4Q$$
$$= -1\cdot05Q^2 + 12\cdot6Q - 30$$
$$dS/dQ = -2\cdot1Q + 12\cdot6$$

When profits are maximized

$$-2\cdot1Q + 12\cdot6 = 0$$

thus

$$Q = 6$$
$$d^2S/dQ^2 = -2\cdot1 < 0$$

Thus profits are maximized when $Q = 6$.   *

(c) With a per unit subsidy profits will increase by $4\cdot2Q$, i.e.

$$S = -1\cdot05Q^2 + 21Q - 30 + 4\cdot2Q$$
$$= -1\cdot05Q^2 + 25\cdot2Q - 30$$
$$dS/dQ = -2\cdot1Q + 25\cdot2$$

When profits are maximized,

$$-2\cdot1Q + 25\cdot2 = 0$$

thus

$$Q = 12$$
$$d^2S/dQ^2 = -2\cdot1 < 0$$

Thus profits are maximized when $Q = 12$.   *

3. (i) $2P - Q_S = 20$
thus

$$P = 10 + 0\cdot5Q_S$$

Since producers adjust the supply function to include the tax this will cause the function to shift, see Chapter 3, Figure 18, and take the form

$$P = 10 + 0\cdot5Q_S + t$$

The demand function is

$$P = 20 - 0\cdot5Q_D$$

At equilibrium $Q_S = Q_D$, thus

$$10 + 0\cdot5Q + t = 20 - 0\cdot5Q$$

i.e.

$$t = 10 - Q$$

t is the tax rate, thus

$$\text{total tax revenue } T = tQ$$

i.e. tax rate multiplied by quantity produced and

$$T = (10 - Q)Q = 10Q - Q^2$$

T is maximized when $dT/dQ = 0$ and $d^2T/dQ^2 < 0$.

$$dT/dQ = 10 - 2Q$$

When total tax revenue T is maximized

$$10 - 2Q = 0 \quad \text{i.e.} \quad Q = 5$$
$$d^2T/dQ^2 = -2 < 0$$

Thus T is maximized when $Q = 5$
  When $Q = 5$,

$$t = 10 - 5 = 5$$

so

$$t = 5 \text{ is the tax rate which maximizes } T \qquad *$$

(ii) $T = tQ = 5(5) = 25$ *

4. $P = 30 - 3Q$
  thus

$$TR = (30 - 3Q)Q = 30Q - 3Q^2$$

A per unit tax will affect the TC function, so that

$$TC = 2Q^2 + 10Q + tQ$$
$$\text{Profits } S = 30Q - 3Q^2 - 2Q^2 - 10Q - tQ$$
$$= -5Q^2 + 20Q - tQ$$

Profits are maximized when $dS/dQ = 0$ and $d^2S/dQ^2 < 0$.

$$dS/dQ = -10Q + 20 - t$$

$$d^2S/dQ^2 = -10 < 0$$

Profits are maximized when $-10Q + 20 - t = 0$.
Expressing Q as a function of t instead of t as a function of Q, as in answer 3 above, gives

$$Q = 2 - 0\cdot1t$$

Since total tax revenue $T = tQ$, then $T = (2 - 0\cdot1t)t$.
T is maximized when $dT/dt = 0$ and $d^2T/dt^2 < 0$.

$$dT/dt = 2 - 0\cdot2t$$

T is at a maximum when $2 - 0\cdot2t = 0$ or $t = 10$

$$d^2T/dt^2 = -0\cdot2 < 0$$

Thus T is at a maximum when $t = 10$. When $t = 10$,

$$Q = 2 - 0\cdot1(10) = 1$$
$$T = tQ = 10$$

i.e. maximum tax revenue is 10.   *
Note that questions 3 and 4 can be answered by expressing T as a function of Q or t.

5. (i) The monopolist will allocate production so that $MC = MR$ in each market, i.e.
$$MC = MR_1 = MR_2$$
Since $TC - 20Q - 20 = 0$ or $TC = 20Q + 20$, then MC is constant at 20.
  In Market I
$$Q_1 = 16 - 0.2P_1 \quad \text{or} \quad P_1 = 80 - 5Q_1$$
$$TR_1 = P_1Q_1 = (80 - 5Q_1)Q_1 = 80Q_1 - 5Q_1^2$$
$$MR_1 = d(TR_1)/dQ_1 = 80 - 10Q_1$$
MR_1 must equal MC, thus
$$80 - 10Q_1 = 20 \quad \text{i.e.} \quad Q_1 = 6$$
When $Q_1 = 6, P_1 = 80 - 30 = 50$. *
  In Market II
$$P_2 = 180 - 20Q_2$$
thus
$$TR_2 = (180 - 20Q_2)Q_2 = 180Q_2 - 20Q_2^2$$
$$MR_2 = 180 - 40Q_2$$
MR_2 must equal MC, thus
$$180 - 40Q_2 = 20 \quad \text{i.e.} \quad Q_2 = 4$$
When $Q_2 = 4$, $P_2 = 180 - 80 = 100$. *

(ii) Profits *with* price discrimination
$$S = (TR_1 + TR_2) - TC$$
$$Q_1 = 6 \quad \text{and} \quad P_1 = 50, \quad \text{so} \quad TR_1 = 300$$
$$Q_2 = 4 \quad \text{and} \quad P_2 = 100, \quad \text{so} \quad TR_2 = 400$$
Thus
$$S = (300 + 400) - [20 + 20(Q_1 + Q_2)]$$
$$= 700 - 20 - 200 = 480 \qquad *$$

Profits *without* price discrimination

To calculate profits without price discrimination it is necessary to find the monopolist's overall market demand function.

If a monopolist splits a market this implies that overall $Q = Q_1 + Q_2$ when the price is the same in both markets
$$Q_1 = 16 - 0.2P_1 \quad \text{Market I}$$
and
$$Q_2 = 9 - 0.05P_2 \quad \text{Market II}$$
$Q = Q_1 + Q_2$ when $P_1 = P_2$, thus
$$Q = 16 - 0.2P + 9 - 0.05P \quad \text{i.e.} \quad Q = 25 - 0.25P$$
The overall demand function is thus
$$Q = 25 - 0.25P \quad \text{or} \quad P = 100 - 4Q$$
so
$$TR = (100 - 4Q)Q = 100Q - 4Q^2$$
or
$$S = 100Q - 4Q^2 - 20Q - 20$$
$$= -4Q^2 + 80Q - 20$$
S is at a maximum when $dS/dQ = 0$ and $d^2S/dQ^2 < 0$.
$$dS/dQ = -8Q + 80$$

S is at a maximum when
$$-8Q + 80 = 0$$
or
$$Q = 10$$
$$\mathrm{d}^2S/\mathrm{d}Q^2 = -8 < 0$$
Thus profits are maximized when $Q = 10$ and
$$S = -4(100) + 80(10) - 20 = 380$$

*

CHAPTER 7

Partial Differentiation

FUNCTIONS OF TWO VARIABLES

The derivative of a function of one variable, i.e. $Y = f(X)$ and the conditions for a maximum or minimum of such a function have been examined. However, most economic problems involve functions of two or more variables. If we have a function of two variables, e.g. $Y = f(X, Z)$, then X is referred to as the first argument of the function and Z as the second argument. $\partial Y/\partial X$ is the partial derivative of Y with respect to the first argument X and is often written as Y_X or f_X or f_1. The partial derivative f_X measures the responsiveness of Y with respect to a change in X, assuming the second argument Z remains constant at a certain level. $\partial Y/\partial Z$ written as Y_Z or f_2 or f_Z is the partial derivative of Y with respect to Z and measures the responsiveness of Y with respect to Z, assuming X is constant at a certain level.

CALCULATION OF PARTIAL DERIVATIVES

First-order Partials

There is no difficulty in calculating partial derivatives as they obey the rules for ordinary differentiation. If $Y = X^\alpha + XZ^\beta$, where α and β are constants, then

$$f_X = \alpha X^{\alpha-1} + Z^\beta$$

since Z is assumed constant and Z^β is the coefficient of X, and

$$f_Z = \beta X Z^{\beta-1}$$

since X is the coefficient of Z^β and the derivative with respect to Z of the constant X^α is zero.

Second-order Partials

The second-order partial gives the rate of change in f_X with respect to X, assuming Z is held constant.

$$\partial^2 Y/\partial X^2 \quad \text{or} \quad f_{XX} = \alpha(\alpha-1)X^{\alpha-2}$$

since the derivative of the constant Z^β is zero.

$$\partial^2 Y / \partial Z^2 \quad \text{or} \quad f_{ZZ} = \beta(\beta - 1)XZ^{\beta - 2}$$

Cross Partials

The second-order cross partial gives the rate of change in f_X as Z varies, assuming X is held constant.

$$\partial(f_X) / \partial Z = f_{ZX} = \frac{\partial(\alpha X^{\alpha - 1} + Z^\beta)}{\partial Z} = \beta Z^{\beta - 1}$$

since the derivative of the constant $\alpha X^{\alpha - 1}$ is zero.

$$\partial(f_Z) / \partial X = f_{XZ} = \frac{\partial(\beta X Z^{\beta - 1})}{\partial X} = \beta Z^{\beta - 1}$$

since Z is constant and $\beta Z^{\beta - 1}$ is the coefficient of X.

Thus $f_{ZX} = f_{XZ} = \beta Z^{\beta - 1}$. It is true for most of the functions used in economics that $f_{XZ} = f_{ZX}$.

MARGINAL PRODUCTS

If we take the Cobb-Douglas production function

$$Q = AL^\alpha K^\beta$$

where Q is output, L is labour, K is capital, A, α, and β are constants, then the first-order partial derivatives f_L and f_K will provide a measure of the marginal product of labour, MPL, and capital, MPK, respectively, Q_L and Q_K can be used to denote the first-order partial with respect to L and K respectively, but we will continue to use f_L and f_K.

$$MPL = f_L = A\alpha L^{\alpha - 1} K^\beta = \text{the rate of change in } Q \text{ with respect to changes in } L, \text{ assuming } K \text{ is held constant}$$

$$MPK = f_K = A\beta L^\alpha K^{\beta - 1} = \text{the rate of change in } Q \text{ with respect to changes in } K, \text{ assuming } L \text{ is held constant}$$

If $\alpha < 1$ and $\beta < 1$ each factor has a diminishing marginal product.

Suppose $\alpha = \beta = 0.5$ or $Q = AL^{0.5} K^{0.5}$; then

$$f_L = A(0.5)L^{-0.5} K^{0.5} = 0.5AK^{0.5}L^{-0.5}$$

As L increases f_L decreases but always remains positive. Therefore, as the quantity of labour increases and the stock of capital is held constant, output increases but at a diminishing rate, i.e. MPL decreases.

$$f_K = 0.5AL^{0.5} K^{-0.5}$$

As K increases, assuming L is constant, f_K decreases but always remains positive, i.e. MPK decreases. $\alpha + \beta \leqslant 1$ is a sufficient, though not necessary, condition for a decreasing MPL and MPK. MPL and MPK will increase if $\alpha > 1$ and $\beta > 1$. MPL will increase and MPK will decrease if $\alpha > 1 > \beta$. The opposite will hold if $\alpha < 1 < \beta$.

MARGINAL UTILITY

Taking a Cobb-Douglas type utility function

$$U = AX^{\alpha}Y^{\beta}$$

where X and Y are two goods, then f_X and f_Y or U_X and U_Y will provide a measure of marginal utility for good X and Y respectively. When $\alpha + \beta = 1$ each of the goods will have diminishing marginal utility. For utility functions it would be implausible to assume $\alpha > 1$ or $\beta > 1$ since this would give increasing marginal utility to X and Y respectively.

RETURNS TO SCALE

A production function is said to have constant returns to scale if multiplying all inputs by the same amount $\lambda > 0$ results in multiplication of output by λ also. If all inputs are multiplied by $\lambda > 1$ then returns to scale are increasing if output is multiplied by more than λ. Returns to scale are decreasing if multiplying all inputs by $\lambda > 1$ multiplies output by less than λ. With a Cobb-Douglas production function

$$Q_0 = AL^{\alpha}K^{\beta}$$

If all inputs are multiplied by λ then

$$Q_1 = A(\lambda L)^{\alpha}(\lambda K)^{\beta} = A\lambda^{\alpha}L^{\alpha}\lambda^{\beta}K^{\beta}$$
$$= \lambda^{\alpha+\beta}(AL^{\alpha}K^{\beta}) = \lambda^{\alpha+\beta}Q_0$$

Output is thus multiplied by $\lambda^{\alpha+\beta}$. The Cobb-Douglas function is thus homogeneous of degree $\alpha + \beta$.

When $\alpha + \beta = 1$ the degree of homogeneity is 1, i.e. $Q_1 = \lambda^1 Q_0$ and returns to scale are constant.

When $\alpha + \beta > 1$ the degree of homogeneity is > 1 and returns to scale are increasing.

EULER'S THEOREM

If a function $Y = f(X, Z)$ is homogeneous of degree n, i.e.

$$f(\lambda X, \lambda Z) = \lambda^n f(X, Z)$$

then Euler's theorem sates that

$$Xf_X + Zf_Z = nf(X, Z)$$

For a proof of this theorem see Appendix 2.

NON-HOMOGENEOUS PRODUCTION FUNCTIONS

Suppose that $Q(L, K) = A(L^{\alpha} + K^{\beta})$, where $\alpha \neq \beta$. Then $Q(\lambda L, \lambda K)$, where $\lambda \neq 1$, is given by

$$Q(\lambda L, \lambda K) = A(\lambda^{\alpha}L^{\alpha} + \lambda^{\beta}K^{\beta})$$
$$= A\lambda^{\alpha}(L^{\alpha} + \lambda^{\beta-\alpha}K^{\beta})$$

Since $\lambda^{\beta-\alpha} \neq 1$ the ratio of $Q(\lambda L, \lambda K)$ to $Q(L, K)$ depends on the values of L and K, and the function is non-homogeneous.

THE COBB–DOUGLAS PRODUCTION FUNCTION

This production function is based on the observation that the wage rate is a constant proportion of output per head. We assume that

$$Q = AL^{\alpha}K^{\beta}$$

If firms are operating in perfectly competitive markets and maximizing profits then the wage rate W is equal to the marginal product of labour f_L, thus

$$W = f_L = A\alpha L^{\alpha-1}K^{\beta} = A\alpha\frac{L^{\alpha}}{L}K^{\beta} = \alpha\frac{Q}{L}$$

so

$$\frac{1}{\alpha}W = \frac{Q}{L} = \text{output per head}$$

The constant elasticity of substitution production function, which will be discussed in Chapter 8, is based on the observation that output per head is a varying proportion of the wage rate.

COMPARATIVE STATICS

Partial Differentiation and Static Macroeconomic Models

Taking a closed economic system with no government activity gives

$$Y = C + I$$

where Y is national income, C is consumption, and I is investment.

Assume I is autonomous at I^* and C is a function of income Y, i.e. $C = C^* + cY$, where C^* is the autonomous part of consumption and c is the marginal propensity to consume.

$$Y = C + I$$

thus

$$Y = C^* + cY + I^*$$

or

$$(1 - c)Y = C^* + I^*$$

and

$$Y = \frac{C^*}{1 - c} + \frac{I^*}{1 - c}$$

so

$$\partial Y/\partial C^* \quad \text{or} \quad f_{C^*} = \frac{1}{1 - c}$$

f_{C*} is the rate of change in Y with respect to changes in C^*; this is the multiplier.

$$\partial Y / \partial I^* = f_{I*} = \frac{1}{1-c}$$

also. Clearly f_{C*} and f_{I*} take a constant value so long as c remains constant. This is a consequence of the linear relations assumed in the model. In each case the partial derivative gives the rate of change in national income with respect to a change in any of the *autonomous component* parts of Y, assuming all other autonomous elements remain constant. If the autonomous part of the consumption function increases by ΔC^* then Y will increase by $\Delta C^*/(1-c)$.

Because it analyses only the effects of changes in autonomous variables on equilibrium positions for the economy, and does not attempt to deal with the process of transition, this type of analysis is known as comparative statics.

A Closed Macroeconomic Model with Government Activity

The identity now becomes

$$Y = C + I + G$$

where G is government expenditure. I and G are assumed to be autonomous and consumption C is assumed to be a function of disposible income Y_d, i.e.

$$C = C^* + cY_d$$

where C^* is the autonomous part of consumption and c is the marginal propensity to consume. Y_d is national income minus taxes. If

$$T = T^* + tY$$

where T^* is that part of taxes which does not depend on income and t is the tax rate, then

$$C = C^* + c(Y - T) = C^* + c(Y - T^* - tY)$$

i.e.

$$C = C^* + c(1 - t)Y - cT^*$$
$$Y = C + I + G$$

thus

$$Y = C^* + c(1 - t)Y - cT^* + I + G$$

so

$$[1 - c(1 - t)]Y = C^* - cT^* + I + G$$

and

$$Y = \frac{C^*}{1 - c(1 - t)} - \frac{cT^*}{1 - c(1 - t)} + \frac{I}{1 - c(1 - t)} + \frac{G}{1 - c(1 - t)}$$

$f_{T*} = -c/[1 - c(1 - t)]$, i.e. the rate of change in income with respect to changes in T^* is constant. Consequently if T^* increases by ΔT^*, national income Y will decrease by

$$\Delta T^* \cdot \frac{c}{1 - c(1 - t)}$$

f_G gives the rate of change in Y with respect to G. If G increases by ΔG then national income Y will increase by

$$\Delta G \cdot f_G = \Delta G \cdot \frac{1}{1 - c(1 - t)}$$

If $\Delta G = \Delta T^*$, i.e. an increase in expenditure is financed by an equal increase in the autonomous part of taxation, then the overall change in national income will equal

$$- \Delta T^* \cdot \frac{c}{1 - c(1 - t)} + \Delta G \cdot \frac{1}{1 - c(1 - t)} = \frac{\Delta G(1 - c)}{1 - c(1 - t)}$$

since $\Delta T^* = \Delta G$, i.e.

$$\Delta Y = \frac{\Delta G(1 - c)}{1 - c + ct}$$

but

$$\frac{1 - c}{1 - c + ct} < 1$$

since the denominator is greater than the numerator. Consequently the change in income ΔY is less than the change in government expenditure ΔG, i.e.

$$\Delta Y / \Delta G < 1 \quad \text{but} \quad > 0$$

In the above model partial derivatives are used to show the effectiveness of fiscal policy.

An Open Macro Model with Government Activity

If the foreign sector is included in the above model the identity becomes

$$Y = C + I + G + X - M$$

where X is exports and M imports. If exports are constant, but imports M are a function of Y_d, i.e.

$$M = M^* + mY_d$$

where M^* is the autonomous part and m is the marginal propensity to import, then

$$M = M^* + m(Y - T) = M^* + m(Y - T^* - tY)$$
$$= M^* + m(1 - t)Y - mT^*$$
$$Y = C + I + G + X - M$$

thus

$$Y = C^* + c(1 - t)Y - cT^* + I + G + X - M^* - m(1 - t)Y + mT^*$$

or

$$[1 - c(1 - t) + m(1 - t)]Y = C^* - (c - m)T^* + G + I + X - M^*$$
$$1 - c(1 - t) + m(1 - t) = 1 - (1 - t)(c - m)$$

thus

$$Y = \frac{C^* - (c - m)T^* + G + I + X - M^*}{1 - (1 - t)(c - m)}$$

so

$$f_{T^*} = \frac{-(c - m)}{1 - (1 - t)(c - m)}$$

i.e. f_{T^*} is constant, and

$$f_{C^*} = f_G = f_I = f_X = -f_{M^*} = \frac{1}{1 - (1 - t)(c - m)}$$

so that f_{C^*} is also constant. If the autonomous part of imports M^* increases by ΔM^*, assuming all other components of Y remain constant, national income will decrease by

$$\Delta M^* \cdot \frac{1}{1 - (1 - t)(c - m)}$$

If the autonomous part of the tax function T^* increases by ΔT^*, national income will decrease by

$$\Delta T^* \cdot \frac{c - m}{1 - (1 - t)(c - m)}$$

MAXIMIZING OR MINIMIZING FUNCTIONS OF TWO VARIABLES

If a firm sells two products X and Y, profits are then a function of two variables, i.e. $S = f(X, Y)$. This three-dimensional profits surface can be represented by a system of profit contours, see Figure 28. The coordinates of any point on one of these contours, which are drawn as smooth circles for simplicity, represents a combination of X and Y which will yield the same level of profits. There will be some point, A, representing a combination of X and Y at which profits are maximized. Profits increase as one moves from contour 1 towards A.

Changes in One Variable

Profits are maximized at A, where $X = WB$ and $Y = WC$. In Chapter 6 we saw that as a function increases its derivative is positive and as it decreases its derivative is negative. As one moves from F towards A, profits S increase, i.e. S increases as X increases and Y remains constant at WC. Thus

$$\partial S / \partial X \quad \text{or} \quad f_X \quad \text{is} > 0$$

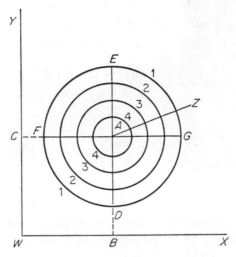

Figure 28   Contours of a function of
two variables

As one moves from $A \to G$, profits S decrease, i.e. S decreases as X increases and Y remains constant. Thus

$$\partial S/\partial X \quad \text{or} \quad f_X \quad \text{is} < 0$$

From $F \to A, f_X > 0$.

From $A \to G, f_X < 0$.

Consequently at $A, f_X = 0$, i.e. the first-order partial is zero. This is the first-order condition for a maximum or a minimum. It is a necessary condition for either.

f_X is positive from F to A, zero at A and negative from A to G. Therefore f_X is decreasing at A. Consequently its derivative f_{XX} is negative.

In the same way if X is kept constant at WB and Y is allowed to vary from $D \to E$, then f_Y will be zero and f_{YY} negative when profits are maximized at A. If $f_{XX} = f_{YY} = 0$ there may be a maximum, but further investigation is necessary. Consequently $f_{XX} \leqslant 0$ and $f_{YY} \leqslant 0$ are necessary conditions for a maximum. To obtain the above conditions for a maximum one of the variables was held constant while the other varied. However, even if the above conditions hold it is still necessary to consider the possibility of changes in both X and Y. Before doing so we will discuss differentials.

DIFFERENTIALS

Previously dY/dX was used as a single symbol. We shall now use it as a ratio of some quantity dY to another dX. dX is the differential of X and dY is the differential of Y. Both refer to infinitesimal changes in X and Y.

$$\frac{\Delta Y}{\Delta X} \equiv \frac{\Delta Y}{\Delta X}$$

thus

$$\Delta Y = \frac{\Delta Y}{\Delta X} \cdot \Delta X$$

In the limit when ΔX and $\Delta Y \to 0$

$$dY = \frac{dY}{dX} \cdot dX$$

If $Y = aX^2 + bX + c$, then

$$dY = (2aX + b)dX$$

i.e. if X changes by an infinitesimal amount then the change in Y or the differential dY is $(2aX + b)$ times the infinitesimal change in X.

TOTAL DIFFERENTIALS

When $S = f(X, Y)$, then the differential dS is the sum of the infinitesimal change in X times the partial derivative of S with respect to X and the infinitesimal change in Y times the partial derivative of S with respect to Y, i.e.

$$dS = f_X dX + f_Y dY$$

Along a curve of equal profits, $dS = 0$ thus

$$f_X dX + f_Y dY = 0$$

or

$$\frac{dX}{dY} = -\frac{f_Y}{f_X} = \text{marginal rate of commodity substitution}$$

If S is a function of n commodities, i.e.

$$S = f(X_1, X_2, X_3, \dots, X_n)$$

then

$$dS = f_{X_1} dX_1 + f_{X_2} dX_2 + \dots + f_{X_n} dX_N$$

In each of the above cases dS is the total differential.

Movement in Any Direction

If X changes from WB and the value of Y changes from WC at the same time, then profits S must decrease if quantity WB of X and WC of Y provide the optimum combination. Let Z represent a movement in any direction, see Figure 28, so that

$$dX = \alpha \, dZ$$

and

$$dY = \beta \, dZ$$

where α and β are not both zero. $dS/dZ = 0$ will then be a necessary condition for a maximum and minimum, and provided that this is satisfied $d^2S/dZ^2 < 0$ will be a sufficient condition for a maximum and $d^2S/dZ^2 > 0$ will be a sufficient condition for a minimum, with respect to Z, see Chapter 6.

First-order Conditions

If $dX = \alpha\, dZ$ and $dY = \beta\, dZ$, then dividing by the differential dZ gives

$$dX/dZ = \alpha \quad \text{and} \quad dY/dZ = \beta$$
$$S = f(X, Y)$$

thus the total differential $dS = f_X dX + f_Y dY$. Dividing by the differential dZ gives

$$\frac{dS}{dZ} = f_X \frac{dX}{dZ} + f_Y \frac{dY}{dZ}$$
$$= f_X \alpha + f_Y \beta$$

If $f_X = f_Y = 0$ then dS/dZ will be zero. At point A we saw that $f_X = f_Y = 0$, i.e. dS/dZ is zero at A.

Second-order Conditions

$$\frac{d^2 S}{dZ^2} = \frac{\partial(f_X \alpha + f_Y \beta)}{\partial X} \cdot \frac{dX}{dZ} + \frac{\partial(f_X \alpha + f_Y \beta)}{\partial Y} \cdot \frac{dY}{dZ}$$
$$= (f_{XX}\alpha + f_{XY}\beta)\alpha + (f_{YX}\alpha + f_{YY}\beta)\beta$$
$$= f_{XX}\alpha^2 + 2f_{XY}\alpha\beta + f_{YY}\beta^2$$

because $f_{XY} = f_{YX}$

$$= \alpha^2\left[f_{XX} + 2f_{XY}\left(\frac{\beta}{\alpha}\right) + f_{YY}\left(\frac{\beta}{\alpha}\right)^2 \right]$$

The bracket is now in the standard form

$$c + bX + aX^2$$

where $c = f_{XX}$, $b = 2f_{XY}$, $a = f_{YY}$ and $X = \beta/\alpha$. We have seen that if

$$c + bX + aX^2 = 0$$

then the parabola will lie in both the positive and negative quadrants if its roots are real and distinct. Consequently we can say that if

$$\alpha^2\left[f_{XX} + 2f_{XY}\left(\frac{\beta}{\alpha}\right) + f_{YY}\left(\frac{\beta}{\alpha}\right)^2 \right] \quad \text{or} \quad \frac{d^2 S}{dZ^2}$$

ever changes sign then it will equal zero for some value of (β/α). To ensure an extreme value of S with respect to variations in both X and Y, we must ensure that $d^2 S/dZ^2 \neq 0$, i.e. that it will always have the same sign. This requires that the equation

$$\alpha^2\left[f_{XX} + 2f_{XY}\left(\frac{\beta}{\alpha}\right) + f_{YY}\left(\frac{\beta}{\alpha}\right)^2 \right] = 0$$

or

$$f_{XX} + 2f_{XY}\left(\frac{\beta}{\alpha}\right) + f_{YY}\left(\frac{\beta}{\alpha}\right)^2 = 0$$

must have no real roots. The latter equation has no real roots when

$$(2f_{XY})^2 - 4f_{XX}f_{YY} < 0$$

i.e.

$$(f_{XY})^2 < f_{XX}f_{YY}$$

If we assume $\alpha = 0$

$$\frac{d^2S}{dZ^2} = f_{XX}\alpha^2 + 2f_{XY}\alpha\beta + f_{YY}\beta^2$$

thus

$$\frac{d^2S}{dZ^2} = 0 + 0 + f_{YY}\beta^2 = f_{YY}\beta^2$$

If $f_{YY} < 0$ then $d^2S/dZ^2 < 0$.
If $f_{YY} > 0$ then $d^2S/dZ^2 > 0$.

SUFFICIENT CONDITIONS FOR A MAXIMUM

We have seen that $f_Y = 0$ and $f_{YY} < 0$ is a sufficient condition for a maximum for movement in the Y-direction alone. If $f_{YY} < 0$, i.e. $d^2S/dZ^2 < 0$ for movements in this direction, then $(f_{XY})^2 < f_{XX}f_{YY}$ will ensure that d^2S/dZ^2 is negative for all α and β, i.e. for a movement in any direction. Consequently sufficient conditions for a maximum are

$$f_X = f_Y = 0$$
$$f_{XX} < 0, f_{YY} < 0$$

and

$$(f_{XY})^2 < f_{XX}f_{YY}$$

SUFFICIENT CONDITIONS FOR A MINIMUM

If $f_{YY} > 0$, i.e. $d^2S/dZ^2 > 0$ for movement in one direction then $(f_{XY})^2 < f_{XX}f_{YY}$ will ensure that d^2S/dZ^2 is positive for all α and β. Consequently the sufficient conditions for a minimum are

$$f_X = f_Y = 0$$
$$f_{XX} > 0, f_{YY} > 0$$

and

$$(f_{XY})^2 < f_{XX}f_{YY}$$

If $f_{XX}f_{YY} = (f_{XY})^2$ there may be a minimum or maximum but further investigation is necessary before we can be certain. If $(f_{XY})^2 > f_{XX}f_{YY}$ then d^2S/dZ^2 would take negative values for some Z and positive values for other Z, which would give us a saddle point, i.e. a point such that for movement in some directions there was a maximum of S but for movement in other directions there was a minimum.

APPLICATION OF THE CONDITIONS FOR A MAXIMUM: A MONOPOLIST WITH TWO PRODUCTS

Consider a monopolist selling two products X and Y with the following linear demand functions

$$P_X = a_1 - b_1 X$$
$$P_Y = a_2 - b_2 Y$$

where a_1, a_2, b_1, and b_2 are positive constants. Assume the joint cost function takes the form

$$TC = X^2 + gXY + Y^2$$

where g is a positive constant.

$$\text{Profits} \quad S = TR - TC$$
$$TR = P_X X + P_Y Y$$

When $P_X = a_1 - b_1 X$, then $P_X X = a_1 X - b_1 X^2$; when $P_Y = a_2 - b_2 Y$, then $P_Y Y = a_2 Y - b_2 Y^2$, thus

$$TR = a_1 X - b_1 X^2 + a_2 Y - b_2 Y^2$$

and

$$S = a_1 X - b_1 X^2 + a_2 Y - b_2 Y^2 - X^2 - gXY - Y^2$$
$$= -(b_1 + 1)X^2 - (b_2 + 1)Y^2 + a_1 X + a_2 Y - gXY$$

S is a quadratic function in two variables and will take a maximum when

$$f_X = f_Y = 0$$
$$f_{XX} \text{ and } f_{YY} \text{ are negative}$$
$$(f_{XY})^2 < f_{XX} f_{YY}$$

Once again the symbols S_X, S_Y, etc. could be used instead of f_X, f_Y, etc.

The first-order derivatives are

$$f_X = -2(b_1 + 1)X + a_1 - gY \tag{1}$$
$$f_Y = -2(b_2 + 1)Y + a_2 - gX \tag{2}$$

When S is at a maximum expressions (1) and (2) will equal zero. This gives two simultaneous equations in X and Y which will provide a solution for X and Y.

$$f_{XX} = -2(b_1 + 1)$$
$$f_{YY} = -2(b_2 + 1)$$

Since b_1 and b_2 are positive constants then f_{XX} and f_{YY} are both negative.

$$f_{XY} = \frac{\partial(f_Y)}{\partial X} = \frac{\partial[-2(b_2 + 1)Y + a_2 - gX]}{\partial X} = -g$$

S will have a maximum if

$$g^2 < [-2(b_1 + 1)][-2(b_2 + 1)]$$

i.e.

$$g^2 < 4(b_1 + 1)(b_2 + 1)$$

IMPLICIT DIFFERENTIATION

Consider a function relating two variables,

$$f(X, Y) = 0$$

If we consider small changes in X and Y, then differentiating both sides of this equation we have

$$f_X dX + f_Y dY = 0$$

thus

$$\frac{dY}{dX} = -\frac{f_X}{f_Y}$$

This is a very convenient method to use when X and Y are connected by an implicit function and it is difficult or impossible to express X or Y as an explicit function of the other. The method is still valid, however, even if this can be done. For example, if

$$aX^2 + bXY + cY^2 = 0,$$
$$(2aX + bY)dX + (bX + 2cY)dY = 0$$

so

$$\frac{dY}{dX} = -\frac{2aX + bY}{bX + 2cY}$$

EXERCISES

1. You are given the following information about an open economy with government activity

   Consumption $C = 50 + 0.9\,Y_d$
   Investment $I = 20 + 0.15\,Y$
   Imports $M = 5 + 0.15\,Y_d$
   Total taxes $T = 10 + 0.2\,Y$ and $Y_d = Y - T$

   Government expenditure G and exports X are autonomous. If the initial level of national income Y is 2000,

   (i) find the value of X, when $G = T$,

   (ii) show how an increase of 45 in the autonomous part of the imports function will affect Y.

   (iii) Starting from the situation in (i), show how an increase of 60 in government expenditure and in the autonomous part of taxation will affect Y.

2. You are given the following information about an open economy with government activity

   Consumption $C = 5 + 0.8\,Y_d$
   Imports $M = 2.5 + 0.1\,Y$

   The only tax is a uniform income tax at a rate of 25 per cent. Investment I, government expenditure G, and exports X, are autonomous. The initial level of national income Y is 100 and trade is initially balanced.

   (i) Find the value of $I + G$.

(ii) Find the change in national income Y if the autonomous part of imports increases to 5.

(iii) Calculate the trade imbalance resulting from (ii).

(iv) Find the new tax rate which would be necessary to restore the trade balance.

3. Given the following production functions say whether returns to scale will be increasing, decreasing, or constant. Q is output, L and K are factor inputs.

(i) $Q = 2L + 4K$

(ii) $Q = L^{0.5} K^{0.5}$

(iii) $Q = 2L^{1/2} K^{1/3}$

(iv) $Q = 4L^{0.75} K^{0.5}$

(v) $Q = 0.2(L^{0.5} + K^{0.25})$

(vi) $Q = 2L^2 + L.K + K^2$

(vii) $Q = 20[\frac{1}{4}L^{-1/4} + \frac{3}{4}K^{-1/4}]^{-4}$

4. A monopolist sells two products, X and Y, and has the following demand functions

$$0.1P_X - 1.2 + 0.2X = 0$$

and

$$10P_Y - 320 + 40Y = 0$$

His joint cost function is

$$TC - X^2 - 2XY - Y^2 = 0$$

Find the price and output for each good which will maximize profits. Check that profits have a maximum at this point.

5. A firm sells two products, X and Y, in related markets, with demand functions given by

$$P_X - 13 + 2X + Y = 0$$

and

$$P_Y - 13 + X + 2Y = 0$$

If $TC = X + Y$ find the price and output for each good which will maximize profits. Check that profits have a maximum at this point.

SOLUTIONS TO EXERCISES

1.
$$Y = C + I + G + X - M$$

Assume that

Consumption $C = C^* + cY_d$

Investment $I = I^* + iY$

Imports $M = M^* + mY_d$

Government expenditure $G = G^*$

Exports $X = X^*$

$Y_d = Y - T$ and $T = T^* + tY$

where C^*, I^*, M^*, G^*, X^*, and T^* are the autonomous parts of the respective functions, c, i, and m are the marginal propensities to consume, invest, and import, and t is the tax rate. Substituting these values into the identity gives

$$Y = C^* + cY_d + I^* + iY + G^* + X^* - M^* - mY_d$$
$$Y_d = Y - T = Y - T^* - tY = (1 - t)Y - T^*$$

thus

$$Y = C^* + c[(1-t)Y - T^*] + I^* + iY + G^* + X^* - M^* - m[(1-t)Y - T^*]$$

so

$$[1 - c(1-t) - i + m(1-t)]Y = C^* - cT^* + I^* + G^* + X^* - M^* + mT^*$$

i.e.

$$[1 - i - (1-t)(c-m)]Y = C^* + I^* + G^* + X^* - M^* - (c-m)T^*$$

or

$$Y = \frac{C^* + I^* + G^* + X^* - M^* - (c-m)T^*}{1 - i - (1-t)(c-m)}$$

thus

$$f_{C^*} = f_{I^*} = f_{G^*} = f_{X^*} = -f_{M^*} = \frac{1}{1 - i - (1-t)(c-m)}$$

and

$$f_{T^*} = -\frac{c-m}{1 - i - (1-t)(c-m)}$$

(i) $Y = 2000$, $C^* = 50$, $I^* = 20$, $M^* = 5$, $T^* = 10$, $c = 0\cdot9$, $i = 0\cdot15$, $m = 0\cdot15$, and $t = 0\cdot2$. Thus

$$2000 = \frac{50 + 20 + G^* + X^* - 5 - (0\cdot9 - 0\cdot15)(10)}{1 - 0\cdot15 - (1 - 0\cdot2)(0\cdot9 - 0\cdot15)}$$

i.e.

$$2000 = \frac{57\cdot5 + G^* + X^*}{0\cdot25} \quad \text{or} \quad G^* + X^* = 442\cdot5$$

G and X are autonomous, i.e.

$$G^* = G \quad \text{and} \quad X^* = X$$

thus

$$G + X = 442\cdot5$$
$$G = T = 10 + 0\cdot2Y = 10 + 0\cdot2(2000) = 410$$

and

$$X = 32\cdot5 \qquad\qquad *$$

(ii)

$$f_{M^*} = -\frac{1}{1 - 0\cdot15 - (1 - 0\cdot2)(0\cdot9 - 0\cdot15)} = -\frac{1}{0\cdot25} = -4$$

Therefore an increase of 45 in M^* will cause Y to decrease by $4(45) = 180$ *

(iii) $f_G = 4$. Therefore if G increases by 60, Y will increase by 240. However,

$$f_{T^*} = -\frac{0\cdot9 - 0\cdot15}{0\cdot25} = -\frac{0\cdot75}{0\cdot25} = -3$$

Consequently an increase of 60 in T^* will cause Y to decrease by 180, thus Y will increase by $240 - 180 = 60$.   *

2. (i) I is now autonomous, thus

$$I^* = I \quad \text{and} \quad i = 0$$
$$G^* = G \quad \text{and} \quad X^* = X$$

There is no autonomous part in the tax function, thus all terms with T^* disappear.

$$M = f(Y) \text{ not } f(Y_d)$$

thus we can omit the term mt. Thus

$$Y = \frac{C^* + I^* + G^* + X^* - M^*}{1 - c(1-t) + m}$$

so

$$100 = \frac{5 + I^* + G^* + X^* - 2{\cdot}5}{1 - 0{\cdot}8(1 - 0{\cdot}25) + 0{\cdot}1} = \frac{2{\cdot}5 + I^* + G^* + X^*}{0{\cdot}5}$$

i.e.

$$I^* + G^* + X^* = 47{\cdot}5$$
$$M = X^* = 2{\cdot}5 + 0{\cdot}1(100) = 12{\cdot}5$$

thus

$$I^* + G^* + 12{\cdot}5 = 47{\cdot}5 \quad \text{or} \quad I + G = 35 \qquad *$$

(ii) $f_{M^*} = -\dfrac{1}{1 - c(1 - t) + m} = -\dfrac{1}{0{\cdot}5} = -2$

Consequently, when the autonomous part of the import function increases by 2·5, Y will decrease by 2(2·5) = 5.

(iii) As a result of (ii), Y decreases to 95, thus

new level of imports $M = 5 + 0{\cdot}1(95) = 14{\cdot}5$

Since X = 12·5 there is now a trade deficit of 14·5 − 12·5 = 2.

(iv) Since X is autonomous at 12·5, M must equal 12·5 if the balance is to be restored, i.e. 5 + 0·1 Y must equal 12·5 or Y must equal 75.
   When Y = 75

$$75 = \frac{5 + 47{\cdot}5 - 5}{1 - 0{\cdot}8(1 - t) + 0{\cdot}1} = \frac{47{\cdot}5}{0{\cdot}3 + 0{\cdot}8t}$$

i.e.

$$22{\cdot}5 + 60t = 47{\cdot}5 \quad \text{or} \quad t = \frac{25}{60} = 41{\cdot}67 \text{ per cent}$$

Thus t = 41·67 per cent when Y = 75. *

3. To calculate returns to scale for each production function let us multiply all inputs by some constant β.
   (i) $Q = 2L + 4K$

$$Q_1 = 2\beta L + 4\beta K = \beta(2L + 4K) = \beta Q$$

Thus returns to scale are constant. *

   (ii) When $Q = AL^a K^b$,

$$Q_1 = A(\beta L)^a (\beta K)^b = \beta^{a+b}(AL^a K^b) = \beta^{a+b}Q$$

Therefore returns to scale are constant when $a + b = 1$, increasing when $a + b > 1$ and decreasing when $a + b < 1$. When $Q = L^{0{\cdot}5}K^{0{\cdot}5}$, $a + b = 1$, thus returns to scale are constant. *

   (iii) When $Q = 2L^{1/2}K^{1/3}$, $a + b = 5/6 < 1$, thus return to scale are decreasing. *
   (iv) When $Q = 4L^{0{\cdot}75}K^{0{\cdot}5}$, $a + b = 1{\cdot}25 > 1$, thus returns to scale are increasing. *
   (v) $Q = 0{\cdot}2(L^{0{\cdot}5} + K^{0{\cdot}25})$

$$Q_1 = 0{\cdot}2(\beta L)^{0{\cdot}5} + 0{\cdot}2(\beta K)^{0{\cdot}25}$$

$Q_1 < \beta Q$ since $\beta^{0{\cdot}5}$ and $\beta^{0{\cdot}25}$ are each $< \beta$ if $\beta > 1$, and $Q_1 > \beta Q$ if $\beta < 1$. Thus returns to scale are decreasing. *

   (vi) $Q = 2L^2 + LK + K^2$

$$Q_1 = 2(\beta L)^2 + (\beta L)(\beta K) + (\beta K)^2 = \beta^2(2L^2 + LK + K^2) = \beta^2 Q$$

Thus returns to scale are increasing.

(vii) $Q = 20[\frac{1}{4}L^{-1/4} + \frac{3}{4}K^{-1/4}]^{-4}$

$$Q_1 = 20[\frac{1}{4}\beta^{-1/4}L^{-1/4} + \frac{3}{4}\beta^{-1/4}K^{-1/4}]^{-4}$$
$$= 20(\beta^{-1/4})^{-4}[\frac{1}{4}L^{-1/4} + \frac{3}{4}K^{-1/4}]^{-4}$$
$$= \beta Q$$

Thus returns to scale are constant.   *

4. Profits $S = TR - TC$
$$= P_X X + P_Y Y - TC$$
$$P_X = 12 - 2X$$

thus

$$P_X X = (12 - 2X)X = 12X - 2X^2$$
$$P_Y = 32 - 4Y$$

thus

$$P_Y Y = (32 - 4Y)Y = 32Y - 4Y^2$$

so

$$S = 12X - 2X^2 + 32Y - 4Y^2 - X^2 - 2XY - Y^2$$
$$= 12X - 3X^2 + 32Y - 5Y^2 - 2XY$$

Profits are at a maximum when $f_X = f_Y = 0$, f_{XX} and f_{YY} are negative, and $(f_{XY})^2 < f_{XX}f_{YY}$.

When $f_X = 0$, $12 - 6X - 2Y = 0$.
When $f_Y = 0$, $32 - 10Y - 2X = 0$.
S is at a maximum when

$$-6X - 2Y = -12 \tag{3}$$

and

$$-2X - 10Y = -32 \tag{4}$$

Multiplying equation (4) by 3 gives

$$-6X - 30Y = -96 \tag{5}$$

Subtracting equation (5) from equation (3) leaves $28Y = 84$ or $Y = 3$. When $Y = 3$, $-6X - 6 = -12$, from equation (3), i.e. $X = 1$.

$$f_{XX} = -6 < 0 \quad \text{and} \quad f_{YY} = -10 < 0$$

$$f_{XY} = \frac{\partial(32 - 10Y - 2X)}{\partial X} = -2$$

$(f_{XY})^2 < f_{XX}f_{YY}$ holds since $(-2)^2 < (-6)(-10)$. Therefore profits are maximized when

$$X = 1 \text{ and } P_X = 10 \qquad *$$
$$Y = 3 \text{ and } P_Y = 20 \qquad *$$

5. $S = TR - TC$
$$= P_X X + P_Y Y - TC$$

$$P_X X = (13 - 2X - Y)X = 13X - 2X^2 - XY$$
$$P_Y Y = (13 - X - 2Y)Y = 13Y - XY - 2Y^2$$
$$S = 13X - 2X^2 - XY + 13Y - XY - 2Y^2 - X - Y$$
$$= 12X - 2X^2 + 12Y - 2XY - 2Y^2$$

Profits S are at a maximum when the same conditions hold as in answer 4. Using

S_X instead of f_X we have

$$S_X = 0 \text{ when } 12 - 4X - 2Y = 0 \text{ or } 4X + 2Y = 12 \tag{6}$$

$$S_Y = 0 \text{ when } 12 - 2X - 4Y = 0 \text{ or } 2X + 4Y = 12 \tag{7}$$

Multiplying equation (7) by 2 gives

$$4X + 8Y = 24 \tag{8}$$

Subtracting equation (6) from equation (8) gives $6Y = 12$ or $Y = 2$. When $Y = 2$, $X = 2$

$$S_{XX} = -4 < 0 \text{ and } S_{YY} = -4 < 0$$
$$S_{XY} = -2$$

$(S_{XY})^2 < S_{XX}S_{YY}$ holds since $(-2)^2 < (-4)(-4)$. Therefore profits are maximized when

$$X = Y = 2 \text{ and } P_X = P_Y = 7 \qquad\qquad *$$

CHAPTER 8

Maxima and Minima
Subject to Constraints

In Chapter 7 partial derivatives were used to find the maxima or minima of functions of more than one variable. If the approach used in the previous chapter is applied to consumers wishing to maximize utility or producers wishing to maximize output, then clearly the assumption is that consumers have no income constraints and producers have no cost constraints. Both assumptions are unrealistic. Suppose a particular consumer's utility is a function of the amount consumed of two goods X and Y, i.e. $U = f(X, Y)$, then utility is maximized when

$$U_X = U_Y = 0$$

Consumption of each good must continue to the point where the marginal utility derived from each good is zero. There may be no finite level of consumption at which this occurs, but even if there is it is hardly likely that the consumer could afford such quantities; i.e. most consumers have an income constraint. The problem then becomes one of maximizing utility subject to a particular constraint. From the production side the problem is usually one of maximizing output subject to a cost constraint or minimizing costs subject to an output constraint.

MAXIMIZING UTILITY SUBJECT TO AN INCOME CONSTRAINT

Consider the Cobb-Douglas type utility function

$$U = AX^\alpha Y^\beta$$

where A, α and β are constants and X and Y are the two commodities in question. If the consumer's income is W and prices are assumed constant then the income constraint is

$$P_X X + P_Y Y \leqslant W$$

Since utility is generally assumed to be an increasing function of consumption, i.e. U_X and U_Y are positive, a utility maximizing consumer will spend all of

his income on consumption and the constraint becomes
$$P_X X + P_Y Y = W$$
The problem is to maximize U subject to the income constraint
$$P_X X + P_Y Y = W$$

METHOD 1: SUBSTITUTION

Since utility is a function of only two variables X and Y the solution to the problem can be obtained by means of substitution.
$$P_X X + P_Y Y = W$$
thus
$$Y = \frac{W - P_X X}{P_Y}$$
Inserting this value for Y into the utility function gives utility as a function of one variable X which can be maximized in the normal way.
$$U = A X^\alpha \left(\frac{W - P_X X}{P_Y} \right)^\beta$$
$$= \frac{A}{P_Y^\beta} X^\alpha (W - P_X X)^\beta$$
$$\frac{A}{P_Y^\beta} = \text{constant}$$
since P_Y is constant. Putting $A/P_Y^\beta = K$ gives
$$U = K X^\alpha (W - P_X X)^\beta$$
U is maximized when $dU/dX = 0$ and $d^2U/dX^2 < 0$. Use of the product rule gives
$$dU/dX = \beta K X^\alpha (W - P_X X)^{\beta - 1}(- P_X) + \alpha K X^{\alpha - 1}(W - P_X X)^\beta$$
When $dU/dX = 0$ the right-hand side must equal zero. Dividing through by $K X^{\alpha - 1}(W - P_X X)^{\beta - 1}$ gives
$$\beta(- P_X)X + \alpha(W - P_X X) = 0$$
i.e.
$$\alpha W = \beta P_X X + \alpha P_X X = (\alpha + \beta)P_X X$$
thus
$$P_X X = \frac{\alpha W}{\alpha + \beta}$$
and
$$X = \frac{\alpha W}{(\alpha + \beta)P_X}$$
$$Y = \frac{W - P_X X}{P_Y}$$

thus

$$Y = \frac{W - \alpha W/(\alpha + \beta)}{P_Y} = \frac{\beta W}{(\alpha + \beta)P_Y}$$

Utility is maximized when

$$X = \frac{\alpha W}{(\alpha + \beta)P_X} \quad \text{and} \quad Y = \frac{\beta W}{(\alpha + \beta)P_Y}$$

When X and Y take these values the income constraint is obeyed;

$$P_X X + P_Y Y = \frac{\alpha W}{\alpha + \beta} + \frac{\beta W}{\alpha + \beta} = \frac{(\alpha + \beta)W}{\alpha + \beta} = W$$

ALTERNATIVE APPROACH: LAGRANGE MULTIPLIERS

For functions of more than two variables solution by means of method 1 becomes very cumbersome. Problems of this nature can be solved using the method of Lagrange multipliers. This technique will now be used to maximize the same utility function $U = AX^\alpha Y^\beta$ subject to the income constraint $P_X X + P_Y Y = W$.

To begin with we consider a new function

$$U^* = AX^\alpha Y^\beta - \lambda(P_X X + P_Y Y - W)$$

λ, the greek lambda, is a new variable and is called the Lagrange multiplier. In this example $\lambda = U_W^*$. Hence the Lagrange multiplier is the change in utility with respect to W, assuming $U^* = U$, or the marginal benefit obtainable from an increase of purchasing power, optimally allocated.

Suppose we know that at (X, Y) the function $f(X, Y)$ has zero first derivatives, i.e. $f_X = f_Y = 0$. The necessary first-order conditions for a maximum or minimum are thus satisfied. If we have not checked whether $f(X, Y)$ does in fact take a maximum or minimum we simply say that $f(X, Y)$ has a stationary value at (X, Y). Let $U^* = U - \lambda[f(X, Y) - K]$. For U^* to have a stationary value the following conditions must hold:

$$U_X^* = U_X - \lambda f_X = 0 \quad \text{thus} \quad U_X = \lambda f_X$$
$$U_Y^* = U_Y - \lambda f_Y = 0 \quad \text{thus} \quad U_Y = \lambda f_Y$$
$$U_\lambda^* = K - f(X, Y) = 0 \quad \text{thus} \quad f(X, Y) = K$$

Thus when U^* has a stationary value the constraint is obeyed and

$$U^* = U - \lambda.0 = U$$
$$dU = U_X dX + U_Y dY = \lambda f_X dX + \lambda f_Y dY = \lambda(f_X dX + f_Y dY)$$

But if $f(X, Y) = K$, $f_X dX + f_Y dY = 0$ so $dU = 0$. Thus when U^* has a stationary value, U subject to $f(X, Y) = K$ has a stationary value also. We therefore seek a stationary value of U^* knowing that when found it will also be a stationary value of U subject to the constraint.

Where $U^* = AX^\alpha Y^\beta - \lambda(P_X X + P_Y Y - W)$,

$$U_X^* = U_X - \lambda P_X = \alpha A X^{\alpha-1} Y^\beta - \lambda P_X = \alpha A \frac{X^\alpha}{X} Y^\beta - \lambda P_X = \frac{\alpha U}{X} - \lambda P_X$$

since $AX^\alpha Y^\beta = U$. When $U_X^* = 0$,

$$\frac{\alpha U}{X} - \lambda P_X = 0 \quad \text{or} \quad \frac{\alpha U}{P_X X} = \lambda \tag{1}$$

$$U_Y^* = U_Y - \lambda P_Y = \beta A X^\alpha Y^{\beta-1} - \lambda P_Y = \beta A X^\alpha \frac{Y^\beta}{Y} - \lambda P_Y = \frac{\beta U}{Y} - \lambda P_Y$$

When $U_Y^* = 0$,

$$\frac{\beta U}{Y} - \lambda P_Y = 0 \quad \text{or} \quad \frac{\beta U}{P_Y Y} = \lambda \tag{2}$$

When $U_\lambda^* = 0$,

$$P_X X + P_Y Y = W \tag{3}$$

From the three equations in three unknowns, X, Y, and λ it is possible to find the X and Y which will give a stationary value for U^*, and therefore U, and obey the income constraint. From equations (1) and (2)

$$\frac{\alpha U}{P_X X} = \frac{\beta U}{P_Y Y} \quad \text{or} \quad \frac{\alpha}{P_X X} = \frac{\beta}{P_Y Y}$$

thus

$$P_Y Y = \frac{\beta}{\alpha} P_X X$$

Substituting this value into equation (3) gives

$$P_X X + \frac{\beta}{\alpha} P_X X = W \quad \text{or} \quad P_X X = \frac{\alpha W}{\alpha + \beta}$$

thus

$$X = \frac{\alpha W}{(\alpha + \beta)P_X}$$

But

$$P_Y Y = \frac{\beta}{\alpha} P_X X = \frac{\beta}{\alpha}\left[\frac{\alpha W}{\alpha + \beta}\right]$$

thus

$$Y = \frac{\beta W}{(\alpha + \beta)P_Y}$$

This gives the same result as that derived using method 1. We have found the values of X and Y which will maximize utility subject to the income constraint. These are the coordinates of the point of tangency between the budget line and the highest possible indifference curve. The level of utility is obtained by inserting the above values for X and Y into the utility function. If we let the level of utility equal $\bar{U}$ then the highest possible indifference curve which is attainable is described by

$$\bar{U} = AX^\alpha Y^\beta$$

Utility is constant at a given level $\bar{U}$ along this indifference curve.

When Utility is Maximized the Marginal Rate of Commodity Substitution MRCS is Equal to Relative Prices

The *MRCS* at any point on an indifference curve is equal to the slope of its tangent. Taking total derivatives as in Chapter 7, we have

$$dU = U_X dX + U_Y dY$$

Utility is constant along an indifference curve, i.e. $dU = 0$. Consequently $U_X dX + U_Y dY = 0$, so

$$\frac{dY}{dX} = -\frac{U_X}{U_Y}$$

U_X = marginal utility of X, U_Y = marginal utility of Y, and $dY/dX = MRCS$. Taking the budget line $P_X X + P_Y Y = W$ gives

$$Y = -\frac{P_X}{P_Y} X + \frac{W}{P_Y}$$

thus

$$\frac{dY}{dX} = -\frac{P_X}{P_Y}$$

where dY/dX gives a measure of relative prices. Hence the *MRCS* is equal to relative prices, i.e. dY/dX on the indifference curve and budget line is the same, when

$$-\frac{U_X}{U_Y} = -\frac{P_X}{P_Y} \quad \text{or} \quad \frac{U_X}{U_Y} = \frac{P_X}{P_Y}$$

Returning to our problem of maximizing utility subject to an income constraint we see that utility is maximized when

$$U_X^* = 0 \quad \text{or} \quad U_X - \lambda P_X = 0$$

i.e. $U_X = \lambda P_X$ and

$$U_Y^* = 0 \quad \text{or} \quad U_Y - \lambda P_Y = 0$$

i.e. $U_Y = \lambda P_Y$, thus

$$\frac{U_X}{U_Y} = \frac{\lambda P_X}{\lambda P_Y} = \frac{P_X}{P_Y}$$

when utility is maximized. Consequently the *MRCS* is equal to relative prices.

ELASTICITY OF SUBSTITUTION AND THE COBB–DOUGLAS UTILITY FUNCTION

Maximizing U subject to the general income constraint gives

$$\frac{\alpha}{P_X X} = \frac{\beta}{P_Y Y}$$

see equations (1) and (2) above, thus

$$\frac{P_Y}{P_X} = \frac{X}{Y} \cdot \frac{\beta}{\alpha}$$

From this is possible to calculate the elasticity of substitution σ.

$$\sigma = \frac{\text{relative change in } X/Y}{\text{relative change in } P_Y/P_X}$$

$$= \frac{\Delta(X/Y)/(X/Y)}{\Delta(P_Y/P_X)/(P_Y/P_X)} = \frac{\Delta(X/Y)}{\Delta(P_Y/P_X)} \cdot \frac{P_Y/P_X}{X/Y}$$

The point elasticity expression will take the form

$$\sigma = \frac{d(X/Y)}{d(P_Y/P_X)} \cdot \frac{(P_Y/P_X)}{(X/Y)} \quad \text{or} \quad \frac{d(Y/X)}{d(P_X/P_Y)} \cdot \frac{(P_X/P_Y)}{(Y/X)}$$

$$\frac{P_Y}{P_X} = \frac{X}{Y} \cdot \frac{\beta}{\alpha}$$

thus

$$\frac{X}{Y} = \frac{\alpha}{\beta} \cdot \frac{P_Y}{P_X} \quad \text{and} \quad \frac{d(X/Y)}{d(P_Y/P_X)} = \frac{\alpha}{\beta}$$

so

$$\sigma = \frac{\alpha}{\beta} \cdot \frac{(X/Y)(\beta/\alpha)}{(X/Y)} = 1$$

i.e. the elasticity of substitution of a Cobb-Douglas type utility function is unity. This means that if we change relative prices by 5 per cent, commodity proportions will change by 5 per cent.

PROPERTIES OF DEMAND CURVES DERIVED FROM COBB–DOUGLAS TYPE UTILITY FUNCTIONS

We have seen that

$$X = \frac{\alpha W}{(\alpha + \beta)P_X} \quad \text{and} \quad Y = \frac{\beta W}{(\alpha + \beta)P_Y}$$

when utility is maximized subject to the income constraint. The above are the general demand functions for goods X and Y respectively.

From $X = \alpha W/(\alpha + \beta)P_X$ it is possible to calculate the quantity of good X which will maximize utility as the price of X varies and income W is held constant. If $\alpha = 0.25$, $\beta = 0.75$, and $W = 100$, then the demand function would take the form

$$X = \frac{0.25(100)}{(0.25 + 0.75)P_X} = \frac{25}{P_X} \quad \text{or} \quad P_X X = 25$$

These demand functions have unit price and income elasticities. Taking the demand function for X

$$X = \frac{\alpha W}{(\alpha + \beta)P_X}$$

and assuming income W is constant gives

$$E_D = X_P P_X \cdot \frac{P_X}{X} = -\frac{\alpha W}{(\alpha + \beta)P_X^2} \cdot \frac{P_X}{\alpha W/(\alpha + \beta)P_X}$$

$$= -\frac{\alpha W}{(\alpha + \beta)P_X^2} \cdot \frac{P_X^2(\alpha + \beta)}{\alpha W} = -1$$

If P_X is assumed constant then income elasticity of demand is

$$X_W \cdot \frac{W}{X} = \frac{\alpha}{(\alpha + \beta)P_X} \cdot \frac{W}{\alpha W/(\alpha + \beta)P_X} = 1$$

The demand function for Y will provide the same results.

Indifference Curves Derived from Cobb–Douglas Type Utility Functions are Convex to the Origin

For an indifference curve to be convex to the origin its first derivative must be negative and its second derivative must be positive. Our indifference curve will take the form

$$Z = AX^\alpha Y^\beta$$

where Z is a constant level of utility, i.e. $U = Z$, thus

$$Y = \left(\frac{Z}{A}\right)^{1/\beta} X^{-\alpha/\beta}$$

If we let the constant $(Z/A)^{1/\beta} = K$ and $\alpha/\beta = m$, then

$$Y = KX^{-m}$$

thus

$$\frac{dY}{dX} = -mKX^{-m-1} = -mK\left(\frac{1}{X^{m+1}}\right) < 0$$

and

$$\frac{d^2 Y}{dX^2} = (-m-1)(-m)KX^{-m-2} = (m+1)mK\left(\frac{1}{X^{m+2}}\right) > 0$$

Consequently the general indifference curve $Z = AX^\alpha Y^\beta$ is convex to the origin.

MAXIMIZATION OF OUTPUT SUBJECT TO A COST CONSTRAINT

Using a Cobb–Douglas production function, i.e. commodities X and Y are now replaced by factor inputs, the problem is to maximize $Q = AL^\alpha K^\beta$, where L is labour and K is capital, subject to the general cost constraint

$P_L L + P_K K = M$. Given that the production and utility functions are exactly the same except that Q replaces U, then output will be maximized when

$$L = \frac{\alpha M}{(\alpha + \beta)P_L} \quad \text{and} \quad K = \frac{\beta M}{(\alpha + \beta)P_K}$$

These are the coordinates of the point of tangency between the isocost line and the highest possible isoquant. The level of output can be obtained by inserting these values into the production function, i.e.

$$Q = A\left(\frac{\alpha M}{(a + \beta)P_L}\right)^\alpha \left(\frac{\beta M}{(\alpha + \beta)P_K}\right)^\beta$$

If we let this level of output equal $\bar{Q}$ then the highest possible isoquant attainable is

$$\bar{Q} = AL^\alpha K^\beta$$

Since the isocost function is tangential to this isoquant when output is maximized subject to the cost constraint this means dL/dK on the isoquant and isocost line is the same, i.e. the marginal rate of technical substitution $MRTS$ is equal to relative input prices at this point or

$$\frac{Q_L}{Q_K} = \frac{P_L}{P_K}$$

where Q_L = marginal product of labour and Q_K = marginal product of capital.

We have shown that $U_X/U_Y = P_X/P_Y$ when utility is maximized subject to an income constraint. When the utility function is replaced by the production function and output is maximized subject to a cost constraint then $Q_L/Q_K = P_L/P_K$ will hold. As with the utility function it can be shown that the isoquants of the Cobb–Douglas production function are convex to the origin and the elasticity of substitution is unity, i.e. a 10 per cent change in the relative prices of the factors will change factor proportions by 10 per cent.

COST MINIMIZATION SUBJECT TO AN OUTPUT CONSTRAINT

Often in economics a firm is faced with the problem of minimizing costs subject to an output constraint, i.e. the firm wishes to produce a certain amount at minimum cost. If the production function is $Q = AL^\alpha K^\beta$ and the firm wishes to produce at a level $Q = N$ then

$$N = AL^\alpha K^\beta$$

is the output constraint. The total cost function will take the form

$$TC = P_L L + P_K K$$

If the Lagrangian method is used we introduce a new function

$$TC^* = P_L L + P_K K - \lambda(AL^\alpha K^\beta - N)$$

When TC^* has a stationary value

$$TC_L^* = 0$$
$$TC_K^* = 0$$

and

$$TC_\lambda^* = 0$$

When $TC_\lambda^* = 0$ then $AL^\alpha K^\beta - N = 0$ or $AL^\alpha K^\beta = N$. Therefore the output constraint is satisfied and

$$TC^* = P_L L + P_K K - \lambda(0) = TC$$

Consequently, when TC^* has a stationary value TC has a stationary value also. The conditions for a stationary value, i.e. $TC_L^* = 0$, $TC_K^* = 0$ and $TC_\lambda^* = 0$, will provide three equations in three unknowns. From these equations it will be possible to find a value for L and K which will minimize costs and obey the output constraint. Here $\lambda = TC_N^*$. Hence the Lagrange Multiplier gives the rate of change in TC with respect to quantity or a measure of marginal cost, assuming $TC^* = TC$, i.e. assuming total costs are minimized.

Alternative Approach: Tangency of Isoquant and Isocost Lines

The output level is constant at N. It is therefore necessary to reach the isoquant $N = AL^\alpha K^\beta$. Since the total cost function is linear and prices are assumed fixed a parallel movement from the origin represents an increase in cost. Costs will therefore be at a minimum when this cost function is brought out from the origin until it touches the isoquant $AL^\alpha K^\beta = N$. Hence the slope of the tangent to the isoquant must equal the slope of the total cost function.

$$AL^\alpha K^\beta = N$$

is the isoquant, thus

$$dN = N_L dL + N_K dK$$

However N is constant along the isoquant, so

$$dN = 0$$

and

$$N_L dL + N_K dK = 0$$

i.e.

$$\frac{dL}{dK} = -\frac{N_K}{N_L}$$

Taking the total cost function $TC = P_L L + P_K K$ gives

$$L = \frac{TC}{P_L} - \frac{P_K}{P_L} K$$

and

$$\frac{dL}{dK} = -\frac{P_K}{P_L}$$

Therefore when costs are minimized $-N_K/N_L$ must equal $-P_K/P_L$, i.e. N_K/N_L must equal P_K/P_L.

$$N_L = \alpha A L^{\alpha-1} K^\beta$$
$$N_K = \beta A L^\alpha K^{\beta-1}$$

thus

$$\frac{N_K}{N_L} = \frac{\beta L}{\alpha K}$$

When costs are minimized subject to the output constraint

$$\frac{\beta L}{\alpha K} = \frac{P_K}{P_L}$$

thus

$$L = \frac{P_K}{P_L} \cdot \frac{\alpha}{\beta} K$$

Substituting this value for L into the output constraint gives a function entirely in K, i.e.

$$N = A \left(\frac{P_K}{P_L} \cdot \frac{\alpha}{\beta} K \right)^\alpha \cdot K^\beta$$

thus

$$K^{\alpha+\beta} = \frac{N}{A} \left(\frac{P_L}{P_K} \cdot \frac{\beta}{\alpha} \right)^\alpha$$

so

$$K = \left[\frac{N}{A} \left(\frac{P_L}{P_K} \cdot \frac{\beta}{\alpha} \right)^\alpha \right]^{1/(\alpha+\beta)}$$

By symmetry or by substituting the value for K into $L = (P_K/P_L) \cdot (\alpha/\beta) K$ we have

$$L = \left[\frac{N}{A} \left(\frac{P_K}{P_L} \cdot \frac{\alpha}{\beta} \right)^\beta \right]^{1/(\alpha+\beta)}$$

The minimum level of total costs is obtained by inserting these values for L and K into the TC function.

CONSTANT ELASTICITY OF SUBSTITUTION PRODUCTION FUNCTIONS

A *CES* production function takes the form

$$Q = A[bL^\alpha + gK^\alpha]^{1/\alpha}$$

or

$$(Q/A)^\alpha = bL^\alpha + gK^\alpha$$

where A, b, and α are the parameters, L and K are the factors of production, and $b + g = 1$.

If we wish to maximize Q subject to a cost constraint it is simpler to work with the maximand $(Q/A)^\alpha$ instead of Q. When $\alpha > 0$ then $(Q/A)^\alpha$ is an increasing function of Q. Therefore maximizing $(Q/A)^\alpha$ means that Q is automatically maximized. When $\alpha < 0$ then $(Q/A)^\alpha$ is a decreasing function of Q, i.e. its value decreases as Q increases. Consequently, maximizing $(Q/A)^\alpha$ will automatically minimize Q. Hence when $\alpha < 0$ minimizing $(Q/A)^\alpha$ will maximize Q.

Suppose we wish to maximize Q subject to the cost constraint

$$P_L L + P_K K = M$$

If $(Q/A)^\alpha$ is maximized Q is automatically maximized, assuming $\alpha > 0$. Using the Lagrangian expression and maximizing $(Q/A)^\alpha$ subject to the constraint $P_L L + P_K K = M$ gives

$$Q^* = (Q/A)^\alpha - \lambda(P_L L + P_K K - M)$$

or

$$Q^* = bL^\alpha + gK^\alpha - \lambda(P_L L + P_K K - M)$$

Q^* has a stationary value when

$$Q_L^* = 0$$
$$Q_K^* = 0$$

and

$$Q_\lambda^* = 0$$

When $Q_\lambda^* = 0$, then $P_L L + P_K K - M = 0$ or $P_L L + P_K K = M$, thus

$$Q^* = (Q/A)^\alpha - \lambda(0) = (Q/A)^\alpha$$

Thus when Q^* has a stationary value, $(Q/A)^\alpha$ is maximized and Q is also maximized. We will therefore continue to seek a stationary value of Q^*

$$Q_L^* = \alpha bL^{\alpha-1} - \lambda P_L$$
$$Q_K^* = \alpha gK^{\alpha-1} - \lambda P_K$$

When $Q_L^* = Q_K^* = 0$, then

and

$$\alpha bL^{\alpha-1} = \lambda P_L$$

$$\alpha gK^{\alpha-1} = \lambda P_K$$

thus

$$\frac{\alpha bL^{\alpha-1}}{\alpha gK^{\alpha-1}} = \frac{\lambda P_L}{\lambda P_K}$$

i.e.

$$\left(\frac{L}{K}\right)^{\alpha-1} = \frac{g}{b}\cdot\frac{P_L}{P_K}$$

or

$$\left(\frac{K}{L}\right)^{1-\alpha} = \frac{g}{b}\cdot\frac{P_L}{P_K}$$

thus

$$\frac{K}{L} = \left(\frac{g}{b}\right)^{1/(1-\alpha)}\left(\frac{P_L}{P_K}\right)^{1/(1-\alpha)} \tag{4}$$

$$\alpha = -\frac{1}{\sigma} + 1$$
$$= \frac{-1+\sigma}{\sigma}$$
$$\sigma = \frac{-1+\sigma}{\alpha}$$
$$\sigma = \frac{\sigma-1}{\alpha}$$
$$\sigma - \sigma = -\frac{1}{\alpha}$$
$$\sigma\left(1-\frac{1}{\alpha}\right) = -\frac{1}{\alpha}$$
$$\sigma\left(\frac{\alpha-1}{\alpha}\right) = -\frac{1}{\alpha}$$
$$\sigma = -\frac{1}{\alpha-1}$$

$$\frac{1}{1+p}$$
$$-\frac{1}{p-1}$$

$$\alpha = -\frac{1}{\sigma}+1$$
$$\sigma = -\frac{1}{\alpha-1}$$

If we let $1/(1-\alpha) = s$, then

$$\frac{K}{L} = \left(\frac{g}{b}\right)^s \left(\frac{P_L}{P_K}\right)^s \quad \text{or} \quad K = \left(\frac{g}{b}\right)^s \left(\frac{P_L}{P_K}\right)^s L$$

To find the L and K which will maximize output subject to the cost constraint we use the condition, $Q_\lambda^* = 0$, and solve for L.

When $Q_\lambda^* = 0$

$$P_L L + P_K K = M$$

thus

$$P_L L + P_K \left[\left(\frac{g}{b}\right)^s \left(\frac{P_L}{P_K}\right)^s\right] L = M$$

i.e.

$$P_L b^s L + P_K^{1-s} P_L^s g^s L = b^s M$$

so

$$P_L^s (P_L^{1-s} b^s + P_K^{1-s} g^s) L = b^s M$$

i.e.

$$L = \frac{b^s}{P_L^s} \left[\frac{M}{P_L^{1-s} b^s + P_K^{1-s} g^s}\right]$$

By symmetry, or by substituting the value for L into $K = (g/b)^s (P_L/P_K)^s L$, we have

$$K = \frac{g^s}{P_K^s} \left[\frac{M}{P_L^{1-s} b^s + P_K^{1-s} g^s}\right]$$

Elasticity of Substitution of the CES Function

If we let $(g/b)^s = Z$ then from equation (4)

$$\frac{K}{L} = Z \left(\frac{P_L}{P_K}\right)^s$$

and

$$\sigma = \frac{d(K/L)}{d(P_L/P_K)} \cdot \frac{(P_L/P_K)}{(K/L)}$$

$$= sZ(P_L/P_K)^{s-1} \cdot \frac{(P_L/P_K)}{Z(P_L/P_K)^s} = s = \frac{1}{1-\alpha}$$

This production function has a constant elasticity of substitution, namely $1/(1-\alpha)$.

Further Comments on Lagrange Multipliers

In all of the above examples one of the functions was linear. However, it is possible to solve problems where both functions are curvilinear by the Lagrange Multiplier method. It should also be added that we have only dealt with the first-order conditions for a maximum or minimum. Since all of the problems considered had only one possible solution the second-order conditions were not essential. However, for many economic problems it is not sufficient to satisfy the first-order condition.

EXERCISES

1. Two consumers have the following utility functions

$$\text{Consumer 1} \quad U = X^{0.4} Y^{0.6}$$
$$\text{Consumer 2} \quad U = 2X^{0.5} Y^{0.25}$$

where X and Y are two goods. If $P_X = 2$ and $P_Y = 3$ and each consumer has an income constraint of 45

(i) Find the quantities of X and Y which will maximize utility, subject to the income constraint, for each consumer.

(ii) Show how a rise in income to 60 will affect the quantity of X and Y which will maximize utility for each consumer.

2. Two producers have the following production functions

$$\text{Producer 1} \quad Q = 50L^{1/3} K^{2/3}$$
$$\text{Producer 2} \quad Q = 2L^{0.75} K^{0.5}$$

L and K are factor inputs which the entrepreneurs purchase in perfectly competitive markets at constant prices. If $P_L = 4$, $P_K = 6$, and each producer has total costs of 72, find the maximum output for each producer.

$\times$ 3. Two producers have the following production functions

(i) $Q = L^{3/8} K^{5/8}$

(ii) $Q = 5L^{1/2} K^{1/3}$

If $P_L = 4$ and $P_K = 5$, calculate the L and K which will produce an output of 200 at minimum costs.

4. If X and Y are different processes producing the same good and the joint cost function is given by

$$TC = X^2 + 2Y^2 - XY$$

Find the X and Y which minimize costs when the prescribed level of output is 16.

$\times$ 5. A firm's production function is

$$Q = 20[\tfrac{3}{4}L^{-1/4} + \tfrac{1}{4}K^{-1/4}]^{-4}$$

If the firm's isocost function is $4L + 3K = 80$ calculate $\surd$

(i) The L and K which maximizes output.

(ii) The marginal rate of technical substitution between L and K, and show that it equals relative input prices at the equilibrium output level. $\surd$

(iii) The L and K which will produce an output of 120 at minimum costs.

SOLUTIONS TO EXERCISES

1. (i) Maximizing a general utility function

$$U = AX^{\alpha} Y^{\beta}$$

subject to a general income constraint

$$P_X X + P_Y Y = W$$

gives

$$X = \frac{\alpha W}{(\alpha + \beta) P_X} \quad \text{and} \quad Y = \frac{\beta W}{(\alpha + \beta) P_Y}$$

Consumer 1

The income constraint is $2X + 3Y = 45$ and $U = X^{0 \cdot 4} Y^{0 \cdot 6}$, i.e. $A = 1$, $\alpha = 0 \cdot 4$, $\beta = 0 \cdot 6$, $P_X = 2$, $P_Y = 3$ and $W = 45$. Thus

$$X = \frac{0 \cdot 4(45)}{(0 \cdot 4 + 0 \cdot 6)(2)} = 9 \qquad *$$

and

$$Y = \frac{0 \cdot 6(45)}{(0 \cdot 4 + 0 \cdot 6)(2)} = 9 \qquad *$$

when utility is maximized.

Consumer 2

The income constraint is the same but $U = 2X^{0 \cdot 5} Y^{0 \cdot 25}$, i.e. $A = 2$, $\alpha = 0 \cdot 5$ and $\beta = 0 \cdot 25$. Thus

$$X = \frac{0 \cdot 5(45)}{(0 \cdot 5 + 0 \cdot 25)(2)} = 15 \qquad *$$

and

$$Y = \frac{0 \cdot 25(45)}{(0 \cdot 5 + 0 \cdot 25)(3)} = 5 \qquad *$$

when utility is miximized.

(ii) W increases to 60, i.e. an increase of a third. Therefore the quantity of X and Y which maximizes utility will also increase by a third since income elasticity of demand is unity.

Consumer 1

$$X = Y = 12$$

Consumer 2

$$X = 20 \quad \text{and} \quad Y = 6\tfrac{2}{3} \qquad *$$

2. We now have a production function and a cost constraint so that factor inputs L and K take the place of commodities X and Y. The income constraint is now replaced by the cost constraint.

Producer 1

Our problem is to maximize $Q = 50 L^{1/3} K^{2/3}$ subject to the cost constraint $4L + 6K = 72$. Output is maximized when

$$L = \frac{72(\tfrac{1}{3})}{(\tfrac{1}{3} + \tfrac{2}{3})4} = 6$$

and

$$K = \frac{72(\tfrac{2}{3})}{(\tfrac{1}{3} + \tfrac{2}{3})6} = 8$$

Consequently, maximum $Q = 50(6)^{1/3}(8)^{2/3}$. $\qquad *$

Producer 2

Maximizing $Q = 2L^{0.75}K^{0.5}$ subject to the cost constraint $4L + 6K = 72$ gives

$$L = \frac{0.75(72)}{(0.75 + 0.5)(4)} = 10.8$$

and

$$K = \frac{0.5(72)}{(0.75 + 0.5)(6)} = 4.8$$

Thus maximum $Q = 2(10.8)^{0.75}(4.8)^{0.5}$.　　　　　　　　　　*

3. Costs will be at a minimum when the isocost function is tangential to the isoquants

$$200 = L^{3/8}K^{5/8} \quad \text{producer (i)}$$
$$200 = 5L^{1/2}K^{1/3} \quad \text{producer (ii)}$$

i.e. when the marginal rate of technical substitution along the relevant isoquant is equal to relative input prices or when dL/dK on the isocost function is equal to dL/dK on the isoquant.

$$Q \text{ is constant at 200, i.e. } dQ = 0$$

thus

$$dQ = Q_L dL + Q_K dK = 0$$

along the isoquant, thus

$$MRTS = \frac{dL}{dK} = -\frac{Q_K}{Q_L}$$

so costs are minimized when

$$-\frac{Q_K}{Q_L} = -\frac{P_K}{P_L}$$

i.e. when

$$\frac{Q_K}{Q_L} = \frac{P_K}{P_L}$$

Producer (i)

$$\frac{Q_K}{Q_L} = (\tfrac{5}{8}L^{3/8}K^{-3/8})/(\tfrac{3}{8}L^{-5/8}K^{5/8})$$

$$= \frac{5}{3}\cdot\frac{L}{K}$$

when costs are at a minimum $5/3.L/K = P_K/P_L = 5/4$ thus

$$4L = 3K \quad \text{or} \quad L = \tfrac{3}{4}K$$

Substituting into the isoquant function gives

$$200 = \left(\frac{3}{4}\right)^{3/8}K^{3/8}K^{5/8} \quad \text{or} \quad K = 200 \bigg/ \left(\frac{3}{4}\right)^{3/8} = 200\left(\frac{4}{3}\right)^{3/8}$$ 　　*

and

$$L = 200\left(\frac{4}{3}\right)^{3/8}\left(\frac{3}{4}\right) = 200\left(\frac{3}{4}\right)^{5/8}$$ 　　*

Producer (ii)

$$\frac{Q_K}{Q_L} = \left(\frac{5}{3}L^{1/2}K^{-2/3}\right)\bigg/\left(\frac{5}{2}L^{-1/2}K^{1/3}\right)$$

$$= \frac{2}{3}\cdot\frac{L}{K}$$

when costs are minimized $2/3.L/K = 5/4$ thus

$$8L = 15K \quad \text{or} \quad L = \frac{15}{8}K$$

Substituting into the isoquant function gives

$$200 = 5.\left(\frac{15}{8}\right)^{1/2} K^{1/2} K^{1/3} \quad \text{or} \quad 40 = \left(\frac{15}{8}\right)^{1/2} K^{5/6}$$

so

$$K = \left[40\left(\frac{8}{15}\right)^{1/2}\right]^{6/5} \qquad *$$

and

$$L = \left(\frac{15}{8}\right)40^{6/5}\left(\frac{8}{15}\right)^{3/5} = 40^{6/5}\left(\frac{15}{8}\right)^{2/5} \qquad *$$

Solution by means of the Lagrange Multiplier

The output constraint of producer (ii) is

$$200 = 5L^{1/2}K^{1/3}$$

and his total cost function is

$$TC = 4L + 5K$$

To minimize TC let us introduce a new function

$$TC^* = 4L + 5K - \lambda(5L^{1/2}K^{1/3} - 200)$$

When TC^* has a stationary value

$$TC_L^* = 0, \quad TC_K^* = 0 \quad \text{and} \quad TC_\lambda^* = 0$$

When $TC_\lambda^* = 0$ then $5L^{1/2}K^{1/3} = 200$, i.e. the output constraint is satisfied and

$$TC^* = 4L + 5K - \lambda(0) = TC$$

Consequently when TC^* has a stationary value TC is minimized.

When $TC_L^* = 0$, then $4 - 5/2.\lambda L^{-1/2}K^{1/3} = 0$, i.e. $\lambda = 8/5.L^{1/2}/K^{1/3}$.

When $TC_K^* = 0$, then $5 - 5/3.\lambda L^{1/2}K^{-2/3} = 0$, i.e. $\lambda = 3K^{2/3}/L^{1/2}$, so

$$\frac{8}{5}.\frac{L^{1/2}}{K^{1/3}} = \frac{3K^{2/3}}{L^{1/2}}.$$

and

$$8L = 15K \quad \text{or} \quad L = \frac{15}{8}K$$

Since $5L^{1/2}K^{1/3} = 200$, i.e. $TC_\lambda^* = 0$, then substituting this value for L will solve for K; see alternative approach.

4. The problem is to minimize the function

$$TC = X^2 + 2Y^2 - XY$$

subject to the constraint $X + Y = 16$.

Using the Lagrangian expression

$$TC^* = X^2 + 2Y^2 - XY - \lambda(X + Y - 16)$$

TC^* will take a stationary value when

$$TC_X^* = 0$$
$$TC_Y^* = 0$$
$$TC_\lambda^* = 0$$

When $TC^*_\lambda = 0$, then $X + Y - 16 = 0$ and
$$TC^* = X^2 + 2Y^2 - XY - \lambda(0) = TC$$
Therefore when TC^* has a stationary value TC is at a minimum.
When $TC^*_X = 0$,
$$2X - Y - \lambda = 0$$
When $TC^*_Y = 0$,
$$4Y - X - \lambda = 0$$
$$\lambda = 2X - Y = 4Y - X$$
thus
$$3X - 5Y = 0 \tag{5}$$
When $TC^*_\lambda = 0$,
$$X + Y = 16 \tag{6}$$
Multiplying equation (6) by 3 and subtracting equation (5) gives
$$8Y = 48 \quad \text{or} \quad Y = 6 \qquad *$$
When $Y = 6$, then $X = 10$. $\qquad *$

5. (i) $Q = 20[\frac{3}{4}L^{-1/4} + \frac{1}{4}K^{-1/4}]^{-4}$
thus
$$\left(\frac{Q}{20}\right)^{-1/4} = \frac{3}{4}L^{-1/4} + \frac{1}{4}K^{-1/4}$$
If $(Q/20)^{-1/4}$ is minimized, then Q is automatically maximized since $(Q/20)^{-1/4}$ decreases as Q increases. Using the Lagrangian expression and minimizing $(Q/20)^{-1/4}$ subject to the constraint $4L + 3K = 80$ gives
$$Q^* = (Q/20)^{-1/4} - \lambda(4L + 3K - 80)$$
or
$$Q^* = \frac{3}{4}L^{-1/4} + \frac{1}{4}K^{-1/4} - \lambda(4L + 3K - 80)$$
Q^* will take a stationary value when
$$Q^*_L = 0$$
$$Q^*_K = 0$$
and
$$Q^*_\lambda = 0$$
When $Q^*_\lambda = 0$, then $4L + 3K - 80 = 0$ and
$$Q^* = \frac{3}{4}L^{-1/4} + \frac{1}{4}K^{-1/4} - \lambda(0) = (Q/20)^{-1/4}$$
Therefore when Q^* has a stationary value, $(Q/20)^{-1/4}$ is at a minimum and Q is maximized,
$$Q^*_L = -\frac{3}{16}L^{-5/4} - 4\lambda$$
$$Q^*_K = -\frac{1}{16}K^{-5/4} - 3\lambda$$
When $Q^*_L = Q^*_K = 0$, then
$$-\frac{3}{16}L^{-5/4} - 4\lambda = 0 \text{ i.e. } -\frac{3}{64}L^{-5/4} = \lambda$$
and
$$-\frac{1}{16}K^{-5/4} - 3\lambda = 0 \text{ i.e. } -\frac{1}{48}L^{-5/4} = \lambda$$

thus
$$\tfrac{3}{4}L^{-5/4} = \tfrac{1}{3}K^{-5/4}$$

i.e.
$$9K^{5/4} = 4L^{5/4}$$

or
$$L = \left(\frac{9}{4}\right)^{4/5} K$$

$4L + 3K = 80$, i.e. $Q_\lambda^* = 0$, thus
$$4\left(\frac{9}{4}\right)^{4/5} K + 3K = 80$$

or
$$K = \frac{80}{4(9/4)^{4/5} + 3} \qquad *$$

and
$$L = \frac{80(9/4)^{4/5}}{4(9/4)^{4/5} + 3} \qquad *$$

(ii) When Q is maximized at a level $\bar{Q}$ then the $MRTS$ is given by the slope of the tangent to the isoquant
$$\bar{Q} = 20[\tfrac{3}{4}L^{-1/4} + \tfrac{1}{4}K^{-1/4}]^{-4}$$

Along this isoquant $(Q/20)^{-1/4}$ is constant since Q is constant at $\bar{Q}$. If we let $(\bar{Q}/20)^{-1/4} = Z$, then
$$dZ = Z_L dL + Z_K dK$$

Along the $\bar{Q}$ isoquant dZ is constant, thus
$$Z_L dL + Z_K dK = 0$$

i.e.
$$\frac{dL}{dK} = -\frac{Z_K}{Z_L} = MRTS$$

Taking the isocost function $4L + 3K = 80$ gives
$$\frac{dL}{dK} = -\frac{3}{4} = -\frac{P_K}{P_L}$$

$MRTS$ will equal relative prices if
$$-\frac{Z_K}{Z_L} = -\frac{P_K}{P_L} \quad \text{i.e.} \quad \frac{Z_K}{Z_L} = \frac{P_K}{P_L}$$

In part (i) we saw that when Q is maximized
$$Z_L - P_L \lambda = 0 \quad \text{or} \quad Z_L = P_L \lambda \quad \text{i.e.} \quad Q_L^* = 0$$

and
$$Z_K - P_K \lambda = 0 \quad \text{or} \quad Z_K = P_K \lambda \quad \text{i.e.} \quad Q_K^* = 0$$

thus
$$\frac{Z_K}{Z_L} = \frac{P_K}{P_L}$$

i.e.
$$MRTS = \text{relative prices}$$

(iii) Costs will be at a minimum when the isocost function is tangential to the isoquant
$$120 = 20[\tfrac{3}{4}L^{-1/4} + \tfrac{1}{4}K^{-1/4}]^{-4}$$

i.e. when the marginal rate of technical substitution along this isoquant is equal to relative input prices. Q is constant at 120, so

$$dQ = Q_L dL + Q_K dK = 0$$

along the isoquant, thus costs are minimized when $-Q_K/Q_L = -P_K/P_L$, i.e.

$$\frac{Q_K}{Q_L} = \frac{P_K}{P_L}$$

$$Q = 20[\tfrac{3}{4}L^{-1/4} + \tfrac{1}{4}K^{-1/4}]^{-4}$$

thus

$$\frac{-80[\ ]^{-5}(-1/16)K^{-5/4}}{-80[\ ]^{-5}(-3/16)L^{-5/4}} = \frac{3}{4}$$

where $[\ ] = \tfrac{3}{4}L^{-1/4} + \tfrac{1}{4}K^{-1/4}$, thus

$$\frac{1}{3}\frac{K^{-5/4}}{L^{-5/4}} = \frac{3}{4}$$

i.e.

$$L = \left(\frac{9}{4}\right)^{4/5} K$$

Substituting into the isoquant function gives

$$120 = 20\left[\frac{3}{4}\left(\frac{9}{4}\right)^{-1/5} K^{-1/4} + \frac{1}{4}K^{-1/4}\right]^{-4}$$

thus

$$K = 6\left[\frac{3}{4}\left(\frac{9}{4}\right)^{-1/5} + \frac{1}{4}\right]^{4} \qquad *$$

and

$$L = 6\left[\frac{3}{4}\left(\frac{9}{4}\right)^{-1/5} + \frac{1}{4}\right]^{4}\left(\frac{9}{4}\right)^{4/5} = 6\left[\frac{3}{4} + \frac{1}{4}\left(\frac{9}{4}\right)^{1/5}\right]^{4} \qquad *$$

when costs are at a minimum.

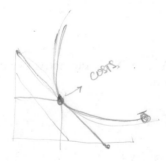

CHAPTER 9

Integration

If we are given the derivative of a function, then integration is the process of deriving the function itself. If $f'(Q)$ is the derivative of $f(Q)$, then $f(Q)$ is the integral of $f'(Q)$ and is written in the form

$$f(Q) = \int f'(Q) \, dQ$$

The sign $\int$ is the integral sign while the function $f'(Q)$ is the integrand, i.e. the function whose integral is to be found. The dQ is added to indicate the variable with respect to which the function $f'(Q)$ is to be integrated. It also acts as an indication of where the function to be integrated finishes.

Integration is thus the reverse of differentiation. Unlike differentiation, however, there is no general rule for integration. Indeed a function need not have an integral. However, most of the functions one finds in economics do have integrals. Knowledge of the results of differentiation is essential if one is to derive the integral of a function.

STANDARD FORM INTEGRALS

Suppose we wish to find the integral of X^2; i.e. X^2 is the derivative of a function and we want to discover the function.

$$X^2 = f'(X)$$
$$f(X) = \int f'(X) \, dX = ?$$

We know that $dX^3/dX = 3X^2$, thus

$$d(X^3/3)/dX = 3X^2/3 = X^2$$

Consequently the function whose derivative is X^2 equals $X^3/3$. $X^3/3$ is not *the* integral of X^2 but an integral. Consider the function

$$\frac{X^3}{3} + k$$

where k is a constant. The derivative of $(X^3/3) + k$, with respect to X is also X^2. Hence the function whose derivative equals X^2 is $(X^3/3) + k$, where k is called the *constant of integration*. Further information is necessary before a definite value can be assigned to k.

146

The function whose derivative is X^{10} is given by

$$f(X) = \int X^{10} dX = \frac{X^{11}}{11} + k$$

where k is the constant of integration.

It is possible to check by taking the derivative of $(X^{11}/11) + k$, with respect to X, e.g.

$$\frac{d[(X^{11}/11) + k]}{dX} = \frac{11 X^{10}}{11} = X^{10}$$

Suppose we wish to find the integral of X^n, i.e.

$$X^n = f'(X)$$

then

$$f(X) = \int X^n dX = \frac{X^{n+1}}{n+1} + k$$

because

$$\frac{d\{[X^{n+1}/(n+1)] + k\}}{dX} = \frac{(n+1)X^n}{n+1} = X^n$$

When $n = 4$, then

$$f(X) = \int X^4 dX = \frac{X^{4+1}}{4+1} + k = \frac{X^5}{5} + k$$

When $n = -2$, then

$$f(X) = \int X^{-2} dX = \frac{X^{-2+1}}{-2+1} + k = \frac{X^{-1}}{-1} + k = -\frac{1}{X} + k$$

This works for all values of n except $n = -1$. This special case will be considered in Chapter 14.

Integral of a Constant

If the constant b is the derivative of a function, i.e. $b = f'(X)$, then

$$f(X) = \int b dX = bX + k$$

because

$$d(bX + k)/dX = b$$

Integral of a Sum

The integral of $X^\alpha + X^\beta + X^\gamma + b$, where α, β, γ, and b are constants is

$$\int X^\alpha dX + \int X^\beta dX + \int X^\gamma dX + \int b dX = \frac{X^{\alpha+1}}{\alpha+1} + \frac{X^{\beta+1}}{\beta+1} + \frac{X^{\gamma+1}}{\gamma+1} + bX + k$$

Integral of a Multiple

The integral of a constant multiple of any integrand is the constant times the integral of that integrand, e.g.

$$\int b(X^\alpha + a)\mathrm{d}X = b\int (X^\alpha + a)\mathrm{d}X = b\left[\frac{X^{\alpha+1}}{\alpha+1} + aX + k\right]$$

where k is the constant of integration.

ECONOMIC APPLICATIONS OF THE INTEGRAL

Marginal and Total Cost

$$MC = \mathrm{d}(TC)/\mathrm{d}Q$$

thus

$$TC = \int MC\mathrm{d}Q$$

Hence if the MC function of a firm is $MC = a + bQ$, where a and b are positive constants, then

$$TC = \int(a + bQ)\mathrm{d}Q = aQ + 0{\cdot}5bQ^2 + k$$

The value of the constant of integration k will depend upon fixed costs, and since these are likely to be positive k will be > 0.

The MPC and the Consumption Function

Given that the marginal propensity to consume is $0{\cdot}8$, i.e.

$$\mathrm{d}C/\mathrm{d}Y = 0{\cdot}8$$

the consumption function will take the form

$$C = \int MPC\mathrm{d}Y = \int 0{\cdot}8\mathrm{d}Y = 0{\cdot}8Y + k$$

The constant of integration k gives the level of consumption when income is zero.

Marginal and Total Revenue

$$MR = \mathrm{d}(TR)/\mathrm{d}Q$$

thus

$$TR = \int MR\mathrm{d}Q$$

If a firm's MR function is

$$MR = a - bQ$$

where a and b are positive constants, then

$$TR = \int(a - bQ)\mathrm{d}Q = aQ - 0{\cdot}5bQ^2 + k$$

k will equal zero because TR is zero when Q is zero.

THE DEFINITE INTEGRAL

When an integral is taken over a definite range it is called a definite integral. Suppose we are asked to find the integral of the MC function

$$MC = a + bQ$$

where a and b are > 0, over the range $Q = m$ to $Q = n$, where m and n are constants, m being the smaller. This is written in the form

$$\int_{Q=m}^{Q=n} (a + bQ)\mathrm{d}Q \quad \text{or} \quad \int_{m}^{n} (a + bQ)\mathrm{d}Q$$

$Q = m$ at the lower limit and $Q = n$ at the upper limit. The indefinite form of the integral is first of all calculated and bracketed and the limits are written outside the bracket. The bracket is then evaluated when $Q = m$, the lower limit, and this value is subtracted from the value of the bracket when $Q = n$, e.g.

$$\int_{m}^{n} (a + bQ)\mathrm{d}Q = [aQ + 0.5bQ^2]_{m}^{n}$$
$$= [an + 0.5bn^2] - [am + 0.5bm^2]$$
$$= a(n - m) + 0.5b(n^2 - m^2)$$

The constant of integration need not be included when dealing with definite integrals. If it were included in the above example it would disappear on subtraction, e.g.

$$\int_{m}^{n} (a + bQ)\mathrm{d}Q = [aQ + 0.5bQ^2 + k]_{m}^{n}$$
$$= an + 0.5bn^2 + k - am - 0.5m^2 - k$$
$$= a(n - m) + 0.5b(n^2 - m^2)$$

This result gives us a measure of total variable costs when output increases from m to n. The definite integral is thus an area under a given curve since the area under the marginal cost curve equals total variable costs.

MEASURING CHANGES IN CAPITAL STOCK

Net investment I is the rate of change in capital stock, i.e. $\mathrm{d}K/\mathrm{d}t = I$. If net investment is a function of time it is possible to calculate the change in capital stock over some period of time by finding the definite integral of I with respect to time.

If $I = at^\beta$, where a and β are positive constants, then the change in capital stock between period $t = 0$ and $t = t^*$ is

$$\int_{0}^{t^*} I\,\mathrm{d}t = \int_{0}^{t^*} at^\beta\,\mathrm{d}t = \left[\frac{at^{\beta+1}}{\beta+1}\right]_{0}^{t^*} = \frac{a(t^*)^{\beta+1}}{\beta+1} - 0 = \frac{a(t^*)^{\beta+1}}{\beta+1}$$

This is also the area under the investment curve from $t = 0$ to $t = t^*$.

THE DEFINITE INTEGRAL: GEOMETRICAL APPROACH

To show that the definite integral measures the area enclosed by a curve over a given range let us take a continuous function, $Y = f'(X)$, and find the area under the curve $Y = f'(X)$ over a certain range.

Figure 29 gives a continuous function $Y = f'(X)$. Suppose we wish to find the area under this curve over the range AB, i.e. the area $ABCD$. Let the

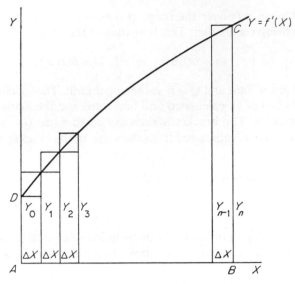

Figure 29   Upper and lower bounds and integration

interval $AB = m$. Subdivide AB into n equal parts each ΔX in length. On each subinterval we have constructed rectangular blocks.

Let S_1 = sum of smaller rectangles, and let S_2 = sum of larger rectangles. Then

$$S_2 > ABCD > S_1$$

As $\Delta X \to 0$, $S_2 - S_1 \to 0$, and $ABCD \to S_1$. In the limit,

$$ABCD = \Delta X \cdot Y_0 + \Delta X \cdot Y_1 + \Delta X \cdot Y_2 + \ldots + \Delta X \cdot Y_{n-2} + \Delta X \cdot Y_{n-1}$$
$$Y = f'(X)$$

When $Y = Y_0$ then $X = 0$ and $Y_0 = f'(0)$

When $Y = Y_1$ then $X = \Delta X$ and $Y_1 = f'(\Delta X)$

When $Y = Y_{n-1}$ then $X = (n-1)\Delta X$ and $Y_{n-1} = f'(\overline{n-1}\Delta X)$

$$ABCD = \Delta X f'(0) + \Delta X f'(\Delta X) + \ldots + \Delta X f'(\overline{n-2}\Delta X) + \Delta X f'(\overline{n-1}\Delta X)$$

The bar over $\overline{n-1}$ is used instead of brackets, i.e.

$$f'(\overline{n-1}\Delta x) \equiv f'[(n-1)\Delta x]$$

When dealing with differentiation

$$Y = f(X)$$
$$Y + \Delta Y = f(X + \Delta X)$$

and

$$\Delta Y = f(X + \Delta X) - f(X)$$

In the limit

$$f'(X) = \frac{f(X + \Delta X) - f(X)}{\Delta X}$$

or

$$f'(X)\Delta X = f(X + \Delta X) - f(X)$$

Applying this to $\Delta X f'(0)$ gives $f(0 + \Delta X) - f(0)$, thus

$$
\begin{aligned}
ABCD &= [f(0 + \Delta X) - f(0)] + [f(\Delta X + \Delta X) - f(\Delta X)] + \dots \\
&\quad + [f(\overline{n-2}\Delta X + \Delta X) - f(\overline{n-2}\Delta X)] + [f(\overline{n-1}\Delta X + \Delta X) \\
&\quad - f(\overline{n-1}\Delta X)] \\
&= [f(\Delta X) - f(0)] + [f(2\Delta X) - f(\Delta X)] + \dots + [f(\overline{n-1}\Delta X) \\
&\quad - f(\overline{n-2}\Delta X)] + [f(n\Delta X) - f(\overline{n-1}\Delta X)] \\
&= f(n\Delta X) - f(0)
\end{aligned}
$$

But $AB = m = n\Delta X$, thus

$$ABCD = f(m) - f(0)$$

$$= [f(X)]_0^m = \int_0^m f'(X)\mathrm{d}X$$

Consequently the definite integral measures the area enclosed by a curve over a given range.

CONSUMER'S SURPLUS

A demand relation gives the price at which any given quantity could be sold. However, in a competitive market the price does not reflect what consumers would be wiliing to pay for each unit of the good rather than go without. In fact the price reflects the valuation they place on the last unit they buy.

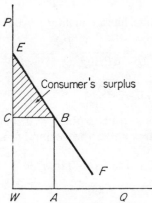

Figure 30   The demand curve
and consumer's surplus

Figure 30 shows a demand function EF, with price on the vertical and quantity demanded on the horizontal axis. At price WC, quantity demanded is WA. $WABC$ is the amount paid by consumers. However, the benefit derived by consumers is the total area under the demand function over the range WA, i.e.

$$WABE = \text{benefit derived by consumers}$$

thus

$$\text{Consumer's surplus} = WABE - WABC = CBE$$

Assuming the demand function is $P = a - bQ$, where a and b are positive constants, then

$$WABE = \int_0^{q^*} (a - bQ)dQ$$

where $q^* = WA$

$$\int_0^{q^*} (a - bQ)dQ = [aQ - 0{\cdot}5bQ^2]_0^{q^*}$$
$$= aq^* - 0{\cdot}5b(q^*)^2 - a.0 - 0{\cdot}5b(0)$$

i.e.

$$WABE = aq^* - 0{\cdot}5b(q^*)^2$$
$$WABC = P.Q = Pq^*$$

but at $B, P = a - bq^*$, thus

$$WABC = (a - bq^*)q^* = aq^* - b(q^*)^2$$
$$\text{Consumer's surplus} = aq^* - 0{\cdot}5b(q^*)^2 - aq^* + b(q^*)^2 = 0{\cdot}5b(q^*)^2$$

Consumer's Surplus and Constant Elasticity of Demand

A demand curve of the form

$$PQ^\beta = a \quad \text{or} \quad P = aQ^{-\beta}$$

where a and β are constants, has a constant price elasticity of demand.

$$E_D = \frac{dQ}{dP} \cdot \frac{P}{Q} = \frac{1}{dP/dQ} \cdot \frac{P}{Q} = \frac{1}{-a\beta Q^{-\beta-1}} \cdot \frac{aQ^{-\beta}}{Q}$$

$$= \frac{-1}{a\beta Q^{-\beta-1}} \cdot aQ^{-\beta-1} = -\frac{1}{\beta} = \text{constant} \quad \checkmark$$

To calculate consumer's surplus from $Q = 0$ to $Q = q^*$ it is necessary to subtract $P.q^*$ from the area under the demand function over this range.

$$\text{Area under demand function from 0 to } q^* = \int_0^{q^*} aQ^{-\beta}dQ$$

$$= \left[\frac{aQ^{-\beta+1}}{-\beta+1} \right]_0^{q^*} = \frac{a(q^*)^{1-\beta}}{1-\beta}$$

only if $\beta < 1$. If $\beta > 1$ or $-E_D < 1$ then this area is not finite. Consequently

β must be < 1 and $-E_D$ must be > 1

$$P.q^* = [a(q^*)^{-\beta}]q^* = a(q^*)^{1-\beta}$$

since $P = a(q^*)^{-\beta}$ when $Q = q^*$. Thus

$$\text{Consumer's surplus} = \frac{a(q^*)^{1-\beta}}{1-\beta} - a(q^*)^{1-\beta}$$

$$= a(q^*)^{1-\beta}\left[\frac{1}{1-\beta} - 1\right]$$

$$= a(q^*)^{1-\beta}\left[\frac{\beta}{1-\beta}\right]$$

But $a(q^*)^{1-\beta} = P.q^* = $ amount spent by consumers. Consequently when a demand function has a constant price elasticity consumer's surplus is a constant, i.e. $\beta/(1-\beta)$ in this case, times the amount spent on the commodity. If the demand function is $P^3Q = 1000$ or $P = 10Q^{-1/3}$ then the amount spent on the commodity when $P = 2$ is 125 and consumer's surplus is

$$125\left[\frac{\frac{1}{3}}{1-\frac{1}{3}}\right] = 62.5$$

since $\beta = \frac{1}{3}$.

PRODUCER'S SURPLUS

In Figure 31 producers will supply quantity WA at price WC. At this price producer's surplus is the area of the rectangle $WABC$ minus the area under the supply curve DE over the range WA,

$$\text{Producer's surplus} = WABC - WABD = DBC$$

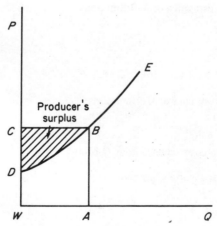

Figure 31 The supply curve and producer's surplus

The supply curve is $P = (a + 2Q)^2$ and $WA = q^*$. We note that

$$\int (a + 2Q)^2 dQ = \frac{(a + 2Q)^3}{6} + k$$

thus

$$WABD = \int_0^{q^*} (a + 2Q)^2 dQ = \left[\frac{(a + 2Q)^3}{6}\right]_0^{q^*}$$

$$= \frac{(a + 2q^*)^3}{6} - \frac{a^3}{6}$$

$$WABC = P.Q = P.q^* = (a + 2q^*)^2 q^*$$

$$\text{Producer's surplus} = q^*(a + 2q^*)^2 - \frac{(a + 2q^*)^3}{6} + \frac{a^3}{6}$$

$$= (a + 2q^*)^2 \left(\frac{4q^* - a}{6}\right) + \frac{a^3}{6}$$

The integral of $(a - 2Q)^2$ with respect to Q is

$$\frac{(a - 2Q)^3}{3(-2)} + k = \frac{(a - 2Q)^3}{-6} + k$$

To check that this is correct let us find the derivative of $\left[(a - 2Q)^3/-6\right] + k$ with respect to Q.

$$\frac{d\{[(a - 2Q)^3/-6] + k\}}{dQ} = \frac{3(a - 2Q)^2(-2)}{-6} = (a - 2Q)^2 = \text{integrand}$$

With integrands of this nature a knowledge of the results of differentiation is essential.

EXERCISES

1. The following are estimates of MR functions
   (i) $MR = 4 - 0.4Q$
   (ii) $MR = 5Q^{-0.5}$
   Q is output.
   Find TR when sales are (a) 4, (b) 9, (c) 16.

2. The following are estimates of MC functions
   (i) $MC = 2 + 0.4Q$
   (ii) $MC = 2 - 4Q + 3Q^2$
   Find total variable costs when output Q is (a) 4, (b) 10.

3. Given the following supply functions
   (i) $Q^2 - 2Q + 6 = 3P$
   (ii) $Q^2 + 4Q + 1 = P$
   (iii) $-(4 + Q)^2 = -P$
   where Q is output and P is price.
   Calculate the producer's surplus when $Q = 3$.

4. Given the following demand functions

    (i) $Q = 10 - P$

    (ii) $Q = 64P^{-2}$

    Find the consumer's surplus when $P = 4$ and $P = 5$.

5. If the demand function is

$$P = 10 - Q - Q^2$$

    and the supply function is

$$P = Q + 2$$

    calculate the consumer's and producer's surplus at the equilibrium price.

6. The demand function of a monopolist is

$$3Q = 60 - 10P$$

    and his *AC* function is

$$AC = \frac{20}{Q} + 1 + 0 \cdot 2Q$$

    If he decides to maximize sales revenue instead of profits show how this will affect consumer's surplus.

SOLUTIONS TO EXERCISES

1. (i) $MR = 4 - 0 \cdot 4Q$, thus

$$TR = \int_0^{q^*} (4 - 0 \cdot 4Q) dQ$$

q^* stands for any value of $Q > 0$

$$TR = \left[4Q - \frac{0 \cdot 4Q^2}{2} \right]_0^{q^*} = [4Q - 0 \cdot 2Q^2]_0^{q^*}$$
$$= [4q^* - 0 \cdot 2(q^*)^2] - [4(0) - 0 \cdot 2(0)]$$
$$= 4q^* - 0 \cdot 2(q^*)^2$$

    (a) When $q^* = 4$, then $TR = 4(4) - 0 \cdot 2(4)^2 = 12 \cdot 8$.   *

    (b) When $q^* = 9$, then $TR = 4(9) - 0 \cdot 2(9)^2 = 19 \cdot 8$.   *

    (c) When $q^* = 16$, then $TR = 4(16) - 0 \cdot 2(16)^2 = 12 \cdot 8$.   *

    (ii) $MR = 5Q^{-0 \cdot 5}$, thus

$$TR = \int_0^{q^*} 5Q^{-0 \cdot 5} dQ$$

$$= \left[\frac{5Q^{0 \cdot 5}}{0 \cdot 5} \right]_0^{q^*} = [10Q^{0 \cdot 5}]_0^{q^*} = 10(q^*)^{0 \cdot 5}$$

    (a) When $q^* = 4$, then $TR = 10(4^{0 \cdot 5}) = 20$.   *

    (b) When $q^* = 9$, then $TR = 10(9^{0 \cdot 5}) = 30$.   *

    (c) When $q^* = 16$, then $TR = 10(16^{0 \cdot 5}) = 40$.   *

2. (i) $MC = 2 + 0 \cdot 4Q$

$$\text{Total variable costs} = TVC = \int_0^{q^*} (2 + 0 \cdot 4Q) dQ$$

$$= \left[2Q + \frac{0 \cdot 4Q^2}{2} \right]_0^{q^*} = [2Q + 0 \cdot 2Q^2]_0^{q^*}$$
$$= 2q^* + 0 \cdot 2(q^*)^2$$

(a) When $q^* = 4$, then $TVC = 2(4) + 0\cdot2(4)^2 = 11\cdot2$.   *

(b) When $q^* = 10$, then $TVC = 2(10) + 0\cdot2(10)^2 = 40$.   *

(ii) $MC = 2 - 4Q + 3Q^2$, thus

$$TVC = \int_0^{q^*} (2 - 4Q + 3Q^2)dQ$$

$$= \left[2Q - \frac{4Q^2}{2} + \frac{3Q^3}{3}\right]_0^{q^*} = [2Q - 2Q^2 + Q^3]_0^{q^*}$$

$$= 2q^* - 2(q^*)^2 + (q^*)^3$$

(a) When $q^* = 4$, then $TVC = 2(4) - 2(4)^2 + (4)^3 = 40$.   *

(b) When $q^* = 10$, then $TVC = 2(10) - 2(10)^2 + (10)^3 = 820$.   *

3. (i) $3P = Q^2 - 2Q + 6$, thus

$$P = \frac{Q^2}{3} - \frac{2Q}{3} + 2$$

$$\text{Area under supply function} = \int_0^{q^*} (\tfrac{1}{3}Q^2 - \tfrac{2}{3}Q + 2)dQ$$

$$= \left[\frac{1}{3}\cdot\frac{Q^3}{3} - \frac{2}{3}\cdot\frac{Q^2}{2} + 2Q\right]_0^{q^*}$$

$$= \frac{1}{9}(q^*)^3 - \frac{1}{3}(q^*)^2 + 2q^*$$

$$= \frac{1}{9}(3)^3 - \frac{1}{3}(3)^2 + 2(3)$$

$$= 3 - 3 + 6 = 6 \quad \text{when } q^* = 3$$

Producer's surplus $= P.q^* -$ area under supply function.

When $q^* = 3$, then $P = 3$ and $P.q^* = 9$.

Thus producer's surplus $= 9 - 6 = 3$.   *

(ii) $P = Q^2 + 4Q + 1$

$$\text{Area under supply curve} = \int_0^{q^*} (Q^2 + 4Q + 1)dQ$$

$$= \left[\frac{Q^3}{3} + \frac{4Q^2}{2} + Q\right]_0^{q^*} = \frac{(q^*)^3}{3} + 2(q^*)^2 + q^*$$

$$= 9 + 18 + 3 = 30 \quad \text{when } q^* = 3$$

When $q^* = 3$, then $P = 22$ and $P.q^* = 66$.

$$\text{Producer's surplus} = P.q^* - \text{area under the supply curve}$$

$$= 66 - 30 = 36$$ *

(iii) $P = (4 + Q)^2$, thus

$$\text{Area under supply function} = \int_0^{q^*} (4 + Q)^2 dQ$$

$$= \left[\frac{(4 + Q)^3}{3}\right]_0^{q^*}$$

$$= \frac{(4 + q^*)^3}{3} - \frac{(4 + 0)^3}{3}$$

$$= \frac{(4 + 3)^3 - 4^3}{3} = \frac{343 - 64}{3} = 93 \quad \text{where } q^* = 3$$

When $q^* = 3$, then $P = 49$ and $P.q^* = 147$.
Thus producer's surplus $= 147 - 93 = 54$.

*

4. (i) $Q = 10 - P$, thus
$$P = 10 - Q$$
$$\text{Area under demand function} = \int_0^{q^*} (10 - Q)\,dQ$$
$$= [10Q - 0{\cdot}5Q^2]_0^{q^*} = 10q^* - 0{\cdot}5(q^*)^2$$

When $P = 4, q^* = 6$ and
$$\text{area under demand function} = 10(6) - 0{\cdot}5(36) = 42.$$
When $P = 5, q^* = 5$ and
$$\text{area under demand function} = 10(5) - 0{\cdot}5(25) = 37{\cdot}5$$
$$\text{Consumer's surplus} = \text{area under demand function} - P.q^*$$
When $q^* = 6$ and $P = 4$, consumer's surplus $= 42 - 24 = 18$.

*

When $q^* = 5$ and $P = 5$, consumer's surplus $= 37{\cdot}5 - 25 = 12{\cdot}5$.

*

(ii) $Q = 64P^{-2} = 64/P^2$, thus

$$P = \sqrt{\frac{64}{Q}} = 8Q^{-0{\cdot}5}$$
$$\text{Area under demand function} = A = \int_0^{q^*} 8Q^{-0{\cdot}5}\,dQ = \left[\frac{8Q^{0{\cdot}5}}{0{\cdot}5}\right]_0^{q^*}$$
$$= 16(q^*)^{0{\cdot}5}$$

When $P = 4, q^* = 4, A = 16(4)^{0{\cdot}5} = 32$ and
$$\text{Consumer's surplus} = A - P.q^* = 32 - 4(4) = 32 - 16$$
$$= 16$$

*

When $P = 5, q^* = 2{\cdot}56, A = 16(2{\cdot}56)^{0{\cdot}5} = 16(1{\cdot}6) = 25{\cdot}6$ and
$$\text{Consumer's surplus} = A - P.q^* = 25{\cdot}6 - 5(2{\cdot}56) = 25{\cdot}6 - 12{\cdot}8$$
$$= 12{\cdot}8$$

*

5. $P = 10 - Q - Q^2$ demand
$P = Q + 2$ supply
At equilibrium $10 - Q - Q^2 = Q + 2$, thus
$$Q^2 + 2Q - 8 = 0$$
i.e.
$$(Q - 2)(Q + 4) = 0$$
thus
$$Q = 2 \quad \text{or} \quad -4$$
Q cannot take a negative value, thus $Q = 2$.
When $Q = 2, P = 4$.

Consumer's surplus

$$\text{Area under demand curve} = \int_0^2 (10 - Q - Q^2)\,dQ$$
$$= [10Q - \tfrac{1}{2}Q^2 - \tfrac{1}{3}Q^3]_0^2$$
$$= 20 - 2 - \tfrac{8}{3} = 15\tfrac{1}{3}$$
$P.Q = 8$, thus consumer's surplus $= 15\tfrac{1}{3} - 8 = 7\tfrac{1}{3}$.

*

Producer's surplus

$$\text{Area under supply function} = \int_0^2 (Q+2)dQ$$
$$= [0.5Q^2 + 2Q]_0^2 = 2+4 = 6$$

$P.Q = 8$, thus producer's surplus $= 8 - 6 = 2$.   *

6. $3Q = 60 - 10P$, thus
$$P = 6 - 0.3Q$$
$$TR = P.Q = 6Q - 0.3Q^2$$
$$AC = \frac{20}{Q} + 1 + 0.2Q$$

thus
$$TC = 20 + Q + 0.2Q^2$$
and
$$\text{Profits } S = 6Q - 0.3Q^2 - 20 - Q - 0.Q^2$$
$$= 5Q - 0.5Q^2 - 20$$

Profits are at a maximum when
$$dS/dQ = 0 \quad \text{and} \quad d^2S/dQ^2 < 0$$
$$dS/dQ = 0 \quad \text{when} \quad 5 - Q = 0 \text{ i.e. } Q = 5$$

When $Q = 5$, $d^2S/dQ^2 = -1 < 0$, thus profits are maximized when $Q = 5$.
Sales revenue is maximized when $d(TR)/dQ = 0$ and $d^2(TR)/dQ^2 < 0$.
$$d(TR)/dQ = 0 \quad \text{when} \quad 6 - 0.6Q = 0 \quad \text{i.e.} \quad \text{when} \quad Q = 10$$

When $Q = 10$, $d^2(TR)/dQ^2 = -0.6 < 0$, thus sales revenue is maximized when $Q = 10$.

When profits are maximized $Q = 5$, thus
$$P = 6 - 0.3(5) = 4.5 \quad \text{and} \quad P.Q = 22.5$$
$$\text{Area under demand function} = \int_0^5 (6 - 0.3Q)dQ$$
$$= \left[6Q - \frac{0.3Q^2}{2} \right]_0^5 = 30 - 3.75 = 26.25$$

Thus consumer's surplus $= 26.25 - 22.5 = 3.75$.   *
When sales revenue is maximized $Q = 10$, thus
$$P = 6 - 0.3(10) = 3 \quad \text{and} \quad P.Q = 30$$
$$\text{Area under demand function} = \int_0^{10} (6 - 0.3Q)dQ$$
$$= \left[6Q - \frac{0.3Q^2}{2} \right]_0^{10} = 60 - 15 = 45$$

Thus consumer's surplus $= 45 - 30 = 15$.

CHAPTER 10

Linear Algebra; Vectors and Matrices

So far we have considered economic relations expressed as functions where all the variables and coefficients are numbers, or scalars. When we come to deal with large systems, containing many equations in several variables, the manipulation required becomes cumbersome. The use of linear algebra, or matrix algebra, enables us to deal with large systems using a compact and simple notation. Vectors are used to refer to sets of variables or coefficients without the need to write them out in full every time they are mentioned. Matrices are used to allow us to refer to systems of equations without writing them out in full.

As the name suggests, linear algebra is concerned only with linear relations. Although many economic relations are non-linear, they can often be reduced to linear ones, or it may be possible to obtain a good approximation, at least locally, using a linear function instead of the true non-linear one. This chapter deals with the algebra of vectors and matrices.

VECTORS

A vector is a single symbol used to refer to a list of variables or coefficients. For example, a consumer will spend money on several different goods. If the amounts of these goods purchased are denoted as x_1, x_2, etc. up to x_n where there are n goods, the vector $\mathbf{x}$ is defined as a list of these, e.g. the 'row vector'

$$\mathbf{x} = [x_1 \quad x_2 \quad x_3] \quad \text{where } n = 3$$

x_1, x_2, and x_3 are called the elements of the vector $\mathbf{x}$. They may represent the amounts of the goods purchased algebraically, as above, or if we know them we may insert the actual quantities purchased, e.g.

$$\mathbf{x} = [1 \quad 2 \quad 5]$$

assuming that $x_1 = 1$, $x_2 = 2$, and $x_3 = 5$.

Another vector can be used to represent the prices of the various goods, e.g.

$$\mathbf{p} = [p_1 \quad p_2 \quad p_3]$$

or the inputs of various factors of production used by a firm, e.g.

$$\mathbf{q} = [q_1 \quad q_2]$$

where q_1 is the amount of labour employed and q_2 is the amount of capital used.

The elements of a vector may be positive or negative numbers. In some cases, for example the quantities and prices used above, the economic logic of the situation requires that they be non-negative, but if we considered, for example, a vector of the unit profit levels of various firms then it would make economic sense for some or all elements to be negative.

The number of elements in a vector is called its dimension. Vectors are frequently represented by being printed, as above, in bold-face lower-case type. This notation will be used in this book; the reader should, however, be warned that not all authors using vectors use this notation, nor do all who do employ it do so completely consistently.

Vectors may be written either as rows, as above, or as columns, e.g.

$$\mathbf{x}' = \begin{bmatrix} 1 \\ 2 \\ 5 \end{bmatrix}$$

Each element of $\mathbf{x}'$, counting from the top, is the same as the corresponding element of $\mathbf{x}$, counting from the left. The process of changing from a row to a column vector is known as transposition; the transpose of $\mathbf{x}$ is denoted by adding a prime, to obtain $\mathbf{x}'$. The transpose of $\mathbf{x}$ is $\mathbf{x}'$, and the transpose of $\mathbf{x}'$ is $\mathbf{x}$.

Where the vectors are written out in full they are enclosed in square brackets. Some authors use different types of brackets to distinguish row and column vectors, but this notation will not be used below. Some authors also use a notation which assumes that all vectors are defined as column vectors, so that a prime is always used to distinguish a row vector. This notation again is not adopted here, but it is as well to be able to recognize it. Again, it is a practice which many users do not employ consistently.

The reason for having both row and column vectors is that it is convenient to use different ones for different operations we wish to perform on them, as shown below.

It will be found that we can perform on vectors many of the operations we can perform on ordinary numbers. This is not surprising, since a vector may have any positive number of elements. This means that it is possible to have a vector with only one element. We would not normally bother to define such a thing, but in handling large economic systems it is often con-venient to treat all sets of goods, or prices, alike, and some of these sets may turn out in particular cases to have only one element. Any operation which works on vectors must therefore work on ordinary numbers. There are, however, some special requirements which vectors must meet before various operations can be performed on them. We will now consider some of these operations.

VECTOR EQUALITY

If
$$\mathbf{x} = [x_1 \quad x_2 \quad \ldots \quad x_n]$$
is the vector of goods consumed by Black, and
$$\mathbf{y} = [y_1 \quad y_2 \quad \ldots \quad y_n]$$
is the vector of goods consumed by Bradley, we may want to know whether these vectors are equal. If they are equal,
$$\mathbf{x} = \mathbf{y}$$
This means that for any good i $(i = 1, \ldots n)$,
$$x_i = y_i$$
Vector equality can only hold where the vectors are of the same dimension, in this case n. They are then said to conform.

VECTOR ADDITION

If $\mathbf{x}$ and $\mathbf{y}$ are defined as before, we may wish to know the vector of the total consumption of goods by Black and Bradley combined. This is defined as
$$\mathbf{z} = \mathbf{x} + \mathbf{y}$$
Vector addition is defined only if $\mathbf{x}$ and $\mathbf{y}$ are of the same dimension, in this case n; $\mathbf{z}$ will also have n elements. Each element of $\mathbf{z}$ is equal to the sum of the corresponding elements of $\mathbf{x}$ and $\mathbf{y}$, i.e.
$$z_i = x_i + y_i \quad (i = 1, \ldots n)$$
For example, if
$$\mathbf{x} = [1 \quad 2 \quad 3] \quad \text{and} \quad \mathbf{y} = [2 \quad 0 \quad 2]$$
$$\mathbf{z} = \mathbf{x} + \mathbf{y} = [1+2 \quad 2+0 \quad 3+2] = [3 \quad 2 \quad 5]$$

VECTOR SUBTRACTION

Subtraction of one vector from another may also be performed. Suppose that a shop stocks m different commodities, and that
$$\mathbf{a} = [a_1 \quad a_2 \quad \ldots \quad a_m]$$
is the vector of goods in stock when the shop opens, and
$$\mathbf{b} = [b_1 \quad b_2 \quad \ldots \quad b_m]$$
is the vector of goods sold, then the vector
$$\mathbf{c} = [c_1 \quad c_2 \quad \ldots \quad c_m]$$
of goods still in stock should be
$$\mathbf{c} = \mathbf{a} - \mathbf{b}$$
where each element of $\mathbf{c}$ is found by subtracting the corresponding element

in **b** from that in **a**, i.e.

$$c_i = a_i - b_i \quad (i = 1, \ldots, m)$$

If all the stock has been sold then **b** = **a** and

$$\mathbf{c} = \mathbf{a} - \mathbf{b} = \mathbf{0}$$

0 is called a null vector of dimension m, i.e. it is a list of m zeros,

$$c_i = 0 \quad (i = 1, \ldots, m)$$

MULTIPLICATION BY A SCALAR

A vector may be multiplied by a scalar. For example, suppose that a firm uses a constant returns to scale production function, such that for each unit of output its input requirements are represented by

$$\mathbf{q} = [q_1 \quad q_2 \quad \cdots \quad q_k]$$

where there are k different inputs, i.e. types of labour, fuel, or materials. If there are three different inputs, the input requirements per unit of output could be

$$\mathbf{q} = [4 \quad 2 \quad 10]$$

If it needs to produce 2 units of output using this technique its total input requirements will be given by

$$\mathbf{q}^* = 2\mathbf{q}$$

e.g.

$$\mathbf{q}^* = 2[4 \quad 2 \quad 10] = [8 \quad 4 \quad 20]$$

where each q_i^* is twice q_i. Multiplication of a vector by 2 means multiplying every element by 2.

Similarly, if the firm needs to produce 100 units its input requirements will be given by

$$\mathbf{q}^{**} = 100\mathbf{q}$$

e.g.

$$\mathbf{q}^{**} = 100[4 \quad 2 \quad 10] = [400 \quad 200 \quad 1000]$$

Multiplication of a vector by 100 means multiplying every element by 100 so that

$$q_i^{**} = 100q_i \quad (i = 1, \ldots k)$$

In general, multiplication of a vector by a scalar number means multiplying every element in that vector by the scalar number.

MULTIPLICATION OF A VECTOR BY A VECTOR

Although two vectors cannot be multiplied together as simply as two scalar numbers, it is possible to combine two vectors to obtain what is called a scalar product, or inner product. Suppose we have a vector of goods purchased,

$$\mathbf{x} = [x_1 \quad x_2 \quad \dots \quad x_n]$$

and a vector of prices paid for them per unit,

$$\mathbf{p} = [p_1 \quad p_2 \quad \dots \quad p_n]$$

The total amount spent is found by multiplying the quantity of each good bought, x_i, by its price per unit, p_i, and summing over all i's. By a convention whose significance will become clearer when we go on to deal with matrices, we multiply a row vector by a column vector, writing the row first, e.g.

$$\mathbf{x} = [1 \quad 2 \quad 5]$$

and the column vector second, e.g.

$$\mathbf{p} = \begin{bmatrix} 3 \\ 4 \\ 6 \end{bmatrix}$$

The inner product $\mathbf{xp}$ is found by summing $x_i p_i$ over all i, written

$$\sum_i x_i p_i = (1 \times 3) + (2 \times 4) + (5 \times 6)$$
$$= 3 + 8 + 30 = 41$$

The inner product of two vectors can be written in various ways.

$$\mathbf{xp} \quad \text{or} \quad \mathbf{x'p} \quad \text{or} \quad \mathbf{x} \times \mathbf{p} \quad \text{or} \quad \mathbf{x.p}$$

In what follows we will adopt the $\mathbf{xp}$ notation, but other authors use a variety of notations.

LINEAR DEPENDENCE AND INDEPENDENCE OF VECTORS 163-167

Suppose we have a set of vectors $\mathbf{a}, \mathbf{b}, \dots, \mathbf{k}$, all non-zero vectors of dimension n. They are said to be linearly dependent if it is possible to find a set of weights $\lambda_1, \lambda_2, \dots \lambda_n$, not all $= 0$, such that   *their Linear combination = 0*

$$\lambda_1 \mathbf{a} + \lambda_2 \mathbf{b} + \dots + \lambda_n \mathbf{k} = \mathbf{0}$$

If it is not possible to do this then $\mathbf{a}, \mathbf{b}, \dots, \mathbf{k}$ are said to be linearly independent. If it is possible to find a set of λ_i which satisfies this equation, the vectors $\mathbf{a}, \mathbf{b}, \dots, \mathbf{k}$ are said to be linearly dependent, since some of them can be expressed as a linear function of the others, e.g. if there is some λ with $\lambda_1 \neq 0$, such that

$$\lambda_1 \mathbf{a} + \lambda_2 \mathbf{b} + \lambda_3 \mathbf{c} + \dots + \lambda_n \mathbf{k} = 0$$

then

$$\mathbf{a} = -\left(\frac{1}{\lambda_1}\right)[\lambda_2 \mathbf{b} + \lambda_3 \mathbf{c} + \dots + \lambda_n \mathbf{k}]$$

For example, if

$$\mathbf{a} = \begin{bmatrix} 1 \\ 2 \end{bmatrix} \quad \text{and} \quad \mathbf{b} = \begin{bmatrix} 2 \\ 3 \end{bmatrix}$$

then they are linearly independent, since no $\lambda \neq 0$ can be found for which

$$\lambda_1 \begin{bmatrix} 1 \\ 2 \end{bmatrix} + \lambda_2 \begin{bmatrix} 2 \\ 3 \end{bmatrix} = 0$$

For there to be a solution would require

$$1(\lambda_1) + 2\lambda_2 = 0 \quad \text{and} \quad 2\lambda_1 + 3\lambda_2 = 0$$

whose only solution is $\lambda_1 = 0$, $\lambda_2 = 0$. However, if

$$\mathbf{a} = \begin{bmatrix} 1 \\ 2 \end{bmatrix} \quad \text{and} \quad \mathbf{c} = \begin{bmatrix} 2 \\ 4 \end{bmatrix}$$

then $\mathbf{a}$ and $\mathbf{c}$ are linearly dependent, since $\lambda_1 \mathbf{a} + \lambda_2 \mathbf{c} = 0$ will hold for any $\lambda_1 = -2\lambda_2$.

How many linearly independent vectors can exist in n dimensions? We can see at once that n linearly independent row vectors can exist, by considering the set of n-vectors of dimension n, each of which has a single element of 1 and the remainder zeros. The first has 1 as its first element, the second has 1 as its second element, the ith has 1 as its ith element, and so on. These vectors, known as unit vectors, can be written

$$\mathbf{u}_1 = [1 \quad 0 \quad 0 \quad 0 \quad \ldots \quad 0]$$
$$\mathbf{u}_2 = [0 \quad 1 \quad 0 \quad 0 \quad \ldots \quad 0]$$
$$\mathbf{u}_3 = [0 \quad 0 \quad 1 \quad 0 \quad \ldots \quad 0]$$
$$\vdots$$
$$\mathbf{u}_n = [0 \quad 0 \quad 0 \quad 0 \quad \ldots \quad 1]$$

These vectors are clearly all linearly independent, since none could ever be expressed as a linear combination of any of the others. Thus there can be n linearly independent n-vectors.

Equally, any n-vector $\mathbf{a}$ can be expressed as a linear combination of these, by adding a_1 times $\mathbf{u}_1$, a_2 times $\mathbf{u}_2$, and so on. It can thus be shown that there cannot be more than n linearly independent n-vectors.

An exactly similar argument can be used to show that there can be n linearly independent column vectors of dimension n, but not more than n.

Any set of n linearly independent rows or columns is said to be a basis for, or to span, a space of dimension n. This means that any vector of dimension n can be expressed as a linear combination of any n linearly independent rows or columns.

CONVEX COMBINATIONS OF VECTORS

Suppose that the vectors $\mathbf{a}, \mathbf{b}, \ldots, \mathbf{k}$, all of the same dimension m, are linearly independent. It is then possible to define a new vector, $\mathbf{m}$, also of dimension m, such that

$$\mathbf{m} = \mu_a \mathbf{a} + \mu_b \mathbf{b} + \ldots + \mu_k \mathbf{k}$$

where each $\mu_i \geq 0$ and $\sum_i \mu_i = 1$.

m is then said to be a convex combination of $\mathbf{a}, \mathbf{b}, \ldots, \mathbf{k}$, and **m** may be regarded as a weighted average of $\mathbf{a}, \mathbf{b}, \ldots, \mathbf{k}$.

A GEOMETRICAL APPROACH

It is often helpful in thinking about vectors to illustrate general propositions by the special case of the two-dimensional vector, which can conveniently be represented in diagrams. By convention, in the graphical representation of a vector the horizontal axis is used to represent the first element of each vector, and the vertical axis to represent the second. If $\mathbf{a} = \begin{bmatrix} 1 & 2 \end{bmatrix}$ it is represented in Figure 32 either by A, or by $0A$, the line joining A to the origin 0. If $\mathbf{b} = \begin{bmatrix} 3 & 1 \end{bmatrix}$ it is represented by B, or by $0B$. If $\mathbf{c} = \mathbf{a} + \mathbf{b} = \begin{bmatrix} 4 & 3 \end{bmatrix}$ it is represented by C, or by $0C$. It can be seen that BC is equal and parallel to $0A$ and that AC is equal and parallel to $0B$. Thus where $0C$ represents the addition of the vectors represented by $0A$ and $0B$, $0ACB$ is a parallelogram. This helps to make the point that in the addition of vectors, it does not matter whether you add $\mathbf{b}$ to $\mathbf{a}$, i.e. get from 0 to C via A, or add $\mathbf{a}$ to $\mathbf{b}$, i.e. get from 0 to C via B. Either way you get to C.

If $\mathbf{a}$ is multiplied by a constant λ, the vector will change from $0A$ to $0A'$ when $\lambda' > 1$, to $0A''$ when $1 > \lambda'' > 0$, or to A''' where $\lambda''' < 0$; i.e. multiplying a vector by a negative constant shifts it to the opposite quadrant of the diagram, as shown in Figure 33.

If we let $\mathbf{b} = \mathbf{a}$, then vector subtraction gives $\mathbf{b} - \mathbf{a} = \mathbf{0}$, where $\mathbf{0}$, the null vector, is represented by the origin 0.

All possible convex combinations of $\mathbf{a}$ and $\mathbf{b}$ can be represented by the points on the line segment AB. This is best seen as follows, using Figure 3.4. Let $\mathbf{d}$ be the convex combination of $\mathbf{b}$ and $\mathbf{a}$ given by

$$\mathbf{d} = \lambda \mathbf{b} + (1 - \lambda)\mathbf{a} \quad \text{where} \quad 0 \leqslant \lambda \leqslant 1$$

Figure 32   Graphical representation of vectors

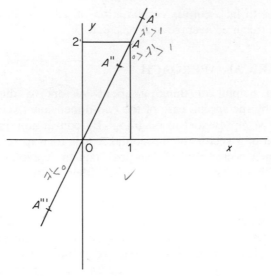

Figure 33   Multiplication of a vector by a scalar

c will be represented by $0C$, where CB is equal and parallel to $0A$. $\lambda\mathbf{c}$ will be represented by points on $0C$, such as E.

$$\mathbf{d} = \lambda\mathbf{b} + (1 - \lambda)\mathbf{a}$$
$$= \mathbf{a} + \lambda(\mathbf{b} - \mathbf{a})$$
$$= \mathbf{a} + \lambda\mathbf{c}$$

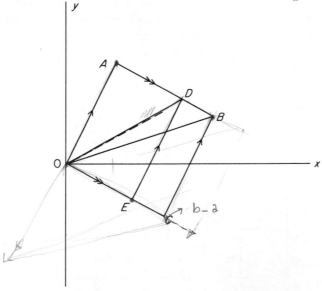

Figure 34   A convex linear combination of vectors

Convex combination of $a \& b$: $\quad d = \mu_a a + \mu_b b \quad$ s.t $\quad \mu_a, \mu_b \geq 0 \quad$ i.e non-negative

$$= d = \lambda a + (1-\lambda)b \qquad \text{s.to} \leq \lambda \leq 1 \quad \mu_a + \mu_b = 1$$

$$d = \cdot 3 \begin{bmatrix} 1 \\ 2 \end{bmatrix} + \cdot 7 \begin{bmatrix} 3 \\ 1 \end{bmatrix} = \begin{bmatrix} \cdot 3 + (\cdot 7)(3) \\ \cdot 3(2) + (\cdot 7)(1) \end{bmatrix}$$

Thus as λc lies on the line $0C$, d will lie on the line AB, since $0E$ is equal and parallel to AD. Thus B, D, and A lie on a straight line, as shown in Figure 34. For example, if $\mathbf{a} = \begin{bmatrix} 1 & 2 \end{bmatrix}$, $\mathbf{b} = \begin{bmatrix} 3 & 1 \end{bmatrix}$, and $\lambda = 0.3$, — *linearly independent* 2 *vectors*

$$\mathbf{d} = [(0.3(1) + 0.7(3)) \quad (0.3(2) + 0.7(1))]$$
$$= [(0.3 + 2.1) \quad (0.6 + 0.7)]$$
$$= [2.4 \quad 1.3]$$

$\lambda_1 = \lambda_2 = 0$

$$\lambda_1 \begin{bmatrix} 1 \\ 2 \end{bmatrix} + \lambda_2 \begin{bmatrix} 3 \\ 1 \end{bmatrix} = \begin{bmatrix} \lambda_1 + 3\lambda_2 = 0 \\ 2\lambda_1 + \lambda_2 = 0 \end{bmatrix}$$

$$\begin{array}{l} 3 \quad 2\lambda_1 + 6\lambda_2 = 0 \\ \underline{ \quad 2\lambda_1 + \lambda_2 = 0} \\ 5\lambda_2 = 0 \end{array}$$

MATRICES

Matrices are a compact and convenient way of writing down systems of linear relations. Suppose we have the system

$$a_{11}x_1 + a_{12}x_2 + a_{13}x_3 = y_1$$
$$a_{21}x_1 + a_{22}x_2 + a_{23}x_3 = y_2$$

This can be written in matrix notation as

$$\begin{bmatrix} a_{11} & a_{12} & a_{13} \\ a_{21} & a_{22} & a_{23} \end{bmatrix} \begin{bmatrix} x_1 \\ x_2 \\ x_3 \end{bmatrix} = \begin{bmatrix} y_1 \\ y_2 \end{bmatrix}$$

The meaning of this notation will be explained in full below. For the moment we concentrate on the matrix $\mathbf{A}$, written

$$\mathbf{A} = \begin{bmatrix} a_{11} & a_{12} & a_{13} \\ a_{21} & a_{22} & a_{23} \end{bmatrix}$$

For example, if

$$x_1 + 2x_2 + 3x_3 = 4$$
$$5x_1 + 6x_2 + 7x_3 = 8$$

$$\mathbf{A} = \begin{bmatrix} 1 & 2 & 3 \\ 5 & 6 & 7 \end{bmatrix}$$

The $\mathbf{A}$ matrix is an array of the coefficients of the equations, where the row tells us which equation the coefficient belongs to, and the column tells us which variable the coefficient refers to. As with vectors, the elements of a matrix may be algebraic symbols, or numbers. The subscripts of a_{ij} tell us that a_{ij} is the element in the ith row and the jth column of $\mathbf{A}$, e.g. a_{23} is the element of $\mathbf{A}$ in the second row and the third column.

If some of the equations do not contain all the variables, they can be treated as if the missing variables had coefficients of zero. If

$$x_1 + 2x_2 \qquad = 4$$
$$6x_2 + 7x_3 = 8$$

$$\mathbf{A} = \begin{bmatrix} 1 & 2 & 0 \\ 0 & 6 & 7 \end{bmatrix}$$

Thus every row of $\mathbf{A}$ has three elements, and every column has two elements.

Matrices are normally denoted by bold-face capital letters, though as usual the reader should be prepared to meet a variety of notations. $\mathbf{A} = [a_{ij}]$ is also used to show that $\mathbf{A}$ is the matrix whose ijth element is a_{ij}.

OPERATIONS ON MATRICES

With matrices, as with vectors, various algebraic operations can be performed. Where they can be defined, these operations have properties similar to those performed on ordinary numbers, or on vectors. This is not surprising, since a vector can be regarded as a special case of a matrix with only one row, or with only one column; and a scalar number can be regarded as the special case of a matrix with only one row and only one column. While we would not normally bother to define a 1×1 matrix, it is a meaningful concept, and must obey the same rules as other matrices.

As with vectors, there are special conditions matrices must satisfy before certain algebraic operations are possible. Making use of matrix notation, the system

$$\begin{bmatrix} a_{11} & a_{12} & a_{13} \\ a_{21} & a_{22} & a_{23} \end{bmatrix} \begin{bmatrix} x_1 \\ x_2 \\ x_3 \end{bmatrix} = \begin{bmatrix} y_1 \\ y_2 \end{bmatrix} \tag{1}$$

can be written compactly as

$$\mathbf{Ax} = \mathbf{y}$$

Equation (1) should be read as saying that the inner product of the first row of $\mathbf{A}$, $[a_{11} \ a_{12} \ a_{13}]$ and the vector $\mathbf{x}$ is equal to y_1, the first element of $\mathbf{y}$; and that the inner product of the second row of $\mathbf{A}$, $[a_{21} \ a_{22} \ a_{23}]$ and the vector $\mathbf{x}$ is equal to y_2, the second element of $\mathbf{y}$.

It may well be felt that this is all a lot of work to write out a system of two equations in three variables, and so it is. When it comes to dealing with large equation systems, however, the notation is invaluable. For example, if there are seven equations in nine variables, the system can be written as

$$\begin{bmatrix} a_{11} & a_{12} & \cdots & a_{19} \\ a_{21} & a_{22} & \cdots & a_{29} \\ \vdots & & & \vdots \\ a_{71} & a_{72} & \cdots & a_{79} \end{bmatrix} \begin{bmatrix} x_1 \\ x_2 \\ \vdots \\ x_9 \end{bmatrix} = \begin{bmatrix} y_1 \\ y_2 \\ \vdots \\ y_7 \end{bmatrix}$$

where the dots represent omitted terms, rows, and columns; or the whole system can be represented as $\mathbf{Ax} = \mathbf{y}$. To write out such systems in full would be very time consuming—the reader who doubts this is advised to try.

MATRIX EQUALITY

Two matrices are capable of being compared to see if they are equal *iff* (if and only if) they both have the same number of rows, and they both have the same number of columns. If **A** has m rows and n columns, **B** can be compared with **A** for equality only if it too has m rows and n columns. **A** and **B** are then equal *iff* every element of **B** equals the corresponding element of **A**, i.e.

$$\mathbf{A} = \mathbf{B} \text{ iff } [a_{ij}] = [b_{ij}]$$

for all $i = 1, 2, \ldots, m$ and all $j = 1, 2, \ldots, n$.

MATRIX ADDITION

Two matrices **A** and **B** can be added if and only if they are 'conformable for addition', i.e. they are both of the same dimension $m \times n$, with m rows and n columns. If they are conformable, then the matrix $\mathbf{C} = \mathbf{A} + \mathbf{B}$ also has m rows and n columns, and every element of **C** is equal to the sum of the corresponding elements of **A** and **B**, i.e. $c_{ij} = a_{ij} + b_{ij}$ for all i, j.

For example, if

$$\mathbf{A} = \begin{bmatrix} 1 & 2 & 3 \\ 1 & 0 & -1 \end{bmatrix} \quad \text{and} \quad \mathbf{B} = \begin{bmatrix} 2 & 3 & 4 \\ 2 & -3 & 0 \end{bmatrix}$$

$$\mathbf{C} = \mathbf{A} + \mathbf{B} = \begin{bmatrix} 1+2 & 2+3 & 3+4 \\ 1+2 & 0-3 & -1+0 \end{bmatrix}$$

$$= \begin{bmatrix} 3 & 5 & 7 \\ 3 & -3 & -1 \end{bmatrix}$$

The reader may prefer simply to accept this as a definition of the rules of the matrix algebra game, and wait to see what must have motivated mathematicians to define the conventions in this way. Those who prefer to try to understand the reasoning at this stage may, however, like to consider the following argument. Matrices are used to write down equation systems. Suppose that $\mathbf{Ax} = \mathbf{a}$ and $\mathbf{Bx} = \mathbf{b}$ are systems for which **A** and **B** are both $m \times n$, and that the vectors **a** and **b** are both of dimension m. Then $\mathbf{Ax} = \mathbf{a}$ and $\mathbf{Bx} = \mathbf{b}$ each represents a system of m equations in n variables. It will then follow that when vector and matrix addition are defined as above, $\mathbf{Cx} = \mathbf{c}$ where $\mathbf{C} = \mathbf{A} + \mathbf{B}$ and $\mathbf{c} = \mathbf{a} + \mathbf{b}$. This can best be seen by considering one row—we choose the first row as our example—of each of these systems.

The first row of $\mathbf{Ax} = \mathbf{a}$ tells us that

$$a_{11}x_1 + a_{12}x_2 + \ldots + a_{1n}x_n = a_1 \tag{2}$$

and the first row of $\mathbf{Bx} = \mathbf{b}$ tells us that

$$b_{11}x_1 + b_{12}x_2 + \ldots + b_{1n}x_n = b_1 \tag{3}$$

If equations (2) and (3) are both true, then the equation we obtain by adding them together must always be true, i.e.

$$(a_{11} + b_{11})x_1 + (a_{12} + b_{12})x_2 + \ldots + (a_{1n} + b_{1n})x_n = a_1 + b_1$$

but this is simply

$$c_{11}x_1 + c_{12}x_2 + \ldots + c_{1n}x_n = c_1$$

i.e. the first row of $\mathbf{Cx} = \mathbf{c}$ as defined above; a similar proof applies to every row of the systems.

It is clear from the definition of matrix addition that the order of addition does not matter;

$$\mathbf{A} + \mathbf{B} = \mathbf{B} + \mathbf{A} \quad \text{since} \quad a_{ij} + b_{ij} = b_{ij} + a_{ij}$$

MATRIX SUBTRACTION

If $\mathbf{A}$ and $\mathbf{B}$ are both of dimension $m \times n$, and are thus conformable for addition, then we can subtract $\mathbf{B}$ from $\mathbf{A}$ to obtain $\mathbf{D} = \mathbf{A} - \mathbf{B}$, where $d_{ij} = a_{ij} - b_{ij}$ for all i, j. This can be proved to work for the same reasons as matrix addition. If $\mathbf{B} = \mathbf{A}$, then $\mathbf{A} - \mathbf{B} = \mathbf{0}$, where $\mathbf{0}$ is the null matrix, every element of whose m rows and n columns is zero.

MULTIPLICATION OF A MATRIX BY A SCALAR

As with multiplication of a vector by a scalar, multiplication of a matrix by a scalar is defined as multiplication of every element of the matrix by the scalar; e.g. if $\mathbf{B} = \lambda \mathbf{A}$ then $b_{ij} = \lambda a_{ij}$ for all i, j.

The motivation of this can be seen by considering $\mathbf{Ax} = \mathbf{a}$. If each element of $\mathbf{A}$ and $\mathbf{a}$ is multiplied by λ, the resulting equation

$$\mathbf{Bx} = \lambda \mathbf{Ax} = \lambda \mathbf{a} = \mathbf{b}$$

will be true.

MULTIPLICATION OF A MATRIX BY A MATRIX

Suppose we have two matrices, $\mathbf{A}$ and $\mathbf{B}$. What can it mean to multiply them together? To see why we should want to be able to perform operations of this sort, consider the following example: $\mathbf{u}$ is a vector of the profit levels of firms 1 to m; $\mathbf{v}$ is a vector of the profits per unit made in operating processes 1 to n, which are employed by the firms; $\mathbf{w}$ is a vector of the prices of materials 1 to p, which are used as inputs to the processes. Suppose that the profit levels $\mathbf{u}$ are linear functions of the unit profitability of processes, $\mathbf{v}$, i.e. $\mathbf{u} = \mathbf{Av}$. This means that the profits of firm 1, u_1, are given by the inner product of the first row of $\mathbf{A}$, which shows the extent to which firm 1 operates each process, and the vector $\mathbf{v}$ of unit profits for each process. The profitability of the various processes, $\mathbf{v}$, is a linear function of the costs of materials, $\mathbf{w}$, i.e. $\mathbf{v} = \mathbf{Bw}$. This means that the profitability of process 2 is given by the inner product of the second row of $\mathbf{B}$, showing the extent of inputs of each material per unit of the second process, times the vector $\mathbf{w}$ of material prices.

We then want to be able to express the profits $\mathbf{u}$ directly as a function of the material prices, $\mathbf{w}$.

If
$$\mathbf{u} = \mathbf{Av} \quad \text{and} \quad \mathbf{v} = \mathbf{Bw},$$
we can write
$$\mathbf{u} = \mathbf{Cw}$$

where $C = AB$, i.e.

$$u = Av = A(Bw) = (AB)w = Cw$$

Before attempting to create a matrix C, let us consider the relation between u and w that can be found by using ordinary methods of substitution.

Suppose that

$$u_1 = -2v_1 - 3v_2$$
$$u_2 = -4v_1 - v_2$$

i.e.

$$A = \begin{bmatrix} -2 & -3 \\ -4 & -1 \end{bmatrix}$$

and

$$v_1 = 5w_1 + 4w_2$$
$$v_2 = 3w_1 + 6w_2$$

i.e.

$$B = \begin{bmatrix} 5 & 4 \\ 3 & 6 \end{bmatrix}$$

$$\begin{aligned} u_1 &= -2v_1 - 3v_2 = -2(5w_1 + 4w_2) - 3(3w_1 + 6w_2) \\ &= -10w_1 - 8w_2 - 9w_1 - 18w_2 \\ &= -19w_1 - 26w_2 \end{aligned}$$

$$\begin{aligned} u_2 &= -4v_1 - v_2 = -4(5w_1 + 4w_2) - (3w_1 + 6w_2) \\ &= -20w_1 - 16w_2 - 3w_1 - 6w_2 \\ &= -23w_1 - 22w_2 \end{aligned}$$

We will now look at the matrix method of obtaining a similar result.

To perform the operation of creating $C = AB$, which is called premultiplying B by A (or, equivalently, postmultiplying A by B), we form a new matrix C, whose ijth element c_{ij} is the inner product of the ith row of A and the jth column of B. Clearly this is possible *iff* the ith row of A and the jth column of B are conformable for vector multiplication, i.e. are vectors with the same number of elements. If the number of elements in the rows of A and the columns of B are not the same then the matrix operation AB is not defined. To allow AB to be performed requires that if A is $m \times n$, B shall be $n \times p$. The product $C = AB$ is then defined and will be $m \times p$, i.e. C will have m rows and p columns. Given the definitions of A and B above, we have

$$\begin{aligned} C = AB &= \begin{bmatrix} -2 & -3 \\ -4 & -1 \end{bmatrix} \begin{bmatrix} 5 & 4 \\ 3 & 6 \end{bmatrix} \\ &= \begin{bmatrix} -2(5) - 3(3) & -2(4) - 3(6) \\ -4(5) - 1(3) & -4(4) - 1(6) \end{bmatrix} \\ &= \begin{bmatrix} -10 - 9 & -8 - 18 \\ -20 - 3 & -16 - 6 \end{bmatrix} \\ &= \begin{bmatrix} -19 & -26 \\ -23 & -22 \end{bmatrix} \end{aligned}$$

so $\mathbf{u} = \mathbf{Cw}$, i.e.

$$\begin{bmatrix} u_1 \\ u_2 \end{bmatrix} = \begin{bmatrix} -19 & -26 \\ -23 & -22 \end{bmatrix} \begin{bmatrix} w_1 \\ w_2 \end{bmatrix}$$

We can check that this gives us the same answers as we have obtained by ordinary methods of substitution.

THE LOGIC OF MATRIX MULTIPLICATION

To see why matrix multiplication is defined as it is, consider the element c_{12} of $\mathbf{C}$. This is the way in which the profits of firm 1 are affected by a rise in the price of material 2. This may happen via the costs of any process; the total effect of a unit change in the price of material 2 is found by adding the effects of a rise in material 2's price on the costs of process 1, b_{12}, times the effects of a unit rise in the costs of process 1 on the profits of firm 1, a_{11}; the effects of a rise in material 2's price on the costs of process 2, b_{22}, times the effects of a unit rise in the costs of process 2 on the profits of firm 1, a_{12}; and so on through all possible processes. Thus

$$c_{12} = a_{11}b_{12} + a_{12}b_{22} + \dots + a_{1n}b_{n2}$$

Because of this property, that c_{12} traces the connection between the first element of $\mathbf{u}$ and the second element of $\mathbf{w}$ through all possible routes, if we have to multiply together more than two matrices it does not matter in what order the various operations are performed; e.g. if we define $\mathbf{D} = \mathbf{ABC}$ it does not matter whether we use the route

$$\mathbf{ABC} = (\mathbf{AB})\mathbf{C} = \mathbf{EC} = \mathbf{D}$$

defining $\mathbf{E} = \mathbf{AB}$, or the route

$$\mathbf{ABC} = \mathbf{A}(\mathbf{BC}) = \mathbf{AF} = \mathbf{G}$$

defining $\mathbf{F} = \mathbf{BC}$. We will find that $\mathbf{G}$ and $\mathbf{D}$ are the same matrix. This is known as the associative law of multiplication. It can be proved as follows. Suppose that $\mathbf{A}$ is $m \times n$, $\mathbf{B}$ is $n \times p$, and $\mathbf{C}$ is $p \times q$ in dimension. $\mathbf{E} = \mathbf{AB}$ is thus $m \times p$ and $\mathbf{F} = \mathbf{BC}$ is $n \times q$. Thus

$$\mathbf{D} = \mathbf{EC} \text{ is } m \times q \quad \text{and} \quad \mathbf{G} = \mathbf{AF} \text{ is } m \times q$$

Thus $\mathbf{G}$ and $\mathbf{D}$ are of the same dimensions. We will now show that they are the same matrix, i.e. that $g_{ij} = d_{ij}$ for every i, j of $\mathbf{D}$'s $m \times q$ elements.

$$e_{iu} = \sum_{k=1}^{n} a_{ik}b_{ku}$$

$$f_{vj} = \sum_{l=1}^{p} b_{vl}c_{lj}$$

$$d_{ij} = \sum_{u=1}^{p} e_{iu}c_{uj} = \sum_{u=1}^{p} c_{uj} \sum_{k=1}^{n} a_{ik}b_{ku}$$

Rearranging the summation signs leaves this expression unchanged, so

$$d_{ij} = \sum_{k=1}^{n} \sum_{u=1}^{p} a_{ik} b_{ku} c_{uj}$$

$$g_{ij} = \sum_{v=1}^{n} a_{iv} f_{vj} = \sum_{v=1}^{n} a_{iv} \sum_{l=1}^{p} b_{vl} c_{lj}$$

$$= \sum_{v=1}^{n} \sum_{l=1}^{p} a_{iv} b_{vl} c_{lj}$$

Thus $g_{ij} = d_{ij}$ for each i, j since the summations concerned differ only in the subscripts used in denoting them, which, however, each vary over the same range. If $\mathbf{x} = \mathbf{ABCy}$ then both procedures, $\mathbf{ABC} = (\mathbf{AB})\mathbf{C} = \mathbf{A}(\mathbf{BC})$ represent the effects of the various elements of $\mathbf{y}$ on those of $\mathbf{x}$, traced through all possible intermediate routes. For example, suppose that

$$\mathbf{A} = \begin{bmatrix} 1 & 2 \\ 3 & 2 \end{bmatrix}, \quad \mathbf{B} = \begin{bmatrix} 2 & 1 \\ 1 & 3 \end{bmatrix}, \quad \text{and} \quad \mathbf{C} = \begin{bmatrix} 1 & 3 \\ 2 & 1 \end{bmatrix}$$

$$(\mathbf{AB})\mathbf{C} = \begin{bmatrix} 4 & 7 \\ 8 & 9 \end{bmatrix} \begin{bmatrix} 1 & 3 \\ 2 & 1 \end{bmatrix} = \begin{bmatrix} 18 & 19 \\ 26 & 33 \end{bmatrix}$$

and

$$\mathbf{A}(\mathbf{BC}) = \begin{bmatrix} 1 & 2 \\ 3 & 2 \end{bmatrix} \begin{bmatrix} 4 & 7 \\ 7 & 6 \end{bmatrix} = \begin{bmatrix} 18 & 19 \\ 26 & 33 \end{bmatrix}$$

It can be seen from the definition of the operations that in combining matrices, it does not matter whether addition is performed before or after multiplication;

$$(\mathbf{A} + \mathbf{B})\mathbf{C} = \mathbf{AC} + \mathbf{BC}$$

This is known as the distributive law of multiplication.

THE EXISTENCE OF MATRIX PRODUCTS

The fact that $\mathbf{C} = \mathbf{AB}$ is defined gives us no clue as to whether or not the reverse product, $\mathbf{D} = \mathbf{BA}$ is defined. Since $\mathbf{B}$ is $n \times p$ and $\mathbf{A}$ is $m \times n$, $\mathbf{BA}$ will be defined only if $p = m$; if $\mathbf{D} = \mathbf{BA}$ is defined it will be $n \times n$. Even if $\mathbf{D} = \mathbf{BA}$ is defined, it will not, except by a fluke, be the same as $\mathbf{C} = \mathbf{AB}$. This is probably best shown by a simple example. Suppose

$$\mathbf{A} = \begin{bmatrix} 1 & 2 & 3 \\ 4 & 5 & 6 \end{bmatrix} \quad \text{and} \quad \mathbf{B} = \begin{bmatrix} 2 & 3 \\ 4 & 5 \\ 6 & 7 \end{bmatrix}$$

where $\mathbf{A}$ is 2×3, $\mathbf{B}$ is 3×2, so $\mathbf{C} = \mathbf{AB}$ and $\mathbf{D} = \mathbf{BA}$ are both defined; but $\mathbf{C}$ is 2×2 and $\mathbf{D}$ is 3×3 so there is no question of their being the same. Even if $\mathbf{C}$ and $\mathbf{D}$ are of the same dimension, which will be the case only if $\mathbf{A}$ and $\mathbf{B}$ have the same number of rows as columns, $\mathbf{C}$ and $\mathbf{D}$ will only by a fluke be the same; if

$$A = \begin{bmatrix} a_{11} & a_{12} \\ a_{21} & a_{22} \end{bmatrix} \quad \text{and} \quad B = \begin{bmatrix} b_{11} & b_{12} \\ b_{21} & b_{22} \end{bmatrix}$$

$$C = AB = \begin{bmatrix} a_{11} & a_{12} \\ a_{21} & a_{22} \end{bmatrix}\begin{bmatrix} b_{11} & b_{12} \\ b_{21} & b_{22} \end{bmatrix}$$

$$= \begin{bmatrix} a_{11}b_{11} + a_{12}b_{21} & a_{11}b_{12} + a_{12}b_{22} \\ a_{21}b_{11} + a_{22}b_{21} & a_{21}b_{12} + a_{22}b_{22} \end{bmatrix}$$

$$D = BA = \begin{bmatrix} b_{11} & b_{12} \\ b_{21} & b_{22} \end{bmatrix}\begin{bmatrix} a_{11} & a_{12} \\ a_{21} & a_{22} \end{bmatrix}$$

$$= \begin{bmatrix} b_{11}a_{11} + b_{12}a_{21} & b_{11}a_{12} + b_{12}a_{22} \\ b_{21}a_{11} + b_{22}a_{21} & b_{21}a_{12} + b_{22}a_{22} \end{bmatrix}$$

e.g. if

$$A = \begin{bmatrix} 1 & 2 \\ 3 & 4 \end{bmatrix} \quad \text{and} \quad B = \begin{bmatrix} 2 & 3 \\ 4 & 5 \end{bmatrix}$$

$$C = AB = \begin{bmatrix} 1 & 2 \\ 3 & 4 \end{bmatrix}\begin{bmatrix} 2 & 3 \\ 4 & 5 \end{bmatrix}$$

$$= \begin{bmatrix} 1(2) + 2(4) & 1(3) + 2(5) \\ 3(2) + 4(4) & 3(3) + 4(5) \end{bmatrix}$$

$$= \begin{bmatrix} 2+8 & 3+10 \\ 6+16 & 9+20 \end{bmatrix} = \begin{bmatrix} 10 & 13 \\ 22 & 29 \end{bmatrix}$$

while

$$D = BA = \begin{bmatrix} 2 & 3 \\ 4 & 5 \end{bmatrix}\begin{bmatrix} 1 & 2 \\ 3 & 4 \end{bmatrix}$$

$$= \begin{bmatrix} 2(1) + 3(3) & 2(2) + 3(4) \\ 4(1) + 5(3) & 4(2) + 5(4) \end{bmatrix}$$

$$= \begin{bmatrix} 2+9 & 4+12 \\ 4+15 & 8+20 \end{bmatrix} = \begin{bmatrix} 11 & 16 \\ 19 & 28 \end{bmatrix}$$

We thus see that in multiplying several matrices, while the order in which the operations are done makes no difference to the result, the order of the matrices themselves does matter, and any change will make the results, even if they are still defined, completely different, sheer flukes excepted.

MULTIPLICATION OF A MATRIX BY A VECTOR

This is a case of matrix multiplication where one of the matrices has only one row, or only one column. If A is an $m \times n$ matrix, and b is an $n \times 1$ column vector, then Ab is defined, and is an $m \times 1$ column vector. If B is an $m \times n$ matrix and a is a $1 \times m$ row vector, then aB is defined and is a $1 \times n$ row vector.

TRANSPOSITION OF MATRICES

As with vectors, it may be convenient to interchange the rows and columns of a matrix. This is denoted by a prime, so that if $\mathbf{A}$ is $m \times n$, its transpose $\mathbf{A}'$ is $n \times m$, and $a'_{ji} = a_{ij}$. The ith row of $\mathbf{A}$ becomes the ith column of $\mathbf{A}'$, and the jth column of $\mathbf{A}$ becomes the jth row of $\mathbf{A}'$. The same rules imply that the transpose of $\mathbf{A}'$ is $\mathbf{A}$.

SQUARE MATRICES

Special interest often attaches to matrices which have the same number of rows as columns. Such matrices play a special role in solving equation systems. If we have a set of n linear equations, represented by $\mathbf{Ax} = \mathbf{a}$ where $\mathbf{A}$ is $n \times n$, then, provided that the rows of $\mathbf{A}$ are linearly independent, it will be shown in the next chapter that there will be just enough information to allow us to solve for $\mathbf{x}$ for every possible $\mathbf{a}$.

SYMMETRIC MATRICES

Square matrices may have the special characteristic of being symmetric, which means that their ith row and ith column are the same vector, i.e. $a_{ij} = a_{ji}$ for all i, j, so that $\mathbf{A}' = \mathbf{A}$. A symmetric matrix is thus the same as its own transpose.
For example, if

$$\mathbf{A} = \begin{bmatrix} 1 & 2 \\ 2 & 3 \end{bmatrix} \quad \text{and} \quad \mathbf{B} = \begin{bmatrix} 1 & 2 \\ 3 & 4 \end{bmatrix}$$

$$\mathbf{A}' = \begin{bmatrix} 1 & 2 \\ 2 & 3 \end{bmatrix} = \mathbf{A}$$

so $\mathbf{A}$ is symmetric, but

$$\mathbf{B}' = \begin{bmatrix} 1 & 3 \\ 2 & 4 \end{bmatrix} \neq \mathbf{B}$$

so $\mathbf{B}$ is not symmetric.

DIAGONAL MATRICES

A special case of the symmetric matrix is the diagonal matrix. Here all $a_{ij} = 0$ for $i \neq j$, and only the a_{ii}'s take non-zero values. These elements, i.e. $a_{11}, a_{22}, \ldots, a_{nn}$ are referred to as the main diagonal.

THE IDENTITY MATRIX

One special diagonal matrix of particular interest is that where all $a_{ii} = 1$ and all $a_{ij} = 0, i \neq j$. This is called the identity matrix. For any $m \times n$ matrix $\mathbf{A}$, premultiplication by $\mathbf{I}_m$, the identity matrix of dimension m, or postmultiplication by $\mathbf{I}_n$, the identity matrix of dimension n, gives us $\mathbf{A}$ again. We do not usually bother to specify the dimension of the **I** matrix explicitly, but simply write **AI** = **A** or **IA** = **A**. However, only the **I**'s of correct dimension are conformable to **A** for multiplication, so in writing **IA** we are implicitly assuming **I** to be I_m, and in writing **AI** we are implicitly assuming **I** to be I_n. The identity matrix **I** plays a role in matrix algebra similar to that of 1 in ordinary multiplication.

MATRIX DIVISION

The reader may be wondering whether, since matrix addition, subtraction, and multiplication all make sense, there is such a thing as matrix division. It is not in fact possible literally to divide one matrix by another; in the next chapter we will see, however, that provided **A** has certain properties it is possible to find another matrix called the inverse of **A**, and written A^{-1}, which has properties in many ways analogous to division by **A**.

ECONOMIC MODELS IN MATRIX NOTATION

It is possible to express many economic models in matrix notation. Having put any model in this form, we will see in the next chapter how to apply the standard properties of matrices to check whether there is a solution, and to find the solution if one exists. For the present, however, we are only concerned with posing the problem.

A NATIONAL INCOME MODEL

$$Y = C + I + G + X - M$$

where Y = income, C = consumption, I = investment, G = government expenditure, X = exports, M = imports, and where

$$C = C^* + cY$$
$$I = I^* + iY$$
$$G = G^*$$
$$X = X^*$$

and

$$M = M^* + mY$$

This can be rewritten as

$$
\begin{aligned}
Y - C - I + M &= G^* + X^* \\
-cY + C \qquad\qquad &= \qquad C^* \\
-iY \qquad + I \qquad &= \qquad I^* \\
-mY \qquad\qquad + M &= \qquad M^*
\end{aligned}
$$

or as

$$\begin{bmatrix} 1 & -1 & -1 & 1 \\ -c & 1 & 0 & 0 \\ -i & 0 & 1 & 0 \\ -m & 0 & 0 & 1 \end{bmatrix} \begin{bmatrix} Y \\ C \\ I \\ M \end{bmatrix} = \begin{bmatrix} G^* + X^* \\ C^* \\ I^* \\ M^* \end{bmatrix}$$

where all the endogenous variables are included in a vector on the left-hand side, and all the exogenous variables are included in a vector on the right-hand side.

AN INPUT–OUTPUT MODEL

Suppose that total final demand is for a given vector of goods **y**. Production takes place with fixed techniques and constant returns to scale. Each firm produces one good only, and production of each good j requires inputs of a_{ij} units of each good i per unit of good j produced. Thus if **x** is the total vector of goods produced, assuming no good to be produced in excess of the demand for it, either as a final or an intermediate good, we have **y** demand for final use, and **Ax** for intermediate use in the production of **x**; thus

$$\mathbf{x} = \mathbf{Ax} + \mathbf{y}$$

so

$$\mathbf{Ix} - \mathbf{Ax} = \mathbf{y} \quad \text{or} \quad (\mathbf{I} - \mathbf{A})\mathbf{x} = \mathbf{y}$$

We will see in the next chapter how matrix methods can be used to find the **x** corresponding to any given **y**.

EXERCISES

1. Use questions 4 and 5 of Chapter 3 to express market equilibrium in matrix notation.

2. Use question 7 of Chapter 3 to express the joint equilibrium of the goods and money markets in matrix form. Disregard parts (b) and (c) of (ii).

3. Suppose the prices of two goods, P_x and P_y, are related to the quantities demanded of the two goods M and N so that

$$P_x = a_1 M + a_2 N$$
$$P_y = b_1 M + b_2 N$$

If

$$M = \alpha_1 R + \alpha_2 P_m \quad \text{and} \quad N = \beta_1 R + \beta_2 P_m$$

Where R is income and P_m the price of imports, express P_x and P_y as a function of R and P_m in matrix notation.

4. Given the National Income Model

$$Y = C + I + G + X - M$$

where

$$C = 100 + 0.8Y$$
$$I = 20 + 0.1Y$$
$$G = 500$$
$$X = 400$$

and

$$M = 200 + 0.1Y$$

Express the model in matrix form.

5. Suppose a comsumer buys 4 units of X, 3 units of Y and 2 units of Z, and that $P_x = 2$, $P_y = 3$ and $P_z = 4$. Express his expenditure on the goods in vector form, and work out the scalar product.

SOLUTIONS TO EXERCISES

1. Question 4: $\begin{bmatrix} 5 & -1 \\ -1 & 3 \end{bmatrix} \begin{bmatrix} P_1 \\ P_2 \end{bmatrix} = \begin{bmatrix} 13 \\ 17 \end{bmatrix}$ or **Ap = b**   *

   Question 5: $\begin{bmatrix} -4 & 2 & -2 \\ 2 & -3 & 2 \\ -1 & 2 & -2 \end{bmatrix} \begin{bmatrix} P_1 \\ P_2 \\ P_3 \end{bmatrix} = \begin{bmatrix} -50 \\ -20 \\ -35 \end{bmatrix}$ or **Ap = b**   *

2. (i) Without government activity:

$$\begin{bmatrix} 0.2 & 30 \\ 0.25 & -25 \end{bmatrix} \begin{bmatrix} Y \\ r \end{bmatrix} = \begin{bmatrix} 1300 \\ 1125 \end{bmatrix}$$ *

   (ii) With government activity:

$$\begin{bmatrix} 0.4 & 30 \\ 0.25 & -25 \end{bmatrix} \begin{bmatrix} Y \\ r \end{bmatrix} = \begin{bmatrix} 2220 \\ 1125 \end{bmatrix}$$ *

3. $\begin{bmatrix} P_x \\ P_y \end{bmatrix} = \begin{bmatrix} a_1 & a_2 \\ b_1 & b_2 \end{bmatrix} \begin{bmatrix} \alpha_1 & \alpha_2 \\ \beta_1 & \beta_2 \end{bmatrix} \begin{bmatrix} R \\ P_m \end{bmatrix}$

   $= \begin{bmatrix} a_1\alpha_1 + a_2\beta_1 & a_1\alpha_2 + a_2\beta_2 \\ b_1\alpha_1 + b_2\beta_1 & b_1\alpha_2 + b_2\beta_2 \end{bmatrix} \begin{bmatrix} R \\ P_m \end{bmatrix}$ *

4. $\qquad Y - C - I + M = G + X = 900$
   $\qquad -0.8Y + C + 0.I + 0.M = 100$
   $\qquad -0.1Y + 0.C + I + 0.M = 20$
   $\qquad -0.1Y + 0.C + 0.I + M = 200$

   or

$$\begin{bmatrix} 1 & -1 & -1 & 1 \\ -0.8 & 1 & 0 & 0 \\ -0.1 & 0 & 1 & 0 \\ -0.1 & 0 & 0 & 1 \end{bmatrix} \begin{bmatrix} Y \\ C \\ I \\ M \end{bmatrix} = \begin{bmatrix} 900 \\ 100 \\ 20 \\ 200 \end{bmatrix}$$ *

5. $E = \begin{bmatrix} 4 & 3 & 2 \end{bmatrix} \begin{bmatrix} 2 \\ 3 \\ 4 \end{bmatrix} = 8 + 9 + 8 = 25$ *

CHAPTER 11

Inverses and Determinants

THE INVERSE MATRIX

We have seen that a system of linear equations can be written as $\mathbf{y} = \mathbf{Ax}$.

If we know $\mathbf{x}$, we can always calculate $\mathbf{y}$. Under what conditions, however, can we solve the equations, i.e. infer $\mathbf{x}$ from a given $\mathbf{y}$? This can be done if we can find a matrix $\mathbf{B}$ such that $\mathbf{BA} = \mathbf{I}$. If such a $\mathbf{B}$ exists, then

$$\mathbf{By} = \mathbf{B(Ax)} = \mathbf{(BA)x} = \mathbf{Ix} = \mathbf{x} \quad \text{so} \quad \mathbf{x} = \mathbf{By}$$

But

$$\mathbf{y} = \mathbf{Ax} = \mathbf{A(By)} = \mathbf{(AB)y}$$

This implies that $\mathbf{AB} = \mathbf{I}$. Also, if $\mathbf{x} = \mathbf{By}$ for any $\mathbf{x}$, then

$$\mathbf{x} = \mathbf{By} = \mathbf{B(Ax)} = \mathbf{(BA)x}$$

This can be true for all $\mathbf{x}$ only if $\mathbf{BA} = \mathbf{I}$.

We can thus infer $\mathbf{x}$ from $\mathbf{y}$ *iff* there exists a matrix $\mathbf{B}$ such that $\mathbf{AB} = \mathbf{I}$ and $\mathbf{BA} = \mathbf{I}$. If such a matrix $\mathbf{B}$ exists we call it the inverse of $\mathbf{A}$, and write it as $\mathbf{A}^{-1}$. The inverse gives us $\mathbf{I}$ both when premultiplied and when post-multiplied by $\mathbf{A}$, i.e.

$$\mathbf{AA}^{-1} = \mathbf{I} \quad \text{and} \quad \mathbf{A}^{-1}\mathbf{A} = \mathbf{I}$$

It is clear from the definition that the inverse of $\mathbf{A}^{-1}$ is $\mathbf{A}$.

We can show that if $\mathbf{A}^{-1}$ exists, it is unique. Suppose that $\mathbf{B}$ and $\mathbf{C}$ are both related to $\mathbf{A}$ by

$$\mathbf{AB} = \mathbf{I} \qquad \mathbf{BA} = \mathbf{I} \qquad \mathbf{AC} = \mathbf{I} \quad \text{and} \quad \mathbf{CA} = \mathbf{I}$$

Thus

$$\mathbf{B} = \mathbf{IB} = \mathbf{(CA)B} = \mathbf{C(AB)} = \mathbf{CI} = \mathbf{C}$$

so $\mathbf{B}$ and $\mathbf{C}$ are identical.

EXISTENCE OF AN INVERSE

We now show under what conditions on $\mathbf{A}$ an inverse matrix $\mathbf{A}^{-1}$ can exist.

Assume $\mathbf{A}$ is $m \times n$. We seek to show that a matrix $\mathbf{B}$ such that $\mathbf{AB} = \mathbf{I}_m$ and $\mathbf{BA} = \mathbf{I}_n$ can exist if and only if $\mathbf{A}$ is square, i.e. $m = n$, and the rows and columns of $\mathbf{A}$ are all linearly independent.

If the n columns of $\mathbf{A}$ are not linearly independent, then an $\mathbf{x} \neq \mathbf{0}$ exists such that $\mathbf{Ax} = \mathbf{0}$. This $\mathbf{x}$ can be regarded as providing a set of weights for combining the various columns of $\mathbf{A}$.

Now if $\mathbf{B}$ exists such that $\mathbf{BA} = \mathbf{I}_n$,

$$\mathbf{x} = \mathbf{I}_n\mathbf{x} = (\mathbf{BA})\mathbf{x} = \mathbf{B}(\mathbf{Ax}) = \mathbf{B}(\mathbf{0}) = \mathbf{0}$$

Thus $\mathbf{x} = \mathbf{0}$, contrary to assumption; postulating that $\mathbf{B}$ exists in this case leads to a contradiction, so $\mathbf{B}$ cannot exist.

Similarly, if the m rows of $\mathbf{A}$ are not linearly independent, then a $\mathbf{z} \neq \mathbf{0}$ must exist such that $\mathbf{zA} = \mathbf{0}$. If then any $\mathbf{B}$ exists such that $\mathbf{AB} = \mathbf{I}_m$,

$$\mathbf{z} = \mathbf{zI}_m = \mathbf{z}(\mathbf{AB}) = (\mathbf{zA})\mathbf{B} = (\mathbf{0})\mathbf{B} = \mathbf{0}$$

so $\mathbf{z} = \mathbf{0}$, contrary to assumption, and $\mathbf{B}$ cannot exist in this case either.

Thus the only conditions under which $\mathbf{B} = \mathbf{A}^{-1}$ can exist are that all $\mathbf{A}$'s m rows and n columns are linearly independent.

We next show that this can only be true if $\mathbf{A}$ is square. $\mathbf{A}$ has m rows and n columns. If $m > n$ the rows cannot all be linearly independent, for each row has only n elements, and it was shown in the previous chapter that there cannot be more than n linearly independent vectors of dimension n. Similarly, if $m < n$, the columns cannot all be linearly independent, since each column has only m elements, and there cannot be more than m linearly independent vectors of dimension m.

Thus the rows and columns of $\mathbf{A}$ can only be linearly independent if $\mathbf{A}$ is square, i.e. if $m = n$. We have so far only shown that it is not impossible for $\mathbf{A}^{-1}$ to exist. Our method of demonstrating that $\mathbf{A}^{-1}$ actually does exist if $\mathbf{A}$ is square and has linearly independent rows and columns is to give a constructive proof; we show that $\mathbf{A}^{-1}$ can actually be calculated and thus proved to exist.

As a test for the independence of the rows and columns of $\mathbf{A}$ we use the determinant of $\mathbf{A}$, defined below, and written $|\mathbf{A}|$. The rows and columns of $\mathbf{A}$ are linearly independent *iff* $|\mathbf{A}| \neq 0$. Thus the inverse matrix $\mathbf{A}^{-1}$ will exist *iff* $|\mathbf{A}| \neq 0$. If $|\mathbf{A}| \neq 0$, $\mathbf{A}$ is said to be non-singular. $\mathbf{A}$ is also said to be of full rank, or of rank n, where the rank of a matrix is the largest number of linearly independent rows or columns which can be found in it (these can be shown to be the same). If $|\mathbf{A}| = 0$, $\mathbf{A}$ is said to be singular, and its rank is less than n.

We provide below constructive methods of claculating both determinants and inverse matrices. It will be seen that the process for calculating an inverse works *iff* the process for calculating a determinant shows that $|\mathbf{A}| \neq 0$. If $|\mathbf{A}| = 0$ then the inverse matrix $\mathbf{A}^{-1}$ cannot be calculated.

DETERMINANTS

Suppose that $\mathbf{A}$ is a square matrix. The determinant of $\mathbf{A}$, written as $|\mathbf{A}|$, or sometimes as $\text{Det}(\mathbf{A})$, is a number found by summing a number of terms each calculated from the elements of $\mathbf{A}$. Each term has as factors one and only one element from each row and column of **A**. The sign attached to each term in the summation is found as follows. Arrange each term with its factors in the order of their row subscripts. Now rearrange the factors of each term, by making successive interchanges of adjacent factors, until they are in the order of their column subscripts. If this requires no change, or any even number of interchanges of adjacent factors, the sign of the term is positive. If the rearrangement takes one or any odd number of interchanges of adjacent factors, the sign of the term is negative. The simplest way of following this is by looking at some examples. Suppose **A** is 2×2; then

$$\mathbf{A} = \begin{bmatrix} a_{11} & a_{12} \\ a_{21} & a_{22} \end{bmatrix}$$

There are only two ways of choosing elements, one from each row and one from each column. These are $(a_{11}a_{22})$ and $(a_{12}a_{21})$. In $(a_{11}a_{22})$ the columns are already in natural order, so the sign is positive. In $(a_{12}a_{21})$ one interchange is needed to get the columns into natural order, so the sign is negative, and

$$|\mathbf{A}| = a_{11}a_{22} - a_{12}a_{21}$$

e.g. if

$$\mathbf{A} = \begin{bmatrix} 1 & 2 \\ 3 & 4 \end{bmatrix}$$

$$|\mathbf{A}| = (1)(4) - (2)(3) = 4 - 6 = -2$$

These rules apply whether the a_{ij} themselves are positive or negative, e.g. if

$$\mathbf{B} = \begin{bmatrix} 2 & -3 \\ -4 & -5 \end{bmatrix}$$

$$|\mathbf{B}| = (2)(-5) - (-3)(-4) = -10 - 12 = -22$$

If **A** is 3×3, there are six terms to the expansion.

$$(a_{11}a_{22}a_{33}), \quad (a_{11}a_{23}a_{32}), \quad (a_{12}a_{21}a_{33})$$
$$(a_{12}a_{23}a_{31}), \quad (a_{13}a_{21}a_{32}), \quad \text{and} \quad (a_{13}a_{22}a_{31})$$

Of these, $(a_{11}a_{22}a_{33})$ already has its columns in natural order and thus requires no changes, so its sign is positive; $(a_{11}a_{23}a_{32})$ becomes $(a_{11}a_{32}a_{23})$ with one change and is negative; $(a_{12}a_{21}a_{33})$ becomes $(a_{21}a_{12}a_{33})$ with one change and is negative; $(a_{12}a_{23}a_{31})$ requires two changes, becoming $(a_{12}a_{31}a_{23})$ then $(a_{31}a_{12}a_{23})$, so it is positive; $(a_{13}a_{21}a_{32})$ requires two changes, becoming $(a_{21}a_{13}a_{32})$ then $(a_{21}a_{32}a_{13})$ and is positive; finally, $(a_{13}a_{22}a_{31})$ requires three changes, becoming $(a_{13}a_{31}a_{22})$ then $(a_{31}a_{13}a_{22})$ then $(a_{31}a_{22}a_{13})$ and so is negative. Thus the expression for $|\mathbf{A}|$ is

$$|\mathbf{A}| = a_{11}a_{22}a_{33} - a_{11}a_{23}a_{32} - a_{12}a_{21}a_{33}$$
$$+ a_{12}a_{23}a_{31} + a_{13}a_{21}a_{32} - a_{13}a_{22}a_{31}$$

An expression of this type, writing out the terms of $|\mathbf{A}|$ with their signs, is referred to as the expansion of $|\mathbf{A}|$.

This set of rules looks complicated and arbitrary. It can be shown however, though it will not be shown here, that these rules are not arbitrary, but are imposed by the necessary conditions for the n rows and columns of $\mathbf{A}$ to be linearly independent.

A similar analysis can be carried out for an $n \times n$ matrix.

We can see from the rules on signs that $|\mathbf{A}|$ and the determinant of its transpose, $|\mathbf{A}'|$, must be the same.

The selection rule 'one element only from each row and each column' is not affected by interchanging rows and columns. Rearranging a term whose row subscripts are in order so as to arrange its column subscripts in order is an exact reversal of the steps required to rearrange a term whose column subscripts are in order so as to get its row subscripts in order, and thus requires the same number of interchanges. Therefore the sign of each term in the determinant is not affected by interchanging rows and columns. Hence the determinants of $\mathbf{A}$ and its transpose $\mathbf{A}'$ must be the same, i.e. $|\mathbf{A}| = |\mathbf{A}'|$.

We can if we wish express the determinant of an $n \times n$ matrix $\mathbf{A}$ as the weighted sum of n 'smaller' determinants, each of order $(n-1) \times (n-1)$, where the 'weights' are the n elements of any one row or any one column of $\mathbf{A}$, and each weight multiplies the determinant of the matrix formed from $\mathbf{A}$ by deleting the row and column of the element used as the weight. The terms are signed according to whether the row or column subscripts of their weight sum to an even number, in which case the term has a $+$ sign, or an odd number in which case the term has a $-$ sign. For example, if $\mathbf{A}$ is 3×3,

$$|\mathbf{A}| = a_{11}a_{22}a_{33} - a_{11}a_{23}a_{32} - a_{12}a_{21}a_{33}$$
$$+ a_{12}a_{23}a_{31} + a_{13}a_{21}a_{32} - a_{13}a_{22}a_{31}$$

We could choose to expand $|\mathbf{A}|$ using any row; selecting the second row, we can group the terms of $|\mathbf{A}|$ as follows:

$$|\mathbf{A}| = -a_{21}(a_{12}a_{33} - a_{13}a_{32}) + a_{22}(a_{11}a_{33} - a_{13}a_{31})$$
$$- a_{23}(a_{11}a_{32} - a_{12}a_{31})$$
$$= -a_{21}\begin{vmatrix} a_{12} & a_{13} \\ a_{32} & a_{33} \end{vmatrix} + a_{22}\begin{vmatrix} a_{11} & a_{13} \\ a_{31} & a_{33} \end{vmatrix} - a_{23}\begin{vmatrix} a_{11} & a_{12} \\ a_{31} & a_{32} \end{vmatrix}$$

This property is indeed commonly used as a method of defining determinants; we have preferred the one used above because it makes derivation of the properties of determinants simpler.

THE PROPERTIES OF DETERMINANTS

If $\mathbf{A}$ has a whole row, or a whole column, of zero elements, then $|\mathbf{A}| = 0$, since every term of the expansion of $|\mathbf{A}|$ will have at least one factor equal to zero.

If $\mathbf{A}$ has every element in any one row, or every element in any one column, multiplied by any constant λ, then $|\mathbf{A}|$ is multiplied by λ, since every term in in the expansion of $|\mathbf{A}|$ has one and only one factor multiplied by λ.

If $\mathbf{A}$ has any two rows, or any two columns, the same, its determinant must be zero. This is shown as follows. Consider the effects of interchanging any two adjacent rows of $\mathbf{A}$, to obtain $\mathbf{B}$, which is otherwise exactly the same as $\mathbf{A}$. This will leave us with the same set of terms in the expansion of $|\mathbf{B}|$ as of $|\mathbf{A}|$, but by the sign rules every term will have changed sign. Thus $|\mathbf{B}| = -|\mathbf{A}|$. A similar result would follow if any two adjacent columns were interchanged.

If we now wish to interchange any two rows of $\mathbf{A}$ to obtain $\mathbf{C}$, which apart from this interchange is exactly the same as $\mathbf{A}$, this can be done by an odd number of interchanges of adjacent rows, each of which multiplies the determinant by -1. Thus $|\mathbf{C}| = -|\mathbf{A}|$, since all odd powers of -1 are negative. It always takes an odd number of changes to interchange rows i and k; if it takes c changes to move i next to k by successive interchanges of adjacent rows, it will take one further change to get i into position k, and c changes to get k back to i's original position. The total number of changes needed is thus $(2c + 1)$, which is always odd. Again, a similar result follows if we consider columns; it always takes an odd number of interchanges of adjacent columns to interchange any two columns of $\mathbf{A}$.

If $\mathbf{A}$ has any two rows the same, then interchanging them will leave us with the same matrix $\mathbf{A}$ that we had to start with; but we have just seen that such an interchange changes the sign of the determinant, so $|\mathbf{A}| = -|\mathbf{A}|$. For any number k, if $k = -k$, $k = 0$, so $|\mathbf{A}| = -|\mathbf{A}| = 0$. The same result occurs if $\mathbf{A}$ has any two columns the same.

If $\mathbf{A}$ has any row proportional to another, $|\mathbf{A}| = 0$. If row i is a constant λ times row $k \neq i$, if we multiply row k by λ we multiply $|\mathbf{A}|$ by λ; but we obtain a matrix with two equal rows, whose determinant is zero; if $\lambda |\mathbf{A}| = 0$, $|\mathbf{A}| = 0$. Similarly, any matrix with one column a multiple λ of another will have a zero determinant.

The determinant of $\mathbf{A}$ will be unchanged if we add one row to another. If $\mathbf{B}$ is the same as $\mathbf{A}$ except for the addition of the kth row of $\mathbf{A}$, $\mathbf{a}_k$, to the ith row, $\mathbf{a}_i$, then each term of the expansion of the determinant of $\mathbf{B}$ has one and only one factor of the form $(a_{ij} + a_{kj})$. The expansion can be split into two parts, one with the factor a_{ij} in each term; this simply gives us $|\mathbf{A}|$. The other term will have the factor a_{kj}, but this gives us the determinant of a matrix with two equal rows, $\mathbf{a}_k$, which must be zero. Thus $|\mathbf{B}| = |\mathbf{A}|$. The same applies to the addition of any one column of a matrix to another; the determinant is unaltered.

If we add any multiple λ of one row to another we get a new matrix whose determinant can be split into two parts, one with a factor a_{ij} which is simply $|\mathbf{A}|$, and the other with a factor λa_{kj} which gives the determinant of a matrix with row i λ times row k, which must be zero. Again, the same argument applies if we add λ times any column to any other column; there is no change in the determinant.

These operations can be performed any number of times; we can add or subtract any multiple of any row of **A** to any other row, or any multiple of any column to another column, obtaining a new matrix whose determinant is equal to $|\mathbf{A}|$. We can now use these rules to show how to evaluate any determinant.

PROCEDURE FOR THE EVALUATION OF ANY DETERMINANT

In this we start from any square matrix **A**. By successive row operations we will change this to a succession of new matrices, **A***, choosing the operations so that at each stage the determinant of **A*** is the same as that of **A**. We will use **A*** to refer to the latest transformation of **A** at each stage of the argument, and a_{ij}^{*} to refer to the element of **A*** in its ith row and jth column.

Start with row 1 of **A**. Is $a_{11} = 0$? If $a_{11} \neq 0$ we can proceed straight to the next step. If $a_{11} = 0$, add some subsequent row of **A** to row 1 to obtain **A*** with $a_{11}^{*} \neq 0$, and then proceed. If this cannot be done then every element in column 1 of **A** is zero, so $|\mathbf{A}| = 0$. Addition of one row to another will not alter the determinant, so $|\mathbf{A}^{*}| = |\mathbf{A}|$.

The next step, with $a_{11}^{*} \neq 0$ (or $a_{11} \neq 0$) is to add to each subsequent row a multiple of row 1 chosen so that we get a column of zeros below a_{11}^{*} in the first column of **A***, i.e. we make $a_{i1}^{*} = 0$ for each $i > 1$. Each of these changes will leave $|\mathbf{A}^{*}|$ unaltered.

Now consider $|\mathbf{A}^{*}|$. Each term in its expansion has one and only one factor from column 1; either this must be a_{11}^{*} or it will be zero. As each term has only one factor from row 1, $|\mathbf{A}^{*}|$ must equal a_{11}^{*} times the determinant of the matrix obtained from **A*** by deleting its first row and first column. If **A** was $n \times n$, this matrix will be $(n-1) \times (n-1)$. We then treat this matrix in the same way as we treated **A**, and so on, gradually reducing the number or rows and columns we need to consider.

The whole process must finish in one of two ways. *Either* we get down to a situation where we have reached row and column n. In this case we have a matrix **A*** which has zeros everywhere below the main diagonal. Every term in the expansion of **A*** must have at least one zero factor, except for the product of all the elements on the main diagonal itself, which is thus the determinant of **A***, i.e.

$$|\mathbf{A}^{*}| = a_{11}^{*} a_{22}^{*} \ldots a_{nn}^{*}$$

But all the operations performed on **A** were chosen to keep its determinant unaltered, so $|\mathbf{A}| = |\mathbf{A}^{*}|$. In this case $|\mathbf{A}| \neq 0$, and **A** is said to be non-singular.

Or, at some stage of the process, after k steps, we get to a matrix $(n-k) \times (n-k)$ which has a first column of zeros. Its determinant must thus be zero, and $|A^{*}|$, being a product of k non-zero terms down the main diagonal and then zero, must itself be zero. Thus $|\mathbf{A}| = |\mathbf{A}^{*}| = 0$. In this case **A** is said to be singular.

It is helpful to consider an example of this process. Consider

$$\mathbf{A} = \begin{bmatrix} 1 & 2 & 3 \\ 4 & 5 & 0 \\ 1 & 1 & 1 \end{bmatrix}$$

Subtract 4 × row 1 from row 2 and subtract 1 × row 1 from row 3, to obtain

$$\mathbf{A}^* = \begin{bmatrix} 1 & 2 & 3 \\ 0 & -3 & -12 \\ 0 & -1 & -2 \end{bmatrix}$$

Subtract 1/3 × row 2 from row 3 to obtain

$$\mathbf{A}^* = \begin{bmatrix} 1 & 2 & 3 \\ 0 & -3 & -12 \\ 0 & 0 & 2 \end{bmatrix}$$

Thus

$$|\mathbf{A}| = |\mathbf{A}^*| = (1)(-3)(2) = -6$$

CALCULATION OF THE INVERSE

This can be done by various methods. The method we will use here is closely similar to the method used above to evaluate determinants. In it, we start with

$$\mathbf{Ax} = \mathbf{Iy}$$

This represents a system of n linear equations in $\mathbf{x}$ and $\mathbf{y}$.

Start with row 1. Is $a_{11} = 0$? If $a_{11} \neq 0$, go on to the next step. If $a_{11} = 0$, add some subsequent row to row 1, on both sides of the = sign, to obtain $a_{11}^* \neq 0$ (we are using the same notation as above, by which a_{ij}^* represents the element on the left-hand side in the ith row and jth column); this can always be done, unless the first column of $\mathbf{A}$ contains all zeros, in which case the n rows of $\mathbf{A}$ cannot be linearly independent, so that the inverse will not exist. We now have another system of equations which we know to be true. Divide all of row 1, both sides of the = sign, by a_{11}^* to obtain a new $a_{11}^* = 1$. Then add to every other row, both sides of the = sign, whatever multiple of row 1 is required to obtain a column of zeros below a_{11}^*, which is 1. Then go on to row 2.

If $a_{22}^* = 0$, add some subsequent row to row 2 to obtain $a_{22}^* \neq 0$. Once $a_{22}^* \neq 0$, divide row 2, both sides of the = sign, by a_{22}^* to obtain a new $a_{22}^* = 1$, and add to all rows other than row 2 whatever multiple of row 2 is required to ensure that the second column of $\mathbf{A}^*$ contains all zeros except for $a_{22}^* = 1$ on the main diagonal. Then go on to column 3, and repeat the process.

This goes on until the left-hand side of matrix $\mathbf{A}^*$ simply consists of 1's down the main diagonal and zeros elsewhere, i.e. $\mathbf{A}^*$ has become the identity matrix $\mathbf{I}$. The right-hand side will now have a matrix $\mathbf{B}$, which allows us to read off

$$\mathbf{Ix} = \mathbf{x} = \mathbf{By}$$

The **B** we have calculated is thus the inverse, $\mathbf{A}^{-1}$, that we have been looking for.

The simplest way to see that this process works is through an example. Suppose

$$\begin{bmatrix} 1 & 2 \\ 3 & 4 \end{bmatrix} \mathbf{x} = \begin{bmatrix} 1 & 0 \\ 0 & 1 \end{bmatrix} \mathbf{y}$$

We already have $a_{11} = 1$. Subtract 3 × row 1 of **A** and **I** from row 2 to obtain

$$\begin{bmatrix} 1 & 2 \\ 0 & -2 \end{bmatrix} \mathbf{x} = \begin{bmatrix} 1 & 0 \\ -3 & 1 \end{bmatrix} \mathbf{y}$$

Multiply row 2 both sides by -0.5 to obtain

$$\begin{bmatrix} 1 & 2 \\ 0 & 1 \end{bmatrix} \mathbf{x} = \begin{bmatrix} 1 & 0 \\ 1.5 & -0.5 \end{bmatrix} \mathbf{y}$$

Subtract 2 × row 2 from row 1 to obtain

$$\begin{bmatrix} 1 & 0 \\ 0 & 1 \end{bmatrix} \mathbf{x} = \begin{bmatrix} -2 & 1 \\ 1.5 & -0.5 \end{bmatrix} \mathbf{y}$$

To check that we have the right answer,

$$\begin{bmatrix} 1 & 2 \\ 3 & 4 \end{bmatrix}\begin{bmatrix} -2 & 1 \\ 1.5 & -0.5 \end{bmatrix} = \begin{bmatrix} -2+3 & 1-1 \\ -6+6 & 3-2 \end{bmatrix} = \begin{bmatrix} 1 & 0 \\ 0 & 1 \end{bmatrix}$$

This method will work on a matrix **A** of any dimension $n \times n$. It is not the most accurate method for large-scale computation, but it does have the great merit that as it leaves us with a set of valid equations at every stage of its working, it is easy to see why it is valid. While for large matrices it is very tedious, it is fairly easy to obtain the right answers as very little has to be carried in the head at any one stage.

If there is an inverse $\mathbf{A}^{-1}$ this method will always find it. If we try to apply it to any matrix which in fact has no inverse, the reason why it will fail to produce one is that at some stage we will find that $a_{ii}^{*} = 0$ and no $a_{ii}^{*} \neq 0$ can be found by adding any subsequent row to row i. Thus there is no way of getting a 1 in the ith position on the main diagonal, and no inverse can be calculated. This was, however, exactly the condition which in the process for calculating the determinant gave $|\mathbf{A}| = 0$.

DETERMINANTS AND INVERSES

We can now see the relation of the determinant $|\mathbf{A}|$ to the existence of an inverse matrix $\mathbf{A}^{-1}$. If the rows of **A** are not linearly independent, **A*** will have a whole row of zeros in the nth row, if not earlier, so $|\mathbf{A}|$ is zero. This is the case where there is no inverse. If we can get right through to the nth row of **A***, then the determinant $|\mathbf{A}|$ is non-zero; **A** has an inverse, and is said to be non-singular. Calculation of the inverse $\mathbf{A}^{-1}$ enables us to calculate **x** if we are given $\mathbf{Ax} = \mathbf{y}$.

AN ECONOMIC APPLICATION OF THE INVERSE MATRIX

Suppose we take the market model $Q_D = 20 - 2P$ and $Q_S = -10 + 3P$. Equilibrium in the market is given by

$$\begin{bmatrix} 1 & 2 \\ 1 & -3 \end{bmatrix} \begin{bmatrix} Q^* \\ P^* \end{bmatrix} = \begin{bmatrix} 20 \\ -10 \end{bmatrix}$$

Where P^* and Q^* are the equilibrium values of P and Q. This can be expressed as

$$\begin{bmatrix} 1 & 2 \\ 1 & -3 \end{bmatrix} \begin{bmatrix} Q^* \\ P^* \end{bmatrix} = \begin{bmatrix} 1 & 0 \\ 0 & 1 \end{bmatrix} \begin{bmatrix} 20 \\ -10 \end{bmatrix}$$

To solve for P^* and Q^* we use row operations to reduce $\begin{bmatrix} 1 & 2 \\ 1 & -3 \end{bmatrix}$ to an identity matrix.

Add $-1 \times$ row 1 to row 2 to get

$$\begin{bmatrix} 1 & 2 \\ 0 & -5 \end{bmatrix} \begin{bmatrix} Q^* \\ P^* \end{bmatrix} = \begin{bmatrix} 1 & 0 \\ -1 & 1 \end{bmatrix} \begin{bmatrix} 20 \\ -10 \end{bmatrix}$$

Multiply row 2 by -0.2 to get

$$\begin{bmatrix} 1 & 2 \\ 0 & 1 \end{bmatrix} \begin{bmatrix} Q^* \\ P^* \end{bmatrix} = \begin{bmatrix} 1 & 0 \\ 0.2 & -0.2 \end{bmatrix} \begin{bmatrix} 20 \\ -10 \end{bmatrix}$$

Add $-2 \times$ row 2 to row 1 to get

$$\begin{bmatrix} 1 & 0 \\ 0 & 1 \end{bmatrix} \begin{bmatrix} Q^* \\ P^* \end{bmatrix} = \begin{bmatrix} 0.6 & 0.4 \\ 0.2 & -0.2 \end{bmatrix} \begin{bmatrix} 20 \\ -10 \end{bmatrix} = \begin{bmatrix} 8 \\ 6 \end{bmatrix}$$

Thus

$$\begin{bmatrix} Q^* \\ P^* \end{bmatrix} = \begin{bmatrix} 8 \\ 6 \end{bmatrix}$$

i.e. $Q^* = 8$ and $P^* = 6$.
Once the inverse of

$$\begin{bmatrix} 1 & 2 \\ 1 & -3 \end{bmatrix}, \text{ i.e.} \begin{bmatrix} 0.6 & 0.4 \\ 0.2 & -0.2 \end{bmatrix}$$

has been found, we can easily determine the effects on the equilibrium price and quantity of any shift in the supply or demand functions. Suppose that the constant term in the demand function is changed to 30, i.e. $Q_D = 30 - 2P$. The new equilibrium price and quantity are given by

$$\begin{bmatrix} Q^* \\ P^* \end{bmatrix} = \begin{bmatrix} 0.6 & 0.4 \\ 0.2 & -0.2 \end{bmatrix} \begin{bmatrix} 30 \\ -10 \end{bmatrix} = \begin{bmatrix} 14 \\ 8 \end{bmatrix}$$

i.e. $Q^* = 14$ and $P^* = 8$. Taking a general market model of the form $Q_D + aP = m$ and $Q_S - bP = -n$ we have at equilibrium

$$\begin{bmatrix} 1 & a \\ 1 & -b \end{bmatrix} \mathbf{v} = \begin{bmatrix} 1 & 0 \\ 0 & 1 \end{bmatrix} \mathbf{w}$$

where

$$\mathbf{v} = \begin{bmatrix} Q^* \\ P^* \end{bmatrix} \quad \text{and} \quad \mathbf{w} = \begin{bmatrix} m \\ -n \end{bmatrix}$$

To solve for $\mathbf{v}$ we apply row operations to reduce $\begin{bmatrix} 1 & a \\ 1 & -b \end{bmatrix}$ to an identity matrix. Add -1 times row 1 to row 2 to get

$$\begin{bmatrix} 1 & a \\ 0 & -a-b \end{bmatrix} \mathbf{v} = \begin{bmatrix} 1 & 0 \\ -1 & 1 \end{bmatrix} \mathbf{w}$$

Multiply row 2 by $-1/(a+b)$ to obtain

$$\begin{bmatrix} 1 & a \\ 0 & 1 \end{bmatrix} \mathbf{v} = \begin{bmatrix} 1 & 0 \\ \dfrac{1}{a+b} & \dfrac{-1}{a+b} \end{bmatrix} \mathbf{w}$$

Add $-a \times$ row 2 to row 1 to obtain

$$\begin{bmatrix} 1 & 0 \\ 0 & 1 \end{bmatrix} \mathbf{v} = \begin{bmatrix} \dfrac{b}{a+b} & \dfrac{a}{a+b} \\ \dfrac{1}{a+b} & \dfrac{-1}{a+b} \end{bmatrix} \mathbf{w}$$

The inverse matrix contains the comparative statics multipliers with respect to the elements in $\mathbf{w}$, i.e. the exogenous components in the supply and demand functions. A great deal of time may be saved by using matrix algebra to determine the impact multipliers within a system. This is especially true for larger economic models. In the above model equilibrium is given by

$$\begin{bmatrix} Q^* \\ P^* \end{bmatrix} = \begin{bmatrix} \dfrac{b}{a+b} & \dfrac{a}{a+b} \\ \dfrac{1}{a+b} & \dfrac{-1}{a+b} \end{bmatrix} \begin{bmatrix} m \\ -n \end{bmatrix} = \begin{bmatrix} \dfrac{bm-an}{a+b} \\ \dfrac{m+n}{a+b} \end{bmatrix}$$

Any shift in the demand function due to a change Δm in the exogenous component results in a change $\Delta Q^* = [b/(a+b)]\Delta m$ in Q^* and $\Delta P^* = [1/(a+b)]\Delta m$ in P^*, assuming n remains constant.

SOLUTION OF EQUATIONS BY THE USE OF DETERMINANTS: CRAMER'S RULE

Suppose that $\mathbf{Ax} = \mathbf{y}$, where $\mathbf{A}$ is a square $n \times n$ matrix and $|\mathbf{A}| \neq 0$. It can then be shown that it is always true that, if $\mathbf{A}_j$ is the matrix obtained from $\mathbf{A}$ by substituting $\mathbf{y}$ for the jth column of $\mathbf{A}$,

$$x_j = \frac{|\mathbf{A}_j|}{|\mathbf{A}|} \text{ for all } j = 1, \dots n$$

This method is known, from its inventor, as Cramer's Rule.

Proof is tricky (see below); equations can in fact be solved without using it, and it is not an efficient computational method. It is, however, a compact notation, and is frequently used in economics, so that it is as well to recognize it. It is simple to see for the 2×2 case that it gives the same results as can be obtained by ordinary substitution methods. Suppose

$$a_{11}x_1 + a_{12}x_2 = y_1$$
$$a_{21}x_1 + a_{22}x_2 = y_2$$

If we eliminate x_2 by suitable multiplication of these equations, we get

$$a_{11}a_{22}x_1 + a_{12}a_{22}x_2 = a_{22}y_1$$
$$a_{12}a_{21}x_1 + a_{12}a_{22}x_2 = a_{12}y_2$$

so

$$(a_{11}a_{22} - a_{12}a_{21})x_1 = a_{22}y_1 - a_{12}y_2$$

thus

$$x_1 = \frac{a_{22}y_1 - a_{12}y_2}{a_{11}a_{22} - a_{12}a_{21}} = \frac{\begin{vmatrix} y_1 & a_{12} \\ y_2 & a_{22} \end{vmatrix}}{\begin{vmatrix} a_{11} & a_{12} \\ a_{21} & a_{22} \end{vmatrix}}$$

and similarly

$$a_{11}a_{21}x_1 + a_{12}a_{21}x_2 = a_{21}y_1$$
$$a_{11}a_{21}x_1 + a_{11}a_{22}x_2 = a_{11}y_2$$

so that

$$(a_{11}a_{22} - a_{12}a_{21})x_2 = a_{11}y_2 - a_{21}y_1$$

thus

$$x_2 = \frac{a_{11}y_2 - a_{21}y_1}{a_{11}a_{22} - a_{12}a_{21}} = \frac{\begin{vmatrix} a_{11} & y_1 \\ a_{21} & y_2 \end{vmatrix}}{\begin{vmatrix} a_{11} & a_{12} \\ a_{21} & a_{22} \end{vmatrix}}$$

A general proof of Cramer's rule proceeds as follows. If the rule is correct, then we can write the column vector $\mathbf{x}$ as

$$\mathbf{x} = [|\mathbf{A}_j|/|\mathbf{A}|] = \begin{bmatrix} |\mathbf{A}_1|/|\mathbf{A}| \\ |\mathbf{A}_2|/|\mathbf{A}| \\ \vdots \\ |\mathbf{A}_n|/|\mathbf{A}| \end{bmatrix} \tag{1}$$

and the original equation can be written as

$$\mathbf{A}[|\mathbf{A}_j|/|\mathbf{A}|] = \mathbf{y}$$

where $[|\mathbf{A}_j|]$ is a column vector of dimension n, which is equivalent to $\mathbf{A}[|\mathbf{A}_j|] = |\mathbf{A}|\mathbf{y}$. If $\mathbf{a}_i$ is the ith row of $\mathbf{A}$, this in turn is equivalent to saying that for all i,

$$\mathbf{a}_i[|\mathbf{A}_j|] = |\mathbf{A}|y_i$$

i.e.

$$[a_{i1} \quad a_{i2} \ldots a_{in}][|\mathbf{A}_j|] = |\mathbf{A}|y_i$$

i.e.

$$a_{i1}|\mathbf{A}_1| + a_{i2}|\mathbf{A}_2| + \ldots + a_{in}|\mathbf{A}_n| = |\mathbf{A}|y_i \tag{2}$$

We will now show that this last proposition must be correct. Consider first the terms on the left-hand side which have y_i as a factor. Each $\mathbf{A}_j$ contains a column $\mathbf{y}$ in which y_i appears. Thus each term on the left-hand side contains y_i multiplied by

$$a_{ij}y_i(-1)^{i+j}|\mathbf{C}_{ij}|$$

where $\mathbf{C}_{ij}$ is a matrix obtained from $\mathbf{A}$ by deleting its ith row and jth column, called a minor of $\mathbf{A}$, and $(-1)^{i+j}|\mathbf{C}_{ij}|$ is called the cofactor of a_{ij}. The sum of terms is

$$y_i\sum_j a_{ij}(-1)^{i+j}|\mathbf{C}_{ij}| = y_i|\mathbf{A}|$$

Consider next the terms on the left-hand side having as a factor y_k for any $k \neq i$. Each term is equal to

$$a_{ij}y_k(-1)^{i+k}|\mathbf{C}_{kj}|$$

thus their sum is

$$y_k\sum_j a_{ij}(-1)^{i+k}|\mathbf{C}_{kj}|$$

The expression after the summation sign is the determinant of a matrix $\mathbf{B}$ which is the same as $\mathbf{A}$ except that $\mathbf{a}_i$, the ith row of $\mathbf{A}$, has been substituted for $\mathbf{a}_k$, the kth row of $\mathbf{A}$. Thus $\mathbf{B}$ has two rows identical so that $|\mathbf{B}| = 0$.

We have thus shown that the term on the LHS of equation (2) with y_i as a factor equals $y_i|\mathbf{A}|$, while the other terms on the LHS are zero. Thus (2) is correct. But equation (1) is equivalent to equation (2), so if equation (2) is correct, equation (1) must be true. This proves Cramer's rule.

Equilibrium in the market model with $Q_D + aP = m$ and $Q_S - bP = n$ is given by

$$\begin{bmatrix} 1 & a \\ 1 & -b \end{bmatrix}\begin{bmatrix} Q^* \\ P^* \end{bmatrix} = \begin{bmatrix} m \\ -n \end{bmatrix}$$

Solving by Cramer's rule gives

$$Q^* = \frac{\begin{vmatrix} m & a \\ -n & -b \end{vmatrix}}{\begin{vmatrix} 1 & a \\ 1 & -b \end{vmatrix}} = \frac{bm - an}{a + b}$$

because Q^* is the first variable in the vector so in the numerator the first column of the coefficients matrix is replaced by the vector $\begin{bmatrix} m \\ -n \end{bmatrix}$. Similarly, replacing the second column,

$$P^* = \frac{\begin{vmatrix} 1 & m \\ 1 & -n \\ 1 & a \\ 1 & -b \end{vmatrix}}{} = \frac{m+n}{a+b}$$

EXERCISES

1. Use Cramer's rule to find the equilibrium prices and quantities of the three-commodity market model given in question 5 of Chapter 3.

2. Answer question 6 of Chapter 3 using the row operations method for finding the inverse.

3. Use matrix algebra to find equilibrium Y and C for a closed macroeconomic model when
   (i) $C = C^* + cY$ and $I = I^*$
   (ii) $C = C^* + cY_d$, $T = tY$ and $I = I^* + iY$
   Comment on the value of the multiplier in (ii).

4. Given the following information on a closed economy:
$$C = C^* + cY$$
$$I = I^* - ar$$
where r is the rate of interest and $a > 0$;
$$M_D = K + \beta Y - gr$$
is the demand for money function
   where $K > 0, \beta > 0$, and $g > 0$;
$$M_S = M^*$$
i.e. money supply is constant.
   Use matrix algebra to find the equilibrium interest rate r and comment on the value of the multipliers.

SOLUTIONS TO EXERCISES

1. Expressing the equations in matrix notation,
$$\begin{bmatrix} 4 & -2 & 2 \\ -2 & 3 & -2 \\ 1 & -2 & 2 \end{bmatrix} \begin{bmatrix} P_1 \\ P_2 \\ P_3 \end{bmatrix} = \begin{bmatrix} 50 \\ 20 \\ 35 \end{bmatrix}$$

$$P_1 = \frac{\begin{vmatrix} 50 & -2 & 2 \\ 20 & 3 & -2 \\ 35 & -2 & 2 \end{vmatrix}}{\begin{vmatrix} 4 & -2 & 2 \\ -2 & 3 & -2 \\ 1 & -2 & 2 \end{vmatrix}}$$

$$= \frac{50 \begin{vmatrix} 3 & -2 \\ -2 & 2 \end{vmatrix} + 2 \begin{vmatrix} 20 & -2 \\ 35 & 2 \end{vmatrix} + 2 \begin{vmatrix} 20 & 3 \\ 35 & -2 \end{vmatrix}}{4 \begin{vmatrix} 3 & -2 \\ -2 & 2 \end{vmatrix} + 2 \begin{vmatrix} -2 & -2 \\ 1 & 2 \end{vmatrix} + 2 \begin{vmatrix} -2 & 3 \\ 1 & -2 \end{vmatrix}}$$

$$= \frac{50(2) + 2(110) + 2(-145)}{4(2) + 2(-2) + 2(1)}$$

$$= \frac{100 + 220 - 290}{8 - 4 + 2}$$

$$= \frac{30}{6} = 5$$

*

$$P_2 = \frac{1}{6} \begin{vmatrix} 4 & 50 & 2 \\ -2 & 20 & -2 \\ 1 & 35 & 2 \end{vmatrix}$$

$$= \frac{1}{6} \left[4 \begin{vmatrix} 20 & -2 \\ 35 & 2 \end{vmatrix} - 50 \begin{vmatrix} -2 & -2 \\ 1 & 2 \end{vmatrix} + 2 \begin{vmatrix} -2 & 20 \\ 1 & 35 \end{vmatrix} \right]$$

$$= \frac{1}{6} [4(110) - 50(-2) + 2(-90)]$$

$$= \frac{1}{6} (440 + 100 - 180)$$

$$= \frac{360}{6} = 60$$

*

$$P_3 = \frac{1}{6} \begin{vmatrix} 4 & -2 & 50 \\ -2 & 3 & 20 \\ 1 & -2 & 35 \end{vmatrix}$$

$$= \frac{1}{6} \left[4 \begin{vmatrix} 3 & 20 \\ -2 & 35 \end{vmatrix} + 2 \begin{vmatrix} -2 & 20 \\ 1 & 35 \end{vmatrix} + 50 \begin{vmatrix} -2 & 3 \\ 1 & -2 \end{vmatrix} \right]$$

$$= \frac{1}{6} [4(145) + 2(-90) + 50(1)]$$

$$= \frac{1}{6} (580 - 180 + 50)$$

$$= \frac{450}{6} = 75$$

*

2. The equations are

$$Q_A - 0 \cdot 2Q_A - 0 \cdot 1Q_I = Y_A$$
$$Q_I - 0 \cdot 3Q_A - 0 \cdot 4Q_I = Y_I$$

which may be written in matrix notation as

$$\begin{bmatrix} 0 \cdot 8 & -0 \cdot 1 \\ -0 \cdot 3 & 0 \cdot 6 \end{bmatrix} \begin{bmatrix} Q_A \\ Q_I \end{bmatrix} = \begin{bmatrix} 1 & 0 \\ 0 & 1 \end{bmatrix} \begin{bmatrix} Y_A \\ Y_I \end{bmatrix}$$

or, letting

$$\mathbf{q} = \begin{bmatrix} Q_A \\ Q_I \end{bmatrix} \quad \text{and} \quad \mathbf{y} = \begin{bmatrix} Y_A \\ Y_I \end{bmatrix}$$

$$\begin{bmatrix} 0 \cdot 8 & -0 \cdot 1 \\ -0 \cdot 3 & 0 \cdot 6 \end{bmatrix} \mathbf{q} = \begin{bmatrix} 1 & 0 \\ 0 & 1 \end{bmatrix} \mathbf{y}$$

To solve for $\mathbf{q}$ we find the inverse of

$$\begin{bmatrix} 0\cdot8 & -0\cdot1 \\ -0\cdot3 & 0\cdot6 \end{bmatrix}$$

by row operations.

Divide row 1 by $0\cdot8$ to get

$$\begin{bmatrix} 1 & -0\cdot125 \\ -0\cdot3 & 0\cdot6 \end{bmatrix}\mathbf{q} = \begin{bmatrix} 1\cdot25 & 0 \\ 0 & 1 \end{bmatrix}\mathbf{y}$$

Add $0\cdot3 \times$ row 1 to row 2 to get

$$\begin{bmatrix} 1 & -0\cdot125 \\ 0 & 0\cdot5625 \end{bmatrix}\mathbf{q} = \begin{bmatrix} 1\cdot25 & 0 \\ 0\cdot375 & 1 \end{bmatrix}\mathbf{y}$$

Divide row 2 by $0\cdot5625$ to get

$$\begin{bmatrix} 1 & -0\cdot125 \\ 0 & 1 \end{bmatrix}\mathbf{q} = \begin{bmatrix} 1\cdot25 & 0 \\ 2/3 & 16/9 \end{bmatrix}\mathbf{y}$$

Add $0\cdot125 \times$ row 2 to row 1 to get

$$\begin{bmatrix} 1 & 0 \\ 0 & 1 \end{bmatrix}\mathbf{q} = \begin{bmatrix} 4/3 & 2/9 \\ 2/3 & 16/9 \end{bmatrix}\mathbf{y}$$

Substituting values for $\mathbf{y}$ we get

(a) $\begin{bmatrix} Q_A \\ Q_I \end{bmatrix} = \begin{bmatrix} 4/3 & 2/9 \\ 2/3 & 16/9 \end{bmatrix}\begin{bmatrix} 200 \\ 600 \end{bmatrix} = \begin{bmatrix} 400 \\ 1200 \end{bmatrix}$ *

(b) $\begin{bmatrix} Q_A \\ Q_I \end{bmatrix} = \begin{bmatrix} 4/3 & 2/9 \\ 2/3 & 16/9 \end{bmatrix}\begin{bmatrix} 100 \\ 300 \end{bmatrix} = \begin{bmatrix} 200 \\ 600 \end{bmatrix}$ *

(c) $\begin{bmatrix} Q_A \\ Q_I \end{bmatrix} = \begin{bmatrix} 4/3 & 2/9 \\ 2/3 & 16/9 \end{bmatrix}\begin{bmatrix} 300 \\ 900 \end{bmatrix} = \begin{bmatrix} 600 \\ 1800 \end{bmatrix}$ *

:. (i) At equilibrium,

$$\begin{aligned} Y - C &= I^* \\ -cY + C &= C^* \end{aligned} \quad \text{or} \quad \begin{bmatrix} 1 & -1 \\ -c & 1 \end{bmatrix}\begin{bmatrix} Y \\ C \end{bmatrix} = \begin{bmatrix} 1 & 0 \\ 0 & 1 \end{bmatrix}\begin{bmatrix} I^* \\ C^* \end{bmatrix}$$

To solve for Y and C we find the inverse of

$$\begin{bmatrix} 1 & -1 \\ -c & 1 \end{bmatrix}$$

by row operations. Add c times row 1 to row 2 to get

$$\begin{bmatrix} 1 & -1 \\ 0 & 1-c \end{bmatrix}\begin{bmatrix} Y \\ C \end{bmatrix} = \begin{bmatrix} 1 & 0 \\ c & 1 \end{bmatrix}\begin{bmatrix} I^* \\ C^* \end{bmatrix}$$

Multiply row 2 by $1/(1-c)$ to obtain

$$\begin{bmatrix} 1 & -1 \\ 0 & 1 \end{bmatrix}\begin{bmatrix} Y \\ C \end{bmatrix} = \begin{bmatrix} 1 & 0 \\ \dfrac{c}{1-c} & \dfrac{1}{1-c} \end{bmatrix}\begin{bmatrix} I^* \\ C^* \end{bmatrix}$$

Add row 2 to row 1 to get

$$\begin{bmatrix} 1 & 0 \\ 0 & 1 \end{bmatrix}\begin{bmatrix} Y \\ C \end{bmatrix} = \begin{bmatrix} \dfrac{1}{1-c} & \dfrac{1}{1-c} \\ \dfrac{c}{1-c} & \dfrac{1}{1-c} \end{bmatrix}\begin{bmatrix} I^* \\ C^* \end{bmatrix}$$

i.e.

$$\begin{bmatrix} Y \\ C \end{bmatrix} = \begin{bmatrix} \dfrac{I^* + C^*}{1 - c} \\ \dfrac{cI^* + C^*}{1 - c} \end{bmatrix}$$

           *

(ii) The equations are

$$Y = C + I = C + I^* + iY \quad \text{or} \quad (1 - i)Y - C = I^*$$

and

$$C = C^* + cY_d = C^* + c(1 - t)Y \quad \text{or} \quad c(1 - t)Y - C = - C^*$$

In matrix notation

$$\begin{bmatrix} 1 - i & -1 \\ c(1 - t) & -1 \end{bmatrix} \begin{bmatrix} Y \\ C \end{bmatrix} = \begin{bmatrix} 1 & 0 \\ 0 & 1 \end{bmatrix} \begin{bmatrix} I^* \\ -C^* \end{bmatrix}$$

Divide row 1 by $1 - i$ to obtain

$$\begin{bmatrix} 1 & \dfrac{-1}{1 - i} \\ c(1 - t) & -1 \end{bmatrix} \begin{bmatrix} Y \\ C \end{bmatrix} = \begin{bmatrix} \dfrac{1}{1 - i} & 0 \\ 0 & 1 \end{bmatrix} \begin{bmatrix} I^* \\ -C^* \end{bmatrix}$$

Subtract $c(1 - t)$ times row 1 from row 2 to to obtain

$$\begin{bmatrix} 1 & \dfrac{-1}{1 - i} \\ 0 & -\dfrac{1 - i - c(1 - t)}{1 - i} \end{bmatrix} \begin{bmatrix} Y \\ C \end{bmatrix} = \begin{bmatrix} \dfrac{1}{1 - i} & 0 \\ -\dfrac{c(1 - t)}{1 - i} & 1 \end{bmatrix} \begin{bmatrix} I^* \\ -C^* \end{bmatrix}$$

Multiply row 2 by $-\alpha(1 - i)$ where $\alpha = 1/[1 - i - c(1 - t)]$ to get

$$\begin{bmatrix} 1 & \dfrac{-1}{1 - i} \\ 0 & 1 \end{bmatrix} \begin{bmatrix} Y \\ C \end{bmatrix} = \begin{bmatrix} \dfrac{1}{1 - i} & 0 \\ \alpha c(1 - t) & -\alpha(1 - i) \end{bmatrix} \begin{bmatrix} I^* \\ -C^* \end{bmatrix}$$

Add $1/(1 - i)$ times row 2 to row 1 to get

$$\begin{bmatrix} 1 & 0 \\ 0 & 1 \end{bmatrix} \begin{bmatrix} Y \\ C \end{bmatrix} = \begin{bmatrix} \alpha & -\alpha \\ \alpha c(1 - t) & -\alpha(1 - i) \end{bmatrix} \begin{bmatrix} I^* \\ -C^* \end{bmatrix}$$

Thus

$$Y = \frac{I^* + C^*}{1 - i - c(1 - t)} \quad \text{and} \quad C = \frac{c(1 - t)I^* + (1 - i)C^*}{1 - i - c(1 - t)}$$

           *

The inverse matrix contains the multipliers. I^* and C^* have the same multiplier effect on Y, i.e. $1/[1 - i - c(1 - t)]$; the multiplier effect on C of a change in I^* is $c(1 - t)/[1 - i - c(1 - t)]$; and the effect on C of a change in C^* is $(1 - i)[1 - i - c(1 - t)]$.

4. The *IS* schedule is given by $Y = C + I$, i.e.

$$Y = C^* + cY + I^* - ar \quad \text{or} \quad (1 - c)Y + ar = C^* + I^*$$

The *LM* schedule is given by $M_D = M_S$, i.e.

$$K + \beta Y - gr = M^* \quad \text{or} \quad \beta Y - gr = M^* - K$$

In matrix notation the equations are

$$\begin{bmatrix} 1-c & a \\ \beta & -g \end{bmatrix}\begin{bmatrix} Y \\ r \end{bmatrix} = \begin{bmatrix} C^* + I^* \\ M^* - K \end{bmatrix}$$

By Cramer's Rule,

$$r = \frac{\begin{vmatrix} 1-c & C^* + I^* \\ \beta & M^* - K \end{vmatrix}}{\begin{vmatrix} 1-c & a \\ \beta & -g \end{vmatrix}} = \frac{\beta(C^* + I^*) - (1-c)(M^* - K)}{(1-c)g + a\beta}$$

*

Any change ΔM^* in the money supply will cause a change of

$$\frac{-(1-c)}{(1-c)g + a\beta}\Delta M^*$$

in r. Since $1 - c > 0$, the rate of interest and the supply of money are inversely related. Here, r is an increasing function of K, C^*, and I^*; the multiplier for r from C^* and I^* is $\beta/[(1-c)g + a\beta]$; and the multiplier from changes in K is $(1-c)/[(1-c)g + a\beta]$.

CHAPTER 12

Linear Programming

LINEAR ALGEBRA AND INEQUALITIES

Many relations in economic models take the form not of equations but rather of inequalities. For example, if there are a_i units of good i available, and the usage of good i is x_i, then it is a constraint on the usage of good i that $x_i \leqslant a_i$. If $\mathbf{a}$ is a vector of available quantities of goods $1, \ldots, n$ and $\mathbf{x}$ is a vector of the usages of the same goods, then it is a constraint on total usage that $\mathbf{x} \leqslant \mathbf{a}$, i.e. that $x_i \leqslant a_i$ for $i = 1, \ldots, n$. These inequalities may be satisfied either as strict inequalities, in which $x_i < a_i$, or as equalities where $x_j = a_j$. In the long run, goods tend to be produced only if they are wanted, so that usually we would expect $x_j = a_j$, but in the short run the effect of accidents and mistakes in planning means that we commonly find $x_i < a_i$.

Many economic problems consist of maximizing some function of $\mathbf{x}$, subject to constraints of the form $\mathbf{x} \leqslant \mathbf{a}$. If function $f(\mathbf{x})$, the objective function, is perfectly general in form, we can say little about the nature of the solution. However, if $f(\mathbf{x})$ is an increasing function of each x_i then each a_i will be fully utilized, and we will find that the optimum $\mathbf{x}$ is given by $\mathbf{x} = \mathbf{a}$.

LINEAR PROGRAMMING

The connection between the constraints and the function to be maximized will frequently be less direct. Suppose for example that the objective is to maximize $f(\mathbf{x})$, and that productive techniques for each good have constant returns to scale, no joint products, and unvarying factor proportions, so that production of each good j requires a_{ij} units of input of resource i, and that only b_i units of i are available for all uses. Then for each resource i we have a constraint

$$\sum_j a_{ij} x_j \leqslant b_i \quad j = 1, \ldots n; i = 1, \ldots m$$

In matrix notation these can be written

$$\mathbf{Ax} \leqslant \mathbf{b}$$

According to the nature of the problem there will usually be some further constraint on $\mathbf{x}$; in this case, as $\mathbf{x}$ is output of goods, $\mathbf{x} \geqslant \mathbf{0}$.

The problem is thus

$$\text{Maximize } f(\mathbf{x}) \quad \text{subject to} \quad \mathbf{Ax} \leqslant \mathbf{b} \quad \mathbf{x} \geqslant \mathbf{0}$$

If the function $f(\mathbf{x})$ is completely general, the optimal $\mathbf{x}$ need not involve using any resource i fully. However, if $f(\mathbf{x})$ is increasing in each x_j, then such a position cannot be optimal, and we can be sure that the optimal $\mathbf{x}$ will be found to have at least some resource i fully utilized, so that increasing any x_j involves an opportunity cost of decreasing some other x_k. In particular, if the objective function is linear, i.e. the objective is to maximize the total value of $\mathbf{x}$ using a set of values $\mathbf{v}$,

$$f(\mathbf{x}) = \sum_j v_j x_j = \mathbf{v}\mathbf{x}$$

then it always pays to use some resource(s) fully. What we need is a method of finding out which resources i should be fully used, and which it pays, given the objective function, to leave under-utilized. The problem becomes

$$\text{Maximize } \mathbf{v}\mathbf{x} \quad \text{subject to} \quad \mathbf{Ax} \leqslant \mathbf{b}, \quad \mathbf{x} \geqslant \mathbf{0}$$

This is known as a linear programming problem, since both the objective function and the constraints are linear in form. It is also possible to have linear programming problems where the objective is to minimize a linear function of $\mathbf{x}$; however, for the moment we will stick to the maximizing problem.

BASIC SOLUTIONS

Given the problem of maximizing $\mathbf{v}\mathbf{x}$ subject to $\mathbf{Ax} \leqslant \mathbf{b}, \mathbf{x} \geqslant \mathbf{0}$, where $\mathbf{b}$ is of dimension m and $\mathbf{x}$ is of dimension n, we have $m + n$ inequalities in the n variables of $\mathbf{x}$. If we let any n of these $m + n$ inequalities hold as equalities, we can solve the set of n equations this gives us. This provides what is called a basic solution of the system. We can then examine the solution we get to find out whether the remaining m inequalities are satisfied. This is best done by the following device. Setting any of the capacity constraints as an equality amounts to deciding that the resource concerned shall be fully utilized; setting any of the non-negativity constraints on $\mathbf{x}$ as an inequality amounts to deciding that the good concerned shall not be produced. Without loss of generality we can re-order the variables, both in $\mathbf{x}$ and in $\mathbf{b}$, so that the b_i we wish to have fully used come first and are written as the vector $\mathbf{b}^1$; and the remaining b_i come below and are written as the vector $\mathbf{b}^2$. This is known as partitioning vector $\mathbf{b}$. The superscripts are used to distinguish $\mathbf{b}^1$, the first part of vector $\mathbf{b}$, from b_1, the first element in vector $\mathbf{b}$, and $\mathbf{b}^2$, the second part of vector $\mathbf{b}$ from b_2 its second element. The vector $\mathbf{x}$ is also partitioned, the x_j set at zero coming later and being written $\mathbf{x}^2$, and the x_j that have been left to be determined coming first and being written $\mathbf{x}^1$. The matrix $\mathbf{A}$ is also partitioned, the northwest component $\mathbf{A}^{11}$ containing the coefficients of $\mathbf{A}$ relating x_j in $\mathbf{x}^1$ to b_i in $\mathbf{b}^1$; the northeast component $\mathbf{A}^{12}$

relating x_j in $\mathbf{x}^1$ to b_i in $\mathbf{b}^2$, etc. The superscripts are used to distinguish $\mathbf{A}^{11}$, a partition of $\mathbf{A}$, from a_{11}, the first element of $\mathbf{A}$. Writing the system in partitioned matrix form we have

$$\begin{bmatrix} \mathbf{A}^{11} & \mathbf{A}^{12} \\ \mathbf{A}^{21} & \mathbf{A}^{22} \end{bmatrix}\begin{bmatrix} \mathbf{x}^1 \\ \mathbf{x}^2 \end{bmatrix} \leqslant \begin{bmatrix} \mathbf{b}^1 \\ \mathbf{b}^2 \end{bmatrix} \qquad \begin{bmatrix} \mathbf{x}^1 \\ \mathbf{x}^2 \end{bmatrix} \geqslant \begin{bmatrix} \mathbf{0} \\ \mathbf{0} \end{bmatrix}$$

The number of rows in $\mathbf{A}^{11}$ is the same as the dimension of the vector $\mathbf{b}^1$, and the number of columns in $\mathbf{A}^{11}$ is the same as the dimension of the vector $\mathbf{x}^1$. Let dimension $(\mathbf{x}^1) = k$, where $0 \leqslant k \leqslant n$. Thus dimension $(\mathbf{x}^2) = n - k$. But dimension $(\mathbf{b}^1)$ + dimension $(\mathbf{x}^2) = n$, by assumption since we started with n equalities; thus dimension $(\mathbf{b}^1) = n - (n - k) = k$, and $0 \leqslant k \leqslant m$. Since dimension $(\mathbf{x}^1)$ = dimension $(\mathbf{b}^1) = k$, $\mathbf{A}^{11}$ is a square $k \times k$ matrix.

Since vector $\mathbf{x}^2 = \mathbf{0}$, $\mathbf{A}^{11}\mathbf{x}^1 = \mathbf{b}^1$. We have already seen that $\mathbf{A}^{11}$ is square; assuming it to be non-singular, we solve $\mathbf{A}^{11}\mathbf{x}^1 = \mathbf{b}^1$ to get

$$\mathbf{x}^1 = \left[\mathbf{A}^{11}\right]^{-1}\mathbf{b}^1$$

This gives us a vector $\mathbf{x}$, consisting of the $\mathbf{x}^1$ just obtained and the $\mathbf{x}^2 = \mathbf{0}$ already assumed. This $\mathbf{x}$ is a basic solution. We now apply two tests to it. Does $\mathbf{x}^1$ satisfy $\mathbf{x}^1 \geqslant \mathbf{0}$, i.e. are all the x_j in $\mathbf{x}^1$ non-negative; and does $\mathbf{x}^1$ satisfy $\mathbf{A}^{21}\mathbf{x}^1 \leqslant \mathbf{b}^2$? If $\mathbf{x}$ fails the first test the basic solution has some x_j negative; if $\mathbf{x}$ fails the second test, some resource b_i in $\mathbf{b}^2$ is over-utilized. A solution which passes both these tests is a basic feasible solution.

BASIC FEASIBLE SOLUTIONS

With the linear objective function, maximize $\mathbf{vx}$, we need only look at basic feasible solutions. Clearly, if a solution is basic but not feasible we need not bother with it. It can also be shown, however, that we need not bother with solutions which are feasible but not basic. The proof of this is too complex to be included here, but a sketch of one can be given. In a linear programming problem each constraint b_i imposes a limit on the $\mathbf{x}$ that can be achieved; the shape of this limit is a linear function in up to n dimensions, whose various slopes are given by the coefficients in the ith row of $\mathbf{A}$. Either the slopes of such a function are exactly equal to those of $\mathbf{v}$, in which case all points on the boundary are of equal value, or (much more likely) they do not exactly equal those of $\mathbf{v}$, in which case it pays to increase any x_j whose value in $\mathbf{v}$ is relatively higher than on the boundary of the feasible set, until either only this x_j is produced, or limits to its expansion imposed by other resource constraints $b_k, k \neq i$, are encountered. But these intersections are exactly the basic feasible solutions of the original problem. If a feasible $\mathbf{x}$ is not on the boundary of the feasible set, its value can be improved by increasing any x_j for which $v_j > 0$ until the boundary is reached; points on the boundary can be expressed as convex combinations of basic feasible solutions provided that all x_j are bounded by some constraint, which seems a reasonable restriction.

To find the maximum for $\mathbf{vx}$ we thus evaluate $\mathbf{vx}$ for each basic feasible solution. If $\mathbf{vx}$ is highest equal for any two or more basic feasible solutions then it does not matter whether we adopt any one of them, or any convex combination of them; we will get the same value of $\mathbf{vx}$. If the $\mathbf{vx}$ for one basic feasible solution is higher than for any other, then this is the one to adopt.

The following example helps to make the point. Suppose we have a vector $\mathbf{x}^*$ which is not basic, but which can be expressed as a convex linear combination of two basic feasible vectors $\mathbf{x}^1$ and $\mathbf{x}^2$.

$$\mathbf{x}^* = \lambda\mathbf{x}^1 + (1 - \lambda)\mathbf{x}^2 \quad \text{where} \quad 1 > \lambda > 0$$

Evaluate $\mathbf{x}^*, \mathbf{x}^1$ and $\mathbf{x}^2$;

$$\mathbf{vx}^* = \mathbf{v}[\lambda\mathbf{x}^1 + (1 - \lambda)\mathbf{x}^2] = \lambda\mathbf{vx}^1 + (1 - \lambda)\mathbf{vx}^2$$
$$\mathbf{vx}^* - \mathbf{vx}^1 = (1 - \lambda)[\mathbf{vx}^2 - \mathbf{vx}^1]$$
$$\mathbf{vx}^* - \mathbf{vx}^2 = \lambda[\mathbf{vx}^1 - \mathbf{vx}^2]$$

so if

$$\mathbf{vx}^2 > \mathbf{vx}^1 \qquad \mathbf{vx}^2 > \mathbf{vx}^* > \mathbf{vx}^1$$

if

$$\mathbf{vx}^2 = \mathbf{vx}^1 \qquad \mathbf{vx}^2 = \mathbf{vx}^* = \mathbf{vx}^1$$

if

$$\mathbf{vx}^2 < \mathbf{vx}^1 \qquad \mathbf{vx}^2 < \mathbf{vx}^* < \mathbf{vx}^1$$

A DIAGRAMMATIC ILLUSTRATION

It is simple to express in diagrammatic form the linear programming problem in the case where there are only two goods, 1 and 2. If there are three constraints, the first one,

$$a_{11}x_1 + a_{12}x_2 \leqslant b_1$$

can be depicted by a straight line showing

$$a_{11}x_1 + a_{12}x_2 = b_1$$

the inequality will be satisfied strictly for points below this line (assuming both a_{1j} positive), and weakly as an equality on the line. This is labelled 11 in Figure 35. Similarly, constraint 2, $a_{21}x_1 + a_{22}x_2 \leqslant b_2$ is represented by line 22, and constraint 3, $a_{31}x_1 + a_{32}x_2 \leqslant b_3$ is represented by line 33. In Figure 35 we have five equations, the lines 11, 22, and 33, and the two axes $0x_1$ and $0x_2$. Any pair of these equations can be solved to give a basic solution, represented by the point where the two lines intersect. There are ten basic solutions, labelled in Figure 35; we check them in turn for feasibility.

A has $x_1 < 0$ and is not feasible;
B has excess use of resources 2 and 3;
C has excess use of resource 2;
D, 0, E, and F are feasible;
G and H have excess use of resource 3;
I has excess use of resources 1 and 3.

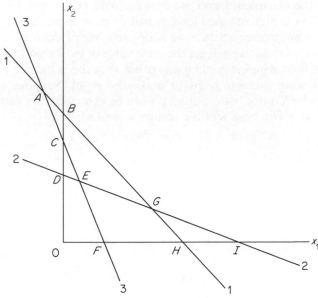

Figure 35   Basic solutions

Thus only $D, 0, E$ and F are feasible. Every feasible vector $\mathbf{x}$ lies in the quadrilateral $0DEF$ and can be expressed as a convex linear combination of the vectors represented by these points. To maximize $\mathbf{vx}$ we need only evaluate it for the four vectors represented by $0, D, E,$ and $F,$ and choose the best of them.

This procedure, of finding all the basic solutions, and eliminating those which are not feasible, rapidly becomes impossible when m and n become large. Thus a method has been evolved which permits of a step-by-step procedure which will locate the best feasible basic solution.

SLACK VARIABLES

In order to apply this method, we define a new set of variables, one for each constraint, called slack variables. The slack is the unused part of each resource, so $\mathbf{s} = \mathbf{b} - \mathbf{Ax}$. In place of the inequalities $\mathbf{Ax} \leqslant \mathbf{b}$ we now have the equations,

$$\mathbf{Ax} + \mathbf{s} = \mathbf{b}$$

The requirement for a feasible solution is now $\mathbf{s} \geqslant \mathbf{0}$. It is convenient to write the equations as

$$\mathbf{Ax} + \mathbf{Is} = \mathbf{b}$$

If we now partition the system as before, to get the x_j not required to be zero first and the rest later, and the resources for which s_i is not required to equal zero to come last, we get

$$\mathbf{A}^{11}\mathbf{x}^1 + \mathbf{A}^{12}\mathbf{x}^2 + \mathbf{s}^1 = \mathbf{b}^1$$
$$\mathbf{A}^{21}\mathbf{x}^1 + \mathbf{A}^{22}\mathbf{x}^2 + \mathbf{s}^2 = \mathbf{b}^2$$

but as $x^2 = 0$ and $s^1 = 0$, this simplifies to

$$A^{11}x^1 \qquad = b^1$$
$$A^{21}x^1 + s^2 = b^2$$

so

$$x^1 = [A^{11}]^{-1}b^1$$

and

$$s^2 = b^2 - A^{21}x^1 = b^2 - A^{21}[A^{11}]^{-1}b^1$$

We require $x^1 \geq 0$ and $s^2 \geq 0$ for this solution to be feasible.

THE SIMPLEX METHOD

The simplex method is an algorithm which has been devised to allow us to start from one basic feasible solution, and shift the basis for it, i.e. the set of variables in x and s which are not constrained to be zero, one variable at a time. If we choose the changes of basis so that at each step we move to another solution which is basic and feasible, and is preferable to the one before it, we will find that in a finite number of steps the process will always get us to the optimal solution, or to one of them if there are several equal best solutions.

To start with we require a basic feasible solution. An obvious one is $x = 0$, which clearly always satisfies both $Ax \leq b$ and $x \geq 0$. As $vx = 0$, it is clearly not optimal.

To decide what change to make in the basis we make use of what is called a simplex tableau. There are variants of this in use; the one described below is chosen to give as close as possible an analogy to the procedure used above for calculating inverses. We start with $(m + n + 1)$ columns, representing the coefficients of the equations

$$Ax + Is = b$$

Labels to show which x_j, s_i are placed above the columns. The s_i form the original basis. The tableau can be regarded as giving s as an explicit function of x, since each s_i appears with a coefficient of 1 in one and only one row, and with zero coefficients in all the other rows. For convenience we include in the tableau a first column which labels the basic variable to which each row refers. The first stage of the simplex tableau will then appear as follows:

The Simplex Tableau: Stage 1

	x_1	x_2	$\ldots$	x_n	s_1	s_2	$\ldots$	s_m	b	
s_1	$a_{11}*$	a_{12}	$\ldots$	a_{1n}	1	0		0	b_1	b_1/a_{11}
s_2	a_{21}	a_{22}	$\ldots$	a_{2n}	0	1	$\ldots$	0	b_2	b_2/a_{21}
$\vdots$	$\vdots$	$\vdots$	$\vdots$	$\vdots$	$\vdots$	$\vdots$	$\vdots$	$\vdots$	$\vdots$	
s_m	a_{m1}	a_{m2}	$\ldots$	a_{mn}	0	0	$\ldots$	1	b_m	b_m/a_{m1}
v	v_1	v_2	$\ldots$	v_n	0	0	$\ldots$	0		

(Dots indicate omitted rows and columns. The final row, and the first and final columns, will be explained below, as will the * on a_{11}.)

There is no unique right rule for which variable in **x** to introduce first into the basis. A common procedure is to start with the x_j for which v_j is greatest (if there is a tie between several x_j for maximum v_j, choose one of the greatest, at random). The final row in stage 1 of the tableau shows the v_j, to aid in this process. Suppose the first x_j chosen for inclusion in the basis is x_1. As x_1 increases, some variable already in the basis, s_i, must fall. The variable to leave the basis must be the first s_i to fall to zero. To find out which one this is, we use the tableau. As x_1 rises, each s_i falls by $a_{i1} x_1$ if $a_{i1} > 0$. The first s_i to reach zero will be that for which (b_i/a_{i1}) has the lowest positive value. To see which one this is, it is useful to enter the values of (b_i/a_{i1}) as an extra column at the extreme right-hand side of the tableau. Having found the i concerned, x_1 must cease to increase when $a_{i1} x_1 = b_i$, since any further increase in x_1 would cause some variable in the basis to become negative. Suppose that s_1 is the first variable to leave the basis.

We now seek to replace the tableau with a new one, in which the new set of basic variables, i.e. $x_1, s_2, s_3, \ldots s_m$ appear with a coefficient of 1 in one and only one equation, and with coefficients of zero in the remaining $(m-1)$ equations. The new basic variables will thus have an identity matrix as their coefficients. This is obtained as follows; consider the element of the old tableau in the x_1 column and the row with minimum positive (b_i/a_{i1}), i.e. a_{11}; this element is called the pivot. In using the tableau it is helpful to label the pivot in some way, e.g. by an asterisk, as is done here. In row 1 each entry is divided by a_{11}. This clearly gives us the coefficients of a new equation, for if

$$a_{11}x_1 + a_{12}x_2 + \ldots + a_{1n}x_n + s_1 = b_1$$

then

$$x_1 + \left(\frac{a_{12}}{a_{11}}\right)x_2 + \ldots + \left(\frac{a_{1n}}{a_{11}}\right)x_n + \left(\frac{1}{a_{11}}\right)s_1 = \left(\frac{b_1}{a_{11}}\right)$$

This row now gives x_1 as a function of the non-basic variables, so it is labelled x_1 in the left-hand column.

We now seek to replace the second equation by one in which x_1 has a coefficient of zero. This can be done by subtracting a_{21} times the *new* row 1 from the *old* row 2. As we are subtracting one linear equation from another the result will clearly be a linear equation. The equation we get expresses s_2 as a function of the new non-basic variables, so the new second row is labelled s_2. A similar procedure is followed to get a new row 3 with a zero coefficient for x_1, subtracting a_{31} times the *new* row 1 from the *old* row 3; this row is labelled s_3. We proceed in this manner for each down to row m. The resulting tableau reads as follows:

The Simplex Tableau; Stage 2

	x_1	x_2	$\cdots$	x_n	s_1	$\cdots$	s_m	b
x_1	1	$\dfrac{a_{12}}{a_{11}}$	$\cdots$	$\dfrac{a_{1n}}{a_{11}}$	$\dfrac{1}{a_{11}}$	$\cdots$	0	$\dfrac{b_1}{a_{11}}$
s_2	0	$\left[a_{22} - \dfrac{a_{21}a_{12}}{a_{11}}\right]$	$\cdots$	$\left[a_{2n} - \dfrac{a_{21}a_{1n}}{a_{11}}\right]$	$-\dfrac{a_{21}}{a_{11}}$	$\cdots$	0	$\left[b_2 - \dfrac{a_{21}}{a_{11}}b_1\right]$
$\vdots$	$\vdots$	$\vdots$	$\vdots$	$\vdots$	$\vdots$		$\vdots$	$\vdots$
s_m	0	$\left[a_{m2} - \dfrac{a_{m1}a_{12}}{a_{11}}\right]$	$\cdots$	$\left[a_{mn} - \dfrac{a_{m1}a_{1n}}{a_{11}}\right]$	$-\dfrac{a_{m1}}{a_{11}}$	$\cdots$	1	$\left[b_m - \dfrac{a_{m1}}{a_{11}}b_1\right]$
z	0	$\left[v_2 - \dfrac{a_{12}}{a_{11}}v_1\right]$	$\cdots$	$\left[v_n - \dfrac{a_{1n}}{a_{11}}v_1\right]$	$-\dfrac{1}{a_{11}}v_1$	$\cdots$	0	

(Dots indicate omitted rows and columns. The last row will be explained below.)

Reading each row as an equation, since all other items in each equation have either zero coefficients if the variable is in the basis and $\neq 0$, or are variables not in the basis and $= 0$ so that their coefficients do not matter, the right-hand column labelled **b** gives the value of each variable in the new basis,

$$x_1 = \left(\frac{b_1}{a_{11}}\right) \qquad s_2 = b_2 - \frac{a_{21}}{a_{11}}b_1, \text{ etc.}$$

We now wish to see whether it will pay to add any further x_j to the basis, for example by including x_2. This will clearly increase **vx** by v_2 per unit of good 2 included, but the increase in x_2 is liable to involve changes in other variables in the basis. In the case of the s_i these changes do not matter since they do not enter the objective function. The change in x_1 must, however, be taken into account. Thus, below the tableau we add a row of z_j, where z_j gives the net change in **vx** consequent on increasing x_j or s_i by one unit.

The first row of the tableau shows that

$$x_1 + \frac{a_{12}}{a_{11}}x_2 + \dots + \frac{a_{1n}}{a_{11}}x_n = \frac{b_1}{a_{11}}$$

where the remaining terms are all either coefficients of variables which by our choice of basis remain at zero, or have zero coefficients. Thus if x_2 increases by a unit, x_1 must fall by a_{12}/a_{11} units, so that

$$z_2 = v_2 - \frac{a_{12}}{a_{11}}v_1$$

Similarly, if x_3 increases by a unit, x_1 must fall by a_{13}/a_{11} units, so that

$$z_3 = v_3 - \frac{a_{13}}{a_{11}}v_1$$

and so on up to z_n.

It pays to introduce any variable into the basis for which $z_j > 0$. If more than one $z_j > 0$, there are various rules for choosing which variable to introduce, of which the commonest is picking the highest z_j (or in the event of a tie choosing at random between the z_j which are highest). Having decided which x_j to introduce next into the basis, we decide which existing member of the basis has to be removed as before, by looking at the coefficients in the column of the latest tableau relating to the x_j to be introduced. If any entry in this column is negative, the variable relating to the row concerned will be increased; this cannot lead to any breach of the non-negativity conditions, so the row can be ignored. If the coefficient is positive, the related variable will fall, and we find how far each variable could fall before coming to zero by dividing the constraint variable for that row in the latest tableau by the coefficient of x_j. This figure for each row can again conveniently be tabulated in the extreme right-hand column of the latest tableau. The row with the lowest positive ratio in this column gives us the next variable to leave the basis, and thus gives us the new pivot (in the event of a tie, again choose at random).

Changes in the rows and columns of the tableau then proceed as before. After each change of basis we calculate the z_j for a change in the value of each non-basic variable. So long as any of these > 0 a change of basis can give us a further increase in the objective function **vx**. When no further change of basis pays, we have found the optimal solution (or one optimal solution, if any others exist).

Once the correct basis has been found, the x_j can be read off from the tableau. Each x_j not in the basis is zero. Each x_j which is in the basis has a unit coefficient in its own column, and zero coefficients in the columns relating to all other basic variables. As all non-basic variables are zero and all basic variables except x_j have unit coefficients, the figure appearing in the constraint column of the tableau, i.e. in the column labelled **b**, gives one the value of x_j for each basic variable. Once **x** is known, **vx** can easily be calculated.

A NUMERICAL EXAMPLE OF THE SIMPLEX METHOD

Let us work through a numerical example. Suppose that the constraints are given as

$$\begin{bmatrix} 1 & 2 & 3 \\ 2 & 2 & 2 \\ 3 & 2 & 1 \end{bmatrix} \begin{bmatrix} x_1 \\ x_2 \\ x_3 \end{bmatrix} \leq \begin{bmatrix} 50 \\ 60 \\ 60 \end{bmatrix}$$

and the objective is to maximize $\mathbf{vx} = 5x_1 + 4x_2 + x_3$. The first tableau will thus appear as follows;

	x_1	x_2	x_3	s_1	s_2	s_3	**b**	
s_1	1	2	3	1	0	0	50	50
s_2	2	2	2	0	1	0	60	30
s_3	3*	2	1	0	0	1	60	20*
v	5*	4	1	0	0	0		

(The extreme right-hand column, and the stars, will be explained below.)

The highest v_j is 5, which is starred. The coefficient above it in each row is divided into the constraint constant in the **b** column for each row, and the ratio entered in the extreme right-hand column. The lowest element in this column is 20, starred; so x_1 must enter the basis and s_3 must leave it. 3, starred, is the pivot.

In stage 2 of the tableau the rows will thus be labelled s_1, s_2, and x_1. The x_1 row is first obtained by dividing each element in the old s_3 row by 3. To get the new s_1 row, subtract the new x_3 row from the old s_1 row. To get the new s_2 row subtract twice the new x_3 row from the old s_2 row. To get the new **z** row subtract from the **v** row 5 (i.e. the price of x_1) × the new x_1 row. The new tableau appears as follows:

	x_1	x_2	x_3	s_1	s_2	s_3	**b**	
s_1	0	1·333*	2·667	1	0	− 0·333	30	22·5*
s_2	0	0·667	1·333	0	1	− 0·667	20	30
x_1	1	0·667	0·333	0	0	0·333	20	30
z	0	0·667*	− 0·667	0	0	− 1·667		

(The extreme right-hand column and the stars will be explained below.)

The highest element in the **z** row is 0·667 (starred). Divide the **b** column by the coefficients above 0·667 and enter the results in the extreme right-hand column. The lowest such element is 22·5 (starred), so x_2 must enter the basis and s_1 must leave it; the pivot, starred, is 1·333.

To get the next stage of the tableau, label the rows x_2, s_2, and x_1. The new x_2 row is the old s_1 row divided by 1·333. To get the new s_2 row subtract 0·667 × the new x_2 row from the old s_2 row. To get the new x_1 row subtract 0·667 × the new x_2 row from the old x_1 row, To get the new **z** row subtract from **v** 4 × the new x_2 row and 5 × the new x_1 row. The result appears as follows:

	x_1	x_2	x_3	s_1	s_2	s_3	**b**
x_2	0	1	2	0·75	0	− 0·25	22·5
s_2	0	0	0	− 0·5	1	− 0·5	5
x_1	1	0	− 1	− 0·5	0	0·5	5
z	0	0	− 2	− 0·5	0	− 1·5	

As no element in the new **z** row is positive, no further change of basis will give any further increase in **vx**. The actual value of **vx** is

$$vx = 5x_1 + 4x_2 + x_3 = 5(5) + 4(22·5) + 0 = 25 + 90 = 115$$

MINIMIZING LINEAR PROGRAMMING PROBLEMS

Suppose we are set the problem,

$$\text{Minimize } \mathbf{vx} \quad \text{subject to} \quad \mathbf{Ax} \geqslant \mathbf{b}, \mathbf{x} \geqslant \mathbf{0}$$

This is clearly a linear programming problem, since the objective function and the constraints are linear. We cannot, however, solve it in quite the same way as the maximizing problem. We can turn the system into a set of equalities by introducing the vector of surplus variables $\mathbf{s}$, so that $\mathbf{Ax} - \mathbf{Is} = \mathbf{b}$, with $\mathbf{s} \geqslant \mathbf{0}$. However, this does not give us any obvious basic feasible solution to start with, unless by chance $\mathbf{Ax} = \mathbf{b}$ happens to be feasible. $-\mathbf{s} = \mathbf{b}$ is of course inconsistent with $\mathbf{s} \geqslant \mathbf{0}$.

To cope with this problem, we introduce a further set of what are known as 'artificial' variables, $\mathbf{q}$, of the same dimension as $\mathbf{b}$ and $\mathbf{s}$. If we set $\mathbf{Ax} - \mathbf{Is} + \mathbf{Iq} = \mathbf{b}$, then the solution $\mathbf{q} = \mathbf{b}$, $\mathbf{x} = \mathbf{0}$, $\mathbf{s} = \mathbf{0}$ satisfies all the constraints, and is basic and feasible. To arrive at a basic feasible solution not involving the $\mathbf{q}$, set a vector of values on $\mathbf{q}$ sufficiently high to ensure that the solution to the problem

$$\text{Minimize } \mathbf{vx} + \mathbf{wq} \quad \text{subject to} \quad \mathbf{Ax} \geqslant \mathbf{b}, \mathbf{x} \geqslant \mathbf{0}, \mathbf{q} \geqslant \mathbf{0}$$

has $\mathbf{q} = \mathbf{0}$ in its solution. The tableau will therefore start off as follows, adding in the extreme left-hand column the values set on the variables in the basis, and taking a numerical example with

$$\mathbf{A} = \begin{bmatrix} 1 & 2 \\ 3 & 1 \end{bmatrix} \quad \text{and} \quad \mathbf{v} = \begin{bmatrix} 1 & 3 \end{bmatrix} \quad \text{and} \quad \mathbf{b} = \begin{bmatrix} 3 \\ 3 \end{bmatrix}$$

We let

$$\mathbf{w} = \begin{bmatrix} 10 & 10 \end{bmatrix}$$

The tableau thus starts off as

		x_1	x_2	s_1	s_2	q_1	q_2	**b**	
10	q_1	1	2	-1	0	1	0	3	3
10	q_2	3*	1	0	-1	0	1	3	1*
	v	1	3	0	0	10	10		
	z	-39*	-27	10	10	0	0		

(The extreme right-hand column, the **z** row and the stars will be explained below).

In the **v** row the v_i and w_i are inserted. The criterion for change is given by the **z** row, which shows v_j minus the coefficients above multiplied by the value of the basic variable concerned. For example, increasing x_1 by 1 unit will mean a fall of 1 unit in q_1 and of 3 units in q_2, so the net change in the objective function is

$$v_1 - w_1 - 3w_2 = 1 - 10 - 30 = -39$$

and similarly for the other entries in the **z** row. As we are seeking to minimize, it will pay to insert into the basis any variable with z_j negative. The negative coefficient with the largest absolute value is z_1, which is -39, starred in the tableau. We now divide the coefficients in the column above -39 into the **b** column for each row, getting the extreme right-hand column. The smallest positive element in this is 1, starred, so x_1 must enter the basis and q_2 must leave it; 3, starred, is the pivot.

To get the new tableau, which will have rows labelled q_1 and x_1, divide every element in the old q_2 row by 3 to get the new x_1 row (except in the price column where v_1 replaces w_2). To get the new q_1 row subtract the new x_1 row from the old q_1 row (except for w_1 which remains the same, as the basic variable q_1 has not changed). To get the new **z** row subtract from **v** the price of each basic variable times the coefficient in its row. As q_2 will never be reintroduced, we omit its column in subsequent stages of the tableau.

		x_1	x_2	s_1	s_2	q_1	**b**	
10	q_1	0	1·667*	-1	0·333	1	2	1·2*
1	x_1	1	0·333	0	$-0·333$	0	1	3
	z	0	$-14*$	10	-3	0		

The largest negative element in the **z** row is now z_2, -14, started. Each element in the column above it is divided into the **b** column to give the extreme right-hand column. The smallest positive element in this column is 1·2, starred, so x_2 must replace q_1 in the basis, and 1·667, starred, is the next pivot. Following the same steps as before, tableau 3 is as follows:

		x_1	x_2	s_1	s_2	**b**	
3	x_2	0	1	$-0·6$	0·2*	1·2	6*
1	x_1	1	0	0·2	$-0·4$	0·6	$-1·5$
	z	0	0	1·6	$-0·2*$		

We now have a basic feasible solution with $\mathbf{q} = \mathbf{0}$; however, the fact that there is still a negative element in the **z** row shows that it is not yet optimal. Here $-0·2$ is starred, and the elements above it are divided into the **b** column, the results being entered in the extreme right-hand column. The minimum positive element in this is 6, starred, so s_2 enters the basis and x_2 leaves it. The new pivot is 0·2, starred. The same operations are now performed to get stage 4 of the tableau, which appears as follows:

		x_1	x_2	s_1	s_2	**b**
0	s_2	0	5	-3	1	6
1	x_1	1	2	-1	0	3
	z	0	1	1	0	

As no element in $\mathbf{z} < 0$, no further reduction in the objective function is possible, and we are left with the optimal basic feasible solution,

$$x_1 = 3 \quad x_2 = 0 \quad s_1 = 0 \quad s_2 = 6$$

so that $\mathbf{vx} = x_1 + 3x_2 = 3$.

THE DUAL

It can be shown that any linear programming problem will have what is called a dual problem. Consider the following two problems:

(1) Choose $\mathbf{x}$ to maximize $\pi_1 = \mathbf{vx}$ subject to $\mathbf{Ax} \leqslant \mathbf{b}, \mathbf{x} \geqslant \mathbf{0}$.
(2) Choose $\mathbf{y}$ to minimize $\pi_2 = \mathbf{yb}$ subject to $\mathbf{yA} \geqslant \mathbf{v}, \mathbf{y} \geqslant \mathbf{0}$.

It should be noted that the $\mathbf{A}$ matrix is the same for each. However the value vector $\mathbf{v}$ in the objective function of problem (1) appears as the constraint vector of problem (2), and the value vector $\mathbf{b}$ in the objective function of problem (2) appears as the constraint vector of problem (1).

To apply the simplex method to problem (2) we would transpose it, making it

$$\text{Min } \mathbf{b'y'} \quad \text{subject to} \quad \mathbf{A'y'} \geqslant \mathbf{v'}, \mathbf{y'} \geqslant \mathbf{0}$$

but it is simpler to state it in discussions of duality without transposing it.

Problem (2) is the dual of problem (1), referred to as the primal problem, and problem (1) is the dual of problem (2).

We can now show that where both problems have feasible solutions, if we can find a pair of feasible solutions $\mathbf{x^*}, \mathbf{y^*}$, such that $\mathbf{vx^*} = \mathbf{y^*b}$, then these will be the optimal solutions to their problems, and $\pi_1 = \mathbf{vx^*} = \mathbf{y^*b} = \pi_2$, i.e. the optimal solutions to the two problems make the values of their objective functions the same. This is shown as follows.

If $\mathbf{x}, \mathbf{y}$ are feasible then $\mathbf{Ax} \leqslant \mathbf{b}$, so $\mathbf{yAx} \leqslant \mathbf{yb}$. This must be true, since $(\mathbf{Ax} - \mathbf{b}) \leqslant \mathbf{0}, \mathbf{y} \geqslant \mathbf{0}$, so every term in their inner product $\leqslant 0$, i.e. $\mathbf{y}(\mathbf{Ax} - \mathbf{b}) \leqslant 0$ so $\mathbf{yAx} \leqslant \mathbf{yb}$. Similarly, $\mathbf{yA} \geqslant \mathbf{v}$ so $\mathbf{yAx} \geqslant \mathbf{vx}$. Thus $\mathbf{yb} \geqslant \mathbf{yAx} \geqslant \mathbf{vx}$. Thus if we can only find $\mathbf{x^*}, \mathbf{y^*}$ where both are feasible and $\mathbf{vx^*} = \mathbf{y^*b}$, these are the optimal $\mathbf{x}, \mathbf{y}$. For every $\mathbf{x}, \mathbf{vx} \leqslant \mathbf{y^*b} = \mathbf{vx^*}$ so $\mathbf{x^*}$ maximizes $\mathbf{vx}$. For every $\mathbf{y}$, $\mathbf{yb} \geqslant \mathbf{vx^*} = \mathbf{y^*b}$ so $\mathbf{y^*}$ minimizes $\mathbf{yb}$.

To see that they can always be found, we make use of the simplex tableau for the maximum problem. (A similar proof can be applied to the simplex for the minimum problem.) Without any loss of generality, supposing we have solved the simplex, we rearrange the order of variables so that those $\mathbf{x}$ which are in the final basis come first, and those which are not in the basis come second, and those slack variables which are not in the final basis come first while those that remain in the final basis come last. The first tableau then reads, in partitioned matrix notation,

$$
\begin{array}{ccccc}
\mathbf{A}^{11} & \mathbf{A}^{12} & \mathbf{I} & \mathbf{0} & \mathbf{b}^1 \\
\mathbf{A}^{21} & \mathbf{A}^{22} & \mathbf{0} & \mathbf{I} & \mathbf{b}^2 \\
\mathbf{v}^1 & \mathbf{v}^2 & \mathbf{0} & \mathbf{0} &
\end{array}
$$

The final tableau is derived from this by a series of row operations. In the actual calculation these are done row by row, but we consider how they appear represented in matrix notation. To get the new first row, $\mathbf{A}^{11}$ must be replaced by $\mathbf{I}$, so the old first row is premultiplied by $[\mathbf{A}^{11}]^{-1}$ to obtain the new first row. In the new second row, we wish to replace the first matrix,

A^{21}, by 0. To obtain the new second row, we take the old second row and subtract from it the new first row premultiplied by A^{21}. To obtain the new third row, i.e. the final z vector, we wish to replace v^1 by 0, so the old third row has subtracted from it the new first row premultiplied by v^1. The tableau then reads as follows:

$$
\begin{array}{ccccc}
I & [A^{11}]^{-1}A^{12} & [A^{11}]^{-1} & 0 & [A^{11}]^{-1}b^1 \\
0 & A^{22} - A^{21}[A^{11}]^{-1}A^{12.} & -A^{21}[A^{11}]^{-1} & I & b^2 - A^{21}[A^{11}]^{-1}b^1 \\
0 & v^2 - v^1[A^{11}]^{-1}A^{12} & -v^1[A^{11}]^{-1} & 0 &
\end{array}
$$

We can read off the solution to the primal problem as

$$x^* = \begin{bmatrix} [A^{11}]^{-1}b^1 \\ 0 \end{bmatrix}$$

If we now take as our y^* the negative of the entry for the slack variables in the third row, we get

$$y^* = [v^1[A^{11}]^{-1} \quad 0]$$

We will find that by construction we have a y^* which satisfies the constraints $Ay \geqslant v, y \geqslant 0$. We can see this, since the simplex procedure made the z row in the final tableau non-positive. Thus by the simplex rules,

$$v^2 - v^1[A^{11}]^{-1}A^{12} \leqslant 0$$

thus

$$v^1[A^{11}]^{-1}A^{12} \geqslant v^2 \quad \text{and} \quad v^1[A^{11}]^{-1} \geqslant 0$$

But

$$y^*A = [v^1[A^{11}]^{-1} \quad 0]\begin{bmatrix} A^{11} & A^{12} \\ A^{21} & A^{22} \end{bmatrix}$$

so

$$y^*A \geqslant v \quad \text{if} \quad [v^1 \quad v^1[A^{11}]^{-1}A^{12}] \geqslant [v^1 \quad v^2]$$

and

$$y^* \geqslant 0 \quad \text{if} \quad [v^1[A^{11}]^{-1} \quad 0] \geqslant 0$$

Both these conditions are clearly satisfied. Also,

$$vx^* = v^1[A^{11}]^{-1}b^1$$

and

$$y^*b = v^1[A^{11}]^{-1}b^1$$

so we have a pair of vectors x, y for which $vx = yb$.

SLACK AND SURPLUS VARIABLES AND DUAL SOLUTIONS

In the maximum problem, the condition $Ax \leqslant b$ can be written using a slack vector s as $Ax + s = b, s \geqslant 0$. In the minimum problem, the condition $yA \geqslant v$ can be written, using a surplus vector r, as $yA - r = v, r \geqslant 0$. For the solution vectors x^* and y^* we have $y^*b = y^*Ax^* = vx^*$. Thus as

$$y^*Ax^* - rx^* = vx^* = y^*Ax^*, \quad rx^* = 0$$

and
$$y^*Ax^* + y^*s = y^*b = y^*Ax^*, \quad y^*s = 0$$

Thus each optimal vector places a zero value on any variable which appears in the other problem as a slack or surplus variable not set equal to zero. If $y^* \geqslant 0$ and $s \geqslant 0$, $y^*s = 0$ implies $y_i s_i = 0$ for every i; this is known as the complementary slackness theorem. Either resources are fully used and their slack variable is zero, in which case they may have a positive value, or they are under-utilized and have a zero value; if they were increased the objective function could not be increased.

AN ECONOMIC INTERPRETATION OF DUALITY

We can in fact put an economic interpretation on duality in linear programmes, in a way analogous to that we found in dealing with Lagrange multipliers. Suppose a firm is seeking to maximize its profits, where v is a vector of profit margins to be obtained in various activities, and $Ax \leqslant b$ describes the constraints put on its various activities by shortages of inputs such as machines or skilled labour. The dual solution y^* shows how profits would change if any slack variable not in the final basis were increased. In the final simplex tableau for the maximizing problem, any rise in a non-basic slack variable would reduce profits. If the available resource vector b increased, then it would become possible to reduce the slack vectors and raise profits. y^* shows how profits would rise if b were increased. If any resource which is fully used were to increase, $y_i > 0$ and $s_i = 0$. If any resource in excess supply were to increase, the slack variable would rise but profits would not, i.e. $s_i > 0$ but $y_i = 0$. The y^* vector is a set of shadow prices on the various scarce inputs.

A similar interpretation can be put on the solution to the dual problem. The problem can be regarded as one of putting the minimum valuations on resources consistent with no process using them yielding a net profit. The vector $v - yA = -r$ can be regarded as the vector of net profit levels on each activity. If $r_j > 0$, activity j makes a loss, but $x_j = 0$, so cannot be further decreased. If $x_j > 0$, $r_j = 0$, so activity j just breaks even and profit cannot be increased by altering x_j.

AN EXAMPLE OF THE DUAL

Suppose we consider the dual of the linear programming problem on p. 206 which seeks to minimize $\pi = vx$ subject to $Ax \geqslant b$, $x \geqslant 0$. The dual of this problem can be written as

$$\text{Maximize } \pi^* = by \quad \text{subject to} \quad A'y \leqslant v, y \geqslant 0$$

i.e. maximize $(3y_1 + 3y_2)$ subject to

$$\begin{bmatrix} 1 & 3 \\ 2 & 1 \end{bmatrix} \begin{bmatrix} y_1 \\ y_2 \end{bmatrix} \leqslant \begin{bmatrix} 1 \\ 3 \end{bmatrix}$$

with y_1 and y_2 non-negative. The first tableau for this problem will take the form, using **r** for the slack variable, and noting that the matrix is the transpose of that p. 206,

		y_1	y_2	r_1	r_2	**v**	
0	r_1	1*	3	1	0	1	1*
0	r_2	2	1	0	1	3	1·5
	b	3*	3	0	0		

There is a tie for the largest entry in the **b** row; deciding to insert y_1 to the basis, 3 is starred, and the entries above it divided into the **v** column and entered on the extreme right. The smallest positive entry in this column is 1, starred, so r_1 must leave the basis, and the pivot is 1, starred. The next stage tableau will then appear as follows:

		y_1	y_2	r_1	r_2	**v**
3	y_1	1	3	1	0	1
0	r_2	0	− 5	− 2	1	1
	z	0	− 6	− 3	0	

As there is no positive entry in the **z** row, this is an optimal solution, and gives $y_1 = 1, y_2 = 0, r_1 = 0$ and $r_2 = 1$. The maximized value of π^* is $\pi^* = 3(1) + 3(0) = 3$. This is the same result as for the minimum problem dealt with above; it comes out in fewer steps, largely because it avoids the need for eliminating the **q** variables introduced into the minimization problem to enable us to find a basic feasible solution.

As seen in the general discussion of duality, the tableau for either problem in fact allows us to read off the solution to the other. The values of the surplus variables in the solution to the minimum problem above, on p. 207, $s_1 = 0$ and $s_2 = 6$, are in fact the negatives of the entries in the **z** row of the final stage tableau of the maximum problem for y_1 and y_2. Similarly, the values of the slack variables in the solution to the maximization problem, $r_1 = 0$ and $r_2 = 1$ correspond to the values of x_1 and x_2 found in the solution to the minimization problem.

We can check that the complementary slackness relation holds between the solutions. Multiplying the solution vector to the minimization problem, **x**, by the slack variables vector to the maximization problem, **r**, we see that **rx** = 0, since

$$r_1 = 0 \qquad r_2 = 1 \qquad x_1 = 3 \text{ and } x_2 = 0$$

Multiplying the solution vector to the maximization problem, **y**, by the surplus variables vector of the minimization problem, **s**, we get **ys** = 0, since

$$y_1 = 1 \quad y_2 = 0 \quad s_1 = 0 \text{ and } s_2 = 6$$

EXERCISES

1. Consider a firm with three types of machine, and two outputs x_1 and x_2. Input

requirements compel the firm to obey the constraints

$$x_1 + 2x_2 \leqslant 10$$
$$x_1 + x_2 \leqslant 7$$
$$2x_1 + x_2 \leqslant 12 \quad \text{where } x_1 \geqslant 0 \text{ and } x_2 \geqslant 0$$

(i) Use the simplex algorithm to find the most profitable combination of outputs when the profit margins on producing x_1 and x_2 are 1 and 1·5 respectively.

(ii) Find the optimal solution of the dual of this problem from the optimal tableau of the primal.

2. (i) Take the above problem as the primal, and solve its counterpart, the dual, by means of the simplex algorithm.

(ii) Show that the optimal value of the objective functions in the primal and in the dual are identical.

(ii) Show that the complementary slackness relation between the primal and dual variables holds.

SOLUTIONS TO EXERCISES

1. (i) The profits function is given by

$$\pi = x_1 + 1·5x_2$$

Simplex Tableau: Stage 1

	x_1	x_2	s_1	s_2	s_3	b	b_i/a_{i2}	
s_1	1	2*	1	0	0	10	5*	($=10/2$)
s_2	1	1	0	1	0	7	7	($=7/1$)
s_3	2	1	0	0	1	12	12	($=12/1$)
v	1	1·5	0	0	0			

v_2 is the largest element in $\mathbf{v}$, so x_2 enters the basis. Calculate (b_i/a_{i2}); the smallest of these is 5. The pivot is thus 2.

Simplex Tableau: Stage 2

	x_1	x_2	s_1	s_2	s_3	b	b_i/a_{i1}	
x_2	0·5	1	0·5	0	0	5	10	($=5/0·5$)
s_2	0·5*	0	$-0·5$	1	0	2	4*	($=2/0·5$)
s_3	1·5	0	$-0·5$	0	1	7	4·667	($=7/1·5$)
z	0·25*	0	$-0·75$	0	0			

z_1 is positive, so x_1 enters the basis. Calculate (b_i/a_{i1}); the minimum of these is 4, so s_2 leaves the basis, and 0·5 is the pivot.

Simplex Tableau: Stage 3

	x_1	x_2	s_1	s_2	s_3	b
x_2	0	1	1	-1	0	3
x_1	1	0	-1	2	0	4
s_3	0	0	1	-3	1	1
z	0	0	$-0·5$	$-0·5$	0	

As no element in the **z** row is positive, no further change of basis will increase profits. Hence the optimal solution is

$$x_1 = 4, \quad x_2 = 3, \quad s_1 = s_2 = 0, \quad \text{and} \quad s_3 = 1$$
$$\pi = 1(4) + 1\cdot5(3) = 8\cdot5$$

(ii) The solution for the dual problem will be given by the negative of the entry for the slack variables in the final **z** row of the tableau. These solutions will be

$$y_1 = 0\cdot5, \quad y_2 = 0\cdot5, \quad y_3 = 0 \qquad\qquad *$$

2. (i) The dual problem will take the form Minimize $\pi^* = 10y_1 + 7y_2 + 12y_3$ subject to

$$y_1 + y_2 + 2y_3 \geqslant 1$$
$$2y_1 + y_2 + y_3 \geqslant 1\cdot5$$

where $y_1 \geqslant 0, y_2 \geqslant 0, y_3 \geqslant 0$.

If we denote the surplus variables by r_1 and r_2 and the artificial variables by q_1 and q_2, then

$$y_1 + y_2 + 2y_3 - r_1 + q_1 = 1$$
$$2y_1 + y_2 + y_3 - r_2 + q_2 = 1\cdot5$$

Since $\pi^* = \mathbf{by} + \mathbf{wq}$, we assume $\mathbf{w} = [100 \quad 100]$

Simplex Tableau: Stage 1

	y_1	y_2	y_3	r_1	r_2	q_1	q_2	v	v_i/a_{i1}
q_1	1	1	2	-1	0	1	0	1	1 $(=1/1)$
q_2	2*	1	1	0	-1	0	⊦	1·5	0·75* $(=1\cdot5/2)$
b	10	7	12	0	0	100	100		
z	$-290*$	-193	-288	100	100	0	0		

The entries for the **z** row are found as follows: increasing y_1 by 1 unit will change π^* by

$$10 - 1(100) - 2(100) = -290$$

increasing y_2 by 1 unit will change π^* by

$$7 - 100 - 100 = -193$$

increasing y_3 by 1 unit changes π^* by

$$12 - 200 - 100 = -288$$

The negative entry in the **z** row with the highest absolute value is -290, so y_1 enters the basis. Divide v_i by a_{i1}; the smallest such ratio is 0·75, so q_2 leaves the basis, and 2 is the pivot.

Simplex Tableau: Stage 2

	y_1	y_2	y_3	r_1	r_2	q_1	q_2	v	v_i/a_{i3}
q_1	0	0·5	1·5*	-1	0·5	1	$-0\cdot5$	0·25	0·167* $(=0\cdot25/1\cdot5)$
y_1	1	0·5	0·5	0	$-0\cdot5$	0	0·5	0·75	1·5 $(=0\cdot75/0\cdot5)$
z	0	-48	$-143*$	100	-45	0	145		

y_3 thus enters the basis and q_1 leaves it; the pivot is 1·5.

Simplex Tableau: Stage 3

	y_1	y_2	y_3	r_1	r_2	q_1	q_2	v	v_i/a_{i2}
y_3	0	0.333*	1	−0.667	0.333	0.667	−0.333	0.167	0.5* (= 0.167/ 0.333)
y_1	1	0.333	0	0.333	−0.667	−0.333	0.667	0.667	2 (= 0.667/ 0.333)
z	0	−0.333*	0	0.667	2.667	95.333	97.333		

The coefficient of y_2 in the z row is negative, so y_2 enters the basis; using the v_i/a_{i2} test, y_3 leaves the basis and the pivot is 0.333.

Simplex Tableau: Stage 4

	y_1	y_2	y_3	r_1	r_2	q_1	q_2	v
y_2	0	1	3	−2	1	2	−1	0.5
y_1	1	0	−1	1	−1	−1	1	0.5
z	0	0	1	4	3	96	93	

As there is now no negative entry in the z row, no further improvement in the objective function is possible, and the solution is thus

$$y_1 = 0.5, \qquad y_2 = 0.5, \qquad y_3 = 0, \qquad r_1 = r_2 = 0 \qquad *$$

Clearly, $q_1 = q_2 = 0$.

(ii) $\pi^* = 10y_1 + 7y_2 + 12y_3 = 10(0.5) + 7(0.5) = 8.5$ *

This is the same as the result for π in problem 1.

(iii) The primal solution is

$$x_1 = 4, \qquad x_2 = 3$$
$$s_1 = 0, \qquad s_2 = 0, \qquad s_3 = 1$$

The dual solution is

$$y_1 = 0.5, \qquad y_2 = 0.5, \qquad y_3 = 0$$
$$r_1 = 0, \qquad r_2 = 0$$

We note that both the choice variables of the primal, x_1 and x_2, are positive; and both the surplus variables of the dual, r_1 and r_2, are zero. Thus $\mathbf{rx} = 0$.

The first two choice variables of the dual, y_1 and y_2, are positive, while y_3 is zero; the first two slack variables of the primal, s_1 and s_2, are zero, while the third slack variable s_3 is positive; thus $\mathbf{ys} = 0$. Thus the complementary slackness condition holds.

CHAPTER 13

Difference Equations

DYNAMIC MODELS: THE CONSUMPTION FUNCTION

So far we have been concerned with static economic models. Our supply functions have assumed that quantity supplied is a function of price in the same time period. The consumption function has assumed that consumption depends on income for the same period. The economy is not static, however. Quantity supplied in one period may depend on the price in the previous period, or current consumption may depend on past incomes, etc. Consideration of the effects of these lags is essential if changes in the economy over time are to be analysed. Once time lags are introduced into the system we are dealing with economic dynamics as opposed to statics.

First-order linear difference equations are used to analyse changes with respect to time when what happens in one period depends upon what happened in the previous period. Applying this to the consumption function it states that consumption in any time period, say t, depends upon income in period $t - 1$. To denote consumption in period t we write C subscript t and for income in period $t - 1$ we write Y subscript $t - 1$, i.e.

$$C_t = f(Y_{t-1})$$

The notation $C(t)$ and $Y(t - 1)$ is also quite common. However, the subscript notation is used throughout this book.

FIRST-ORDER LINEAR DIFFERENCE EQUATIONS

Taking a closed economic system without government activity we know that

$$Y_t = C_t + I_t$$

where Y_t is national income in period t, C_t is consumption in period t, and I_t is investment in period t. The only difference between this national income identity and the one used in Chapter 7, i.e. $Y = C + I$, is that the time period has been numbered as t. When dealing with the static model it was assumed that only one time period was involved. What has been done now is to call this time period t.

215

Assume that I_t takes a value I^*, which does not change as t varies. If consumption is a function of the previous period's income the consumption function will take the form

$$C_t = C^* + cY_{t-1}$$

where C^* is the autonomous part of consumption and c is the marginal propensity to consume.

$$Y_t = C_t + I_t$$

thus

$$Y_t = C^* + cY_{t-1} + I^*$$

This is a first-order linear difference equation. It is of the first-order since income in period t depends partly on what happens in the previous period. If income in period t was affected by events in period $t - 2$ or $t - 3$, i.e. $Y_t = f(Y_{t-2})$ or $Y_t = f(Y_{t-3})$, we would have a second- or third-order difference equation. The above equation is linear because it does not contain terms such as $c(Y_{t-1})^2$, etc.

SOLUTION OF A FIRST-ORDER LINEAR DIFFERENCE EQUATION

Given $Y_t = C^* + cY_{t-1} + I^*$, what does this tell us about changes in income Y with respect to time? If we let $C^* = 100$, $I^* = 500$, and $c = 0.8$, then

$$Y_t = 100 + 0.8\,Y_{t-1} + 500$$

i.e.

$$Y_t = 600 + 0.8\,Y_{t-1}$$

To discover how Y changes with respect to t it is necessary to know the value of income in some initial period, say Y_0. Let us assume $Y_0 = 2500$. Since

$$Y_t = 600 + 0.8\,Y_{t-1}$$

it is possible to calculate Y_t when $t = 1, 2, 3 \ldots$, etc.

$$Y_1 = 600 + 0.8\,Y_{1-1} = 600 + 0.8\,Y_0 = 600 + 0.8(2500) = 2600$$
$$Y_2 = 600 + 0.8\,Y_{2-1} = 600 + 0.8\,Y_1 = 600 + 0.8(2600) = 2680$$
$$Y_3 = 600 + 0.8\,Y_{3-1} = 600 + 0.8\,Y_2 = 600 + 0.8(2680) = 2744$$
$$Y_4 = 600 + 0.8\,Y_{4-1} = 600 + 0.8\,Y_3 = 600 + 0.8(2744) = 2795.2$$

Figure 36 shows the changes in income Y over the first three periods. It also shows clearly that economic models expressed in difference equation form treat time as a discrete or discontinuous variable. Differential equations, which will be considered in Chapters 14 and 18 treat time as a continuous variable. The above method is quite simple if we wish to calculate income for small values of t, but as t gets larger, i.e. as one moves along the time axis, it becomes tedious. To calculate Y_t as $t \to \infty$ the following approach can be used.

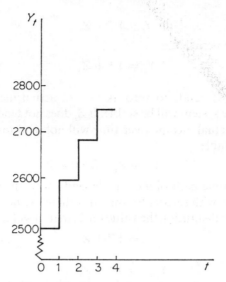

Figure 36   Income changes over time

THE EQUILIBRIUM POSITION

If there is an equilibrium level of Y_t, Y^*, this level is such that if it is once attained it will be repeated, i.e. if

$$Y_{t-1} = Y^* \quad \text{then} \quad Y_t = Y^*$$

Taking the equation

$$Y_t = 600 + 0{\cdot}8\, Y_{t-1} \tag{1}$$

and substituting $Y_t = Y_{t-1} = Y^*$, we obtain

$$Y^* = 600 + 0{\cdot}8\, Y^* \tag{2}$$

i.e.

$$(1 - 0{\cdot}8)Y^* = 600$$

thus

$$Y^* = \frac{600}{1 - 0{\cdot}8} = 3000$$

We now know that a finite equilibrium level of income exists. The next problem is to find out if actual income Y_t approaches the equilibrium and if so, how fast.

DEVIATIONS FROM EQUILIBRIUM

If $Y_t \neq Y^*$ and we let the deviation of actual income in period t from the equilibrium equal Z_t, then

$$Z_t = Y_t - Y^*$$

or
$$Y_t = Y^* + Z_t$$
where t is any time period, e.g.
$$Y_1 = Y^* + Z_1$$
etc.

If the deviation Z_t tends to zero as $t \to \infty$, actual income Y_t will tend towards Y^* and the system will be stable. If Z_t does not tend to zero as $t \to \infty$ the movement of actual income over time will not tend towards Y^* and the system will be unstable.

$$Y_t = Y^* + Z_t = 3000 + Z_t$$

Consequently the time path of income depends only upon the deviation Z_t since Y^* is constant with respect to time. It is therefore necessary to solve for Z_t. This is done by substituting the value for Y_t in terms of Z_t into equation (1).

$$Y_t = Y^* + Z_t$$
thus
$$Y_{t-1} = Y^* + Z_{t-1}$$
and equation (1) will take the form
$$Y^* + Z_t = 600 + 0.8(Y^* + Z_{t-1})$$
i.e.
$$Y^* + Z_t = 600 + 0.8 Y^* + 0.8 Z_{t-1} \qquad (3)$$

Subtracting equation (2) from equation (3) gives
$$Z_t = 0.8 Z_{t-1}$$
Since this is true for all t

$$Z_1 = 0.8 Z_0$$
$$Z_2 = 0.8 Z_1 = 0.8[(0.8)Z_0] = (0.8)^2 Z_0$$
$$Z_3 = 0.8 Z_2 = 0.8[(0.8)^2 Z_0] = (0.8)^3 Z_0 \quad \text{etc.}$$

By repeating the process we get
$$Z_t = (0.8)^t Z_0$$
We have already mentioned that it is necessary to know the value of income in some initial period say Y_0 in order to find Y_t. If $Y_0 = 2500$ then
$$Z_0 = Y_0 - Y^* = 2500 - 3000 = -500$$
With this information
$$Z_t = -500(0.8)^t$$
This enables us to obtain a value for the deviation for any t; e.g. when $t = 100$
$$Z_{100} = -500(0.8)^{100}$$
The time path of actual income Y_t is
$$Y_t = Y^* + Z_t$$
thus
$$Y_t = 3000 - 500(0.8)^t$$

This is the solution to the first-order difference equation; all terms refer to the same time period so that one can immediately find a value for actual income in say period 50, i.e.

$$Y_{50} = 3000 - 500(0.8)^{50}$$

THE STABILITY OF THE SYSTEM

Any constant > 0 and < 1, to the power t, gets smaller as t becomes larger. Consequently any constant > 0 and < 1, to the power t, tends to zero as t tends to infinity. If we take $\frac{1}{2}$ then $(\frac{1}{2})^t$ tends to zero as t tends to infinity, e.g.

$$(\tfrac{1}{2})^1 = \tfrac{1}{2}, \quad (\tfrac{1}{2})^2 = \tfrac{1}{4}, \quad (\tfrac{1}{2})^3 = \tfrac{1}{8}, \quad (\tfrac{1}{2})^4 = \tfrac{1}{16}, \quad \text{etc.}$$

We have found that the time path of actual income is

$$Y_t = 3000 - 500(0.8)^t$$

$(0.8)^t \to 0$ as $t \to \infty$ since $0.8 < 1$, thus $-500(0.8)^t \to 0$ as $t \to \infty$, i.e. $Z_t \to 0$ as $t \to \infty$, and $Y_t \to 3000$ as $t \to \infty$, see Figure 37.

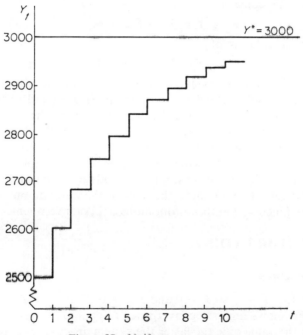

Figure 37   Uniform convergence

Since the deviation tends to zero as t tends to infinity, actual income approaches the equilibrium. This system does not reach equilibrium in finite t. However, for practical purposes we can say that a system is in equilibrium after some t since the deviation becomes negligible. If instead of 0.8 we had

0.6 then $(0.6)^t$ would tend to zero faster as $t \to \infty$. Hence actual income would approach the equilibrium level faster, since the negligible deviations we are willing to ignore for practical purposes occur sooner.

With a linear system stability does not depend upon the initial condition. However, if there is some minimum deviation beyond which we are not interested in minor deviations from equilibrium then a decrease in $|Z_0|$, resulting from a change in Y_0, means that Z_t would reach this minimum sooner. The symbol $|Z_0|$ means the size of Z_0, regardless of sign; i.e. if $Z_0 > 0$, $|Z_0| = Z_0$, or if $Z_0 < 0$, $|Z_0| = -Z_0$. $|Z_0|$ is referred to as the 'absolute value' or 'modulus' of Z_0.

GENERAL SOLUTION OF A FIRST-ORDER DIFFERENCE EQUATION

It should be obvious from the solution of the specific first-order difference equation that given an economic system where

$$Y_t = A + \alpha Y_{t-1}$$

the time path of income is

$$Y_t = Y^* + K\alpha^t$$

which is the general form of

$$Y_t = 3000 - 500(0.8)^t$$

Y^* stands for the equilibrium level of income and its value will depend upon the value of $A > 0$ and α. Y^* will take a finite value if $\alpha \neq 1$ and a positive value if $\alpha < 1$. The case where $\alpha = 1$ will be dealt with later. The value of K will depend upon the initial level of income since it will be the deviation of this level of income from the equilibrium. In a closed system without government activity the values of A and α will depend upon the consumption and investment functions. In the previous example α stood for the marginal propensity to consume while A represented the sum of the autonomous part of the consumption function and the autonomous level of investment.

POSSIBLE TIME PATHS

Uniform Convergence

If $0 < \alpha < 1$, as $t \to \infty$ so $\alpha^t \to 0$ and $K\alpha^t \to 0$. Thus $Y_t \to Y^*$. This system is stable, and shows uniform convergence to equilibrium as in Figure 37. The smaller the value of α, the faster the system approaches equilibrium.

Explosion

If $\alpha > 1$, as $t \to \infty$, $\alpha^t \to \infty$ so $K\alpha^t \to \infty$. Thus the system is unstable, and Y_t departs steadily further from equilibrium, as shown in Figure 38. The larger the value of α the faster the system will explode.

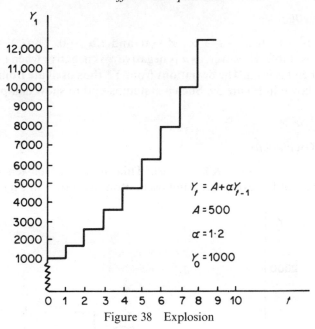

Figure 38   Explosion

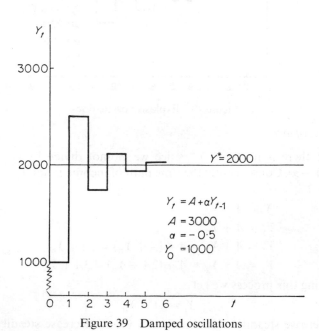

Figure 39   Damped oscillations

Damped Oscillations

If $0 > \alpha > -1$, then as $t \to \infty$, $\alpha^t \to 0$, and $K\alpha^t \to 0$. Thus $Y_t \to Y^*$ and the system is stable. However, as α is negative α^t is negative when t is odd and positive when t is even. The deviations from Y^* thus oscillate from period to period as shown in Figure 39. Such a system is said to show damped oscillations.

Explosive Oscillations

If $-1 > \alpha$, as $t \to \infty$, $K\alpha^t \to \pm\infty$. This system never tends towards equilibrium, but has ever increasing oscillations as shown in Figure 40.

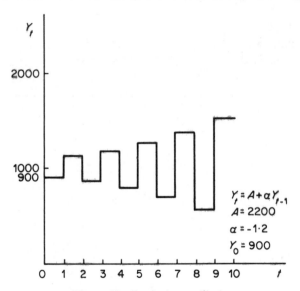

Figure 40   Explosive oscillations

Steady Divergence

If $\alpha = 1$ there is no finite Y^* since we cannot divide by $(1 - \alpha)$ to get $Y^* = A/(1 - \alpha)$. Consequently the time path of income is not $Y_t = Y^* + K\alpha^t$. When $\alpha = 1$

$$Y_t = A + Y_{t-1}$$
$$Y_1 = A + Y_0$$
$$Y_2 = A + Y_1 = A + (A + Y_0) = 2A + Y_0$$
$$Y_3 = A + Y_2 = A + (2A + Y_0) = 3A + Y_0$$

By repeating this process we get

$$Y_t = tA + Y_0$$

Y_t will increase steadily over time if $A > 0$ and decrease steadily if $A < 0$.

A CLOSED ECONOMIC SYSTEM WITH GOVERNMENT ACTIVITY

With the introduction of the government sector the national income identity becomes

$$Y_t = C_t + I_t + G_t$$

If we assume $C_t = C^* + c(Y_d)_{t-1}$, I_t is constant at I^*, G_t is constant at G^* and total taxes $T_t = \tau Y_t$, where τ, Greek letter tau, is the tax rate, then

$$Y_t = C^* + c(Y_d)_{t-1} + I^* + G^*$$
$$(Y_d)_{t-1} = Y_{t-1} - T_{t-1} = Y_{t-1} - \tau Y_{t-1}$$

thus

$$Y_t = C^* + c(Y_{t-1} - \tau Y_{t-1}) + I^* + G^*$$

and

$$Y_t = C^* + I^* + G^* + c(1 - \tau)Y_{t-1}$$

This first-order difference equation is now in the standard form

$$Y_t = A + \alpha Y_{t-1}$$

where $A = C^* + I^* + G^*$ and $\alpha = c(1 - \tau)$. Since the stability of the system depends upon α or $c(1 - \tau)$, this system is stable and free from oscillations if $c(1 - \tau)$ is positive and < 1. This is plausible since it is reasonable to assume that the propensity to consume c and the tax rate τ are both positive and < 1. It would be implausible to assume $c(1 - \tau) < 0$ and hence that oscillations, which arise from a negative α, would occur, given this simple dynamic system.

As a result of the introduction of the government sector the system will approach equilibrium faster since $c(1 - \tau) < c$. The equilibrium level of income Y^* will also change since $Y^* = A/(1 - \alpha)$ and both A and α have changed. The effects of introducing the government sector will depend upon the type of tax and expenditure functions assumed. If we assume that G_t is not autonomous but takes the form $G_t = g Y_{t-1}$ then α would take the value $c(1 - \tau) + g$. This will affect the time it takes for the system to approach equilibrium and also the equilibrium level itself.

DIFFERENCE EQUATIONS APPLIED TO MARKET EQUILIBRIUM

Equilibrium Price and Quantity

Suppose we want to know whether and under what conditions price and quantity will tend towards equilibrium if quantity in period t is a function of price in the same period, i.e.

$$Q_t = a - bP_t \tag{4}$$

but quantity supplied is a function of price in period $t - 1$, i.e.

$$Q_t = c + gP_{t-1} \tag{5}$$

where a, b, c, and g are constants. Suppose there is an equilibrium price P^*

and equilibrium quantity Q^*. The equilibrium price and quantity will continue if once attained. Thus if

$$P_t = P_{t-1} = P^* \quad \text{and} \quad Q_t = Q^*$$

equation (4) implies

$$Q^* = a - bP^* \qquad (6)$$

and equation (5) implies

$$Q^* = c + gP^* \qquad (7)$$

Thus

$$a - bP^* = c + gP^*$$

so

$$P^*(b + g) = a - c \quad \text{i.e.} \quad P^* = \frac{a - c}{b + g}$$

and

$$Q^* = a - b\left(\frac{a - c}{b + g}\right) = \frac{ag + bc}{b + g}$$

Assuming normal demand and supply functions means that $b > 0$ and $g > 0$, i.e. $b + g > 0$. Consequently P^* is positive if $a > c$ and Q^* is positive if $ag + bc > 0$.

Deviations from Equilibrium

If $P_t \neq P^*$ and we let the deviation of actual price in period t from the equilibrium price equal v_t, then

$$v_t = P_t - P^* \quad \text{or} \quad P_t = P^* + v_t$$

If $Q_t \neq Q^*$ and we let the deviation equal w_t, then

$$w_t = Q_t - Q^* \quad \text{or} \quad Q_t = Q^* + w_t'$$

If $v_t \to 0$ as $t \to \infty$, actual price approaches the equilibrium and similarly, if $w_t \to 0$ as $t \to \infty$, actual quantity approaches the equilibrium. Consequently if the system is started off at some arbitrarily chosen P_0 the movement of actual price and quantity with respect to time will depend upon their deviations. It is therefore necessary to solve for v_t and w_t. This is done by substituting the value for P_t in terms of v_t and Q_t in terms of w_t into equations (4) and (5). This will give two equations in two unknowns, v_t and w_t. From equation (4)

$$Q^* + w_t = a - b(P^* + v_t)$$

i.e.

$$Q^* + w_t = a - bP^* - bv_t \qquad (8)$$

Subtracting equation (6) from equation (8) gives

$$w_t = -bv_t \qquad (9)$$

From equation (5)

$$Q^* + w_t = c + g(P^* + v_{t-1})$$

i.e.

$$Q^* + w_t = c + gP^* + gv_{t-1} \tag{10}$$

Subtracting equation (7) from equation (10) gives

$$w_t = gv_{t-1} \tag{11}$$

From equations (9) and (11) we have

$$w_t = -bv_t = gv_{t-1}$$

thus

$$v_t = (-g/b)v_{t-1}$$

Since this is true for all t

$$v_1 = (-g/b)v_0$$
$$v_2 = (-g/b)^2 v_0$$

and

$$v_t = (-g/b)^t v_0$$

v_0 is the deviation of the initial price P_0 from the equilibrium, i.e. $v_0 = P_0 - P^*$

$$P_t = P^* + v_t$$

thus the movement of actual price with respect to time is

$$P_t = P^* + (-g/b)^t v_0$$

Stability of the System

The stability of this system depends upon $(-g/b)^t$. If $(-g/b)^t \to 0$ as $t \to \infty$ then $(-g/b)^t v_0 \to 0$ and $P_t \to P^*$.

We have seen that g/b is positive, assuming normal demand and supply functions. Hence $(-g/b)^t$ will be negative when t is odd and positive when t is even. This will give rise to oscillations around P^*. These oscillations will fade away, however, if $|g/b| < 1$, i.e. if $g < b$, since $(-g/b)^t \to 0$ as $t \to \infty$. If $v_t \to 0$ as $t \to \infty$, then $w_t \to 0$ also, since $w_t = -bv_t$, Consequently $Q_t \to Q^*$ if $|g/b| < 1$. When $|g/b| < 1$ the values of P_t for all $t \geqslant 2$ will lie between P_0 and P_1, e.g. assuming $v_0 > 0$,

$$\begin{aligned}
\text{for } t = 0 \quad & P_0 = P^* + (-g/b)^0 v_0 = P^* + v_0 \\
\text{for } t = 1 \quad & P_1 = P^* + (-g/b)^1 v_0 \\
\text{for } t = 2 \quad & P_2 = P^* + (-g/b)^2 v_0 \\
\text{for } t = 3 \quad & P_3 = P^* + (-g/b)^3 v_0 \\
& P_2 > P_1 \quad \text{but} \quad < P_0 \\
& P_3 > P_1 \quad \text{but} \quad < P_0
\end{aligned}$$

Consequently if we wish to check the economic plausibility of the model, i.e. that P_t and Q_t are positive or zero for all $t \geqslant 0$, this will only involve checking for P_0, P_1, Q_0 and Q_1 if the system is stable. If $|g/b| > 1$, i.e. $g > b$, then $(-g/b)^t \to \pm \infty$ as $t \to \infty$. This will give rise to increasing oscillations around P^* since $v_t \to \pm \infty$ as $t \to \infty$. w_t will also tend to $\pm \infty$ as $t \to \infty$. Such a model will eventually give negative price and quantity for some t. This is economically impossible and implies that the linear system cannot be a good long-term description of the real economy.

EXERCISES

1. In a closed economy without government activity
$$C_t = 200 + 0.75\,Y_{t-1}$$
and
$$I_t = 400$$
Find Y_t when $Y_0 = 2500$. Is the system stable?

2. Consider a closed economy without government activity where
$$C_t = 500 + 0.55\,Y_{t-1}$$
and
$$I_t = 100 + 0.05\,Y_{t-1}$$
Find Y_t when $Y_0 = 1550$, and comment on the stability of the system.

3. In a closed economy
$$C_t = 200 + 0.75(Y_d)_{t-1}$$
and
$$I_t = 100 + 0.25(Y_d)_{t-1}$$
If total taxes and government expenditure depend on current income, i.e.
$$T_t = \tau\,Y_t \quad \text{and} \quad G_t = g\,Y_t$$
(i) Comment on the stability of the system when $Y_0 = 3500$, $\tau = 0.2$ and $g = 0.1$
(ii) If τ increases to 0.25 and g to 0.2, how will this affect the stability of the system?

4. Given the demand function
$$Q_t = 30 - 4P_t$$
and the supply function
$$Q_t = -5 + 3P_{t-1}$$
Find (i) P_t and Q_t,
  (ii) $P_1, Q_1, P_2,$ and Q_2,
when $P_0 = 3$.

5. Given the demand function
$$Q_t = 20 - 2P_t$$
and the supply function
$$Q_t = -5 + 3P_{t-1}$$
Find P_t and Q_t when $P_0 = 4$, and comment on the stability of the system.

SOLUTIONS TO EXERCISES

1. $Y_t = C_t + I_t$
$$= 200 + 0.75\,Y_{t-1} + 400 = 600 + 0.75\,Y_{t-1}$$

If there is an equilibrium level of income Y^* then setting $Y_t = Y_{t-1} = Y^*$ equilibrium is given by

$$Y^* = 600 + 0.75\,Y^*$$

thus

$$Y^* = \frac{600}{1 - 0.75} = 2400$$

Let Z_t be the deviation of Y_t from Y^*, i.e.

$$Z_t = Y_t - Y^*$$

or

$$Y_t = Y^* + Z_t$$
$$Y_t = 600 + 0.75\,Y_{t-1}$$

thus

$$Y^* + Z_t = 600 + 0.75\,Y^* + 0.75 Z_{t-1}$$

but

$$Y^* = 600 + 0.75\,Y^*$$

Subtraction gives

$$Z_t = 0.75 Z_{t-1}$$

thus

$$Z_t = (0.75)^t Z_0$$
$$Z_0 = Y_0 - Y^* = 2500 - 2400 = 100$$

thus

$$Z_t = 100(0.75)^t$$

and

$$Y_t = Y^* + 100(0.75)^t$$
$$= 2400 + 100(0.75)^t \qquad\qquad *$$

$0.75 < 1$ thus $(0.75)^t \to 0$ as $t \to \infty$, so $100(0.75)^t \to 0$ as $t \to \infty$ and $Y_t \to Y^*$. The system is therefore stable.

2. $Y_t = C_t + I_t$
$$= 500 + 0.55\,Y_{t-1} + 100 + 0.05\,Y_{t-1} = 600 + 0.6\,Y_{t-1}$$

If there is an equilibrium level of income Y^* then setting $Y_t = Y_{t-1} = Y^*$ equilibrium is given by

$$Y^* = 600 + 0.6\,Y^* \qquad\qquad (12)$$

Thus

$$Y^* = \frac{600}{1 - 0.6} = 1500$$

Let Z_t be the deviation of Y_t from Y^*, i.e.

$$Z_t = Y_t - Y^*$$

or

$$Y_t = Y^* + Z_t$$
$$Y_t = 600 + 0.6\,Y_{t-1}$$

thus

$$Y^* + Z_t = 600 + 0.6\,Y^* + 0.6 Z_{t-1} \qquad\qquad (13)$$

Subtracting equation (12) from equation (13) gives

$$Z_t = 0.6 Z_{t-1}$$

thus

$$Z_t = (0.6)^t Z_0$$
$$Z_0 = Y_0 - Y^* = 1550 - 1500 = 50$$

thus
$$Z_t = 50(0.6)^t$$
and
$$Y_t = Y^* + 50(0.6)^t$$
$$= 1500 + 50(0.6)^t \qquad *$$
$0.6 < 1$ thus $(0.6)^t \to 0$ as $t \to \infty$, so $50(0.6)^t \to 0$ as $t \to \infty$ and $Y_t \to Y^*$. The system is therefore stable.

3. (i) $Y_t = C_t + I_t + G_t$
$$= 200 + 0.75(Y_{t-1} - 0.2\,Y_{t-1}) + 100 + 0.25(Y_{t-1} - 0.2\,Y_{t-1}) + 0.1\,Y_t$$
$$= 300 + 0.6\,Y_{t-1} + 0.2\,Y_{t-1} + 0.1\,Y_t$$
thus
$$0.9\,Y_t = 300 + 0.8\,Y_{t-1} \qquad (14)$$
Equilibrium is given by
$$0.9\,Y^* = 300 + 0.8\,Y^* \qquad (15)$$
Thus
$$Y^* = 3000$$
Let Z_t be the deviation of Y_t from Y^*, i.e.
$$Y_t = Y^* + Z_t$$
Equation (14) becomes
$$0.9\,Y^* + 0.9\,Z_t = 300 + 0.8\,Y^* + 0.8\,Z_{t-1} \qquad (16)$$
subtracting equation (15) from equation (16) gives
$$0.9\,Z_t = 0.8\,Z_{t-1}$$
or
$$Z_t = 0.89\,Z_{t-1} \text{ approx.}$$
$$= (0.89)^t Z_0$$
$$Z_0 = Y_0 - Y^* = 3500 - 3000 = 500$$
thus
$$Z_t = 500(0.89)^t$$
and
$$Y_t = 3000 + 500(0.89)^t \qquad *$$
$0.89 < 1$, thus $500(0.89)^t \to 0$ as $t \to \infty$, so $Y_t \to Y^*$. The system is therefore stable.
(ii) $Y_t = C_t + I_t + G_t$
$$= 200 + 0.75[Y_{t-1} - 0.25\,Y_{t-1}] + 100 + 0.25[Y_{t-1} - 0.25\,Y_{t-1}] + 0.2\,Y_t$$
$$= 300 + 0.75\,Y_{t-1} + 0.2\,Y_t$$
thus
$$0.8\,Y_t = 300 + 0.75\,Y_{t-1} \qquad (17)$$
Equilibrium is given by
$$0.8\,Y^* = 300 + 0.75\,Y^* \qquad (18)$$
Thus
$$Y^* = 6000$$
Let Z_t be the deviation of Y_t from Y^*, i.e.
$$Y_t = Y^* + Z_t$$
thus
$$0.8\,Y^* + 0.8\,Z_t = 300 + 0.75\,Y^* + 0.75\,Z_{t-1} \qquad (19)$$
Subtracting equation (18) from equation (19) gives
$$0.8\,Z_t = 0.75\,Z_{t-1}$$

thus

$$Z_t = 0.94 Z_{t-1} \quad \text{approx.}$$
$$= (0.94)^t Z_0$$
$$Z_0 = 3500 - 6000 = -2500$$

thus

$$Z_t = -2500(0.94)^t$$

and

$$Y_t = 6000 - 2500(0.94)^t \qquad \qquad \qquad *$$

$0.94 < 1$ thus $-2500(0.94)^t \to 0$ as $t \to \infty$, so $Y_t \to Y^*$. However, the root of the system has increased from 0·89 to 0·94 so that the system, although still stable, converges less rapidly towards its equilibrium.

4. (i) $Q_t = 30 - 4P_t$
   $Q_t = -5 + 3P_{t-1}$

If there is an equilibrium then setting $P_t = P_{t-1} = P^*$ and $Q_t = Q^*$,

$$Q^* = 30 - 4P^* \qquad (20)$$
$$Q^* = -5 + 3P^* \qquad (21)$$

Thus

$$30 - 4P^* = -5 + 3P^*$$

so

$$7P^* = 35$$

and

$$P^* = 5$$

When $P^* = 5$ then $Q^* = 10$.

Let v_t be the deviation of P_t from P^* and w_t be the deviation of Q_t from Q^*,

$$v_t = P_t - P^* \quad \text{or} \quad P_t = P^* + v_t$$
$$w_t = Q_t - Q^* \quad \text{or} \quad Q_t = Q^* + w_t$$

Inserting these values into the demand function $Q_t = 30 - 4P_t$ gives

$$Q^* + w_t = 30 - 4P^* - 4v_t \qquad (22)$$

Subtracting equation (20) from equation (21) gives

$$w_t = -4v_t$$

From the supply function $Q_t = -5 + 3P_{t-1}$

$$Q^* + w_t = -5 + 3P^* + 3v_{t-1} \qquad (23)$$

Subtracting equation (21) from equation (23) gives

$$w_t = 3v_{t-1}$$

thus

$$-4v_t = 3v_{t-1}$$

or

$$v_t = -0.75v_{t-1}$$

so

$$v_t = (-0.75)^t v_0$$

and

$$w_t = -4(-0.75)^t v_0$$
$$v_0 = P_0 - P^* = 3 - 5 = -2$$

thus

$$v_t = -2(-0.75)^t$$

and

$$w_t = 8(-0.75)^t$$

Hence

$$P_t = 5 - 2(-0.75)^t \qquad *$$

and

$$Q_t = 10 + 8(-0.75)^t \qquad *$$

$|0.75| < 1$, thus v_t and $w_t \to 0$ as $t \to \infty$, so $P_t \to P^*$ and $Q_t \to Q^*$.

(ii) $P_1 = 5 - 2(-0.75)^1 = 5 + 1.5 = 6.5 \qquad *$
$\quad Q_1 = 10 + 8(-0.75)^1 = 10 - 6 = 4 \qquad *$
$\quad P_2 = 5 - 2(-0.75)^2 = 5 - 1.125 = 3.875 \qquad *$
$\quad Q_2 = 10 + 8(-0.75)^2 = 10 + 4.5 = 14.5 \qquad *$

5. $Q_t = 20 - 2P_t$
$\quad Q_t = -5 + 3P_{t-1}$

If there is an equilibrium then setting $P_t = P_{t-1} = P^*$ and $Q_t = Q^*$,

$$20 - 2P^* = -5 + 3P^*$$

thus

$$P^* = 5$$

When $P^* = 5$ then $Q^* = 10$.

Let v_t be the deviation of P_t from P^* and let w_t be the deviation of Q_t from Q^*; thus

$$P_t = P^* + v_t = 5 + v_t$$

and

$$Q_t = Q^* + w_t = 10 + w_t$$

Inserting these values for P_t and Q_t into the demand function $Q_t = 20 - 2P_t$ gives

$$w_t = -2v_t \qquad (24)$$

Inserting the values for P_{t-1} and Q_t into the supply function $Q_t = -5 + 3P_{t-1}$ gives

$$w_t = 3v_{t-1} \qquad (25)$$

From equations (24) and (25) we have $-2v_t = 3v_{t-1}$ or $v_t = -1.5v_{t-1}$, thus

$$v_t = (-1.5)^t v_0$$

and

$$w_t = -2(-1.5)^t v_0$$
$$v_0 = P_0 - P^* = 4 - 5 = -1$$

so

$$v_t = -(-1.5)^t$$

and

$$w_t = 2(-1.5)^t$$

thus

$$P_t = 5 - (-1.5)^t \qquad *$$

and

$$Q_t = 10 + 2(-1.5)^t \qquad *$$

but $|1.5| > 1$ thus $(-1.5)^t \to \pm \infty$ as $t \to \infty$ and the system is unstable.

CHAPTER 14

Differential Equations

In many economic problems involving changes over time, it is convenient to assume that change is continuous rather than working in terms of distinct periods of time. This is appropriate, for example, in considering the behaviour of economic aggregates, where there is no central decision-making body which might make decisions at discrete intervals. In particular, it is often desirable to consider the consequence of assuming constant proportional rates of change. This implies seeking a function $y = f(x)$ such that

$$\frac{1}{y}\frac{dy}{dx} = a \quad \text{or} \quad \frac{dy}{dx} = ay$$

for all values of x. This is called a first-order differential equation. We first seek a solution to the equation

$$\frac{dy}{dx} = y$$

for all values of x.

Consider any finite polynomial, i.e. any function which is the sum of a finite number of terms, each a power of x. If the highest powered term in $f(x)$ is cx^n, the highest powered term in the derivative will be ncx^{n-1} and the coefficient of x^n in $f'(x)$ will be zero. Thus no finite polynomial will solve our equation.

A SERIES APPROACH

Let us assume however that there is some infinite series,

$$y = 1 + a_1 x + a_2 x^2 + a_3 x^3 + \dots + a_n x^n + \dots$$

of which the sum of the first n terms is convergent, i.e. tends to a finite limit as $n \to \infty$. We can differentiate this series term by term, to get

$$dy/dx = a_1 + 2a_2 x + 3a_3 x^2 + \dots + na_n x^{n-1} + \dots$$

If we are to have $dy/dx = y$ for all values of x, then these two series must be the same, i.e. the coefficient of x^n must be the same in each series, for all n.

Starting at the beginning, we have

$$a_1 = 1$$

$$2a_2 = a_1 = 1 \quad \text{so} \quad a_2 = \frac{1}{2} = \frac{1}{2!}$$

where $2! = (2)(1)$,

$$3a_3 = a_2 = \frac{1}{2!} \quad \text{so} \quad a_3 = \frac{1}{3} \cdot \frac{1}{2!} = \frac{1}{3!}$$

where $3! = (3)(2)(1)$,

$$4a_4 = a_3 = \frac{1}{3!} \quad \text{so} \quad a_4 = \frac{1}{4} \cdot \frac{1}{3!} = \frac{1}{4!}$$

where $4! = (4)(3)(2)(1)$,

$$\ldots \text{and} \ldots$$

$$na_n = a_{n-1} = \frac{1}{(n-1)!} \quad \text{so} \quad a_n = \frac{1}{n} \cdot \frac{1}{(n-1)!} = \frac{1}{n!}$$

where $n! = (n)(n-1)\ldots(4)(3)(2)(1)$. Thus using these results,

$$y = 1 + x + \frac{x^2}{2!} + \frac{x^3}{3!} + \frac{x^4}{4!} + \ldots + \frac{x^n}{n!} + \ldots$$

Differentiating term by term, we can check that $dy/dx = y$.

CONVERGENCE OF e^x

We can check that this series is convergent by considering the relation of each term to the one before. To get from the term in x^{n-1} to the term in x^n we multiply by x and divide by n. The value of x is fixed, but n increases, so for any finite x there will be some term which is less than the term before since $x/n < 1$, and beyond this the terms decrease ever more rapidly. If the terms beyond this one decreased in a constant proportion $a < 1$, their sum would be finite. This is so because if S_n is the sum of a geometric progression,

$$S_n = 1 + a + a^2 + a^3 + \ldots + a^{n-1}$$

$$aS_n = a + a^2 + a^3 + \ldots + a^{n-1} + a^n$$

and

$$(1-a)S_n = 1 - a^n$$

so

$$S_n = \frac{1 - a^n}{1 - a}$$

If we let $n \to \infty$,

$$S = \text{Lim}_{n \to \infty}(S_n) = \frac{1}{1-a}$$

if $|a| < 1$, because if $|a| < 1$, as $n \to \infty$, $a^n \to 0$.

As its terms approach zero more rapidly than a geometric progression with $|a| < 1$, the series

$$y = 1 + x + \frac{x^2}{2!} + \ldots + \frac{x^n}{n!} + \ldots$$

is convergent, i.e. as more terms are added the sum tends to a finite limit for any finite value of x. This limit is called e^x, which is referred to as the exponential function. The derivative of e^x with respect to x is e^x. The exponential constant e is found by setting $x = 1$, so that

$$e = e^1 = 1 + 1 + \frac{1}{2!} + \ldots + \frac{1}{n!} + \ldots$$

This can easily be evaluated to any required number of places of decimals. To five places, e $= 2 \cdot 71828$.

PROPERTIES OF THE EXPONENTIAL FUNCTION

We can show that functions of the form e^x obey the usual rules for manipulating powers of any number, e.g. a^x. This requires that for any b and any x,

$$(e^x)^b = e^{bx}$$

Define the function $w(x)$ as the ratio of the two functions, $(e^x)^b$ and e^{bx},

$$w(x) = \frac{(e^x)^b}{e^{bx}}$$

If $x = 0$, $e^x = 1$ so $(e^x)^b = 1$ and $e^{bx} = 1$. Thus $w(0) = 1$. By the function of a function rule for derivatives,

$$\frac{d[(e^x)^b]}{dx} = \frac{d[(e^x)^b]}{d(e^x)} \cdot \frac{d(e^x)}{dx} = b \cdot (e^x)^{b-1} \cdot e^x = b \cdot (e^x)^b$$

and

$$\frac{d(e^{bx})}{dx} = \frac{d(e^{bx})}{d(bx)} \cdot \frac{d(bx)}{dx} = e^{bx} \cdot b$$

Thus

$$d[w(x)]/dx = \frac{e^{bx}(b \cdot (e^x)^b) - (e^x)^b \cdot e^{bx} \cdot b}{(e^{bx})^2} = 0$$

so $w(x)$ is the same for all x. Thus $w(x) = 1$ for all x, and

$$(e^x)^b = e^{bx} \quad \text{for all } x \text{ and all } b.$$

When $x = 1$, $(e^1)^b = (e)^b = e^b$. Thus e^x as defined above is the same as the number e raised to the power x.

We can see that if $y = e^x$, as $x \to \infty$, $y \to \infty$. Also, as $dy/dx = y$, e^x increases steadily over the range $x > 0$. If $x < 0$, it is not immediately obvious whether $y \lessgtr 0$; however, let $z = -x > 0$. Then $e^z > 0$ and $e^z \to \infty$ as $z \to \infty$. Thus $e^x = e^{-z} = 1/e^z > 0$ and as $x \to -\infty$, $z \to +\infty$ and

$e^x = 1/e^z \rightarrow 0$. Thus $e^x > 0$ over the whole range from $x \rightarrow -\infty$ to $x \rightarrow +\infty$, and e^x increases steadily over this range, see Figure 41.

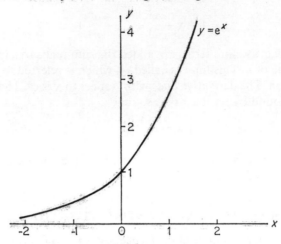

Figure 41   The exponential function

CONSTANT PROPORTIONAL GROWTH RATES

If we have $z = e^m$, where $m = f(x)$, then by the function of a function rule

$$dz/dx = \frac{dz}{dm}\cdot\frac{dm}{dx} = e^m f'(x) = e^{f(x)} f'(x)$$

For example if $z = e^{ax}$,

$$dz/dx = a.e^{ax}$$

This means that the proportional rate of increase of z as x increases is

$$\frac{1}{z}\cdot\frac{dz}{dx} = \frac{ae^{ax}}{e^{ax}} = a$$

so that the proportional rate of increase of z as x increases is the constant a.

If we consider z as a function of time, then if the proportional rate of growth of z_t is a, this is satisfied by a function of the form

$$z_t = ke^{at}$$

so that

$$dz/dt = \dot{z}_t = ake^{at}$$

and

$$\frac{1}{z_t}\cdot\frac{dz_t}{dt} = \frac{\dot{z}_t}{z_t} = a$$

The use of a dot over a variable to denote its rate of change over time is common in economic texts, and is adopted below.

Alternative Bases of Exponential Functions

It is in fact possible to have exponential functions to any positive base, other than 1, e.g.

$$y = c^x$$

The simplest way to treat these is to let $c = e^b$, so that

$$y = c^x = e^{bx}$$

and

$$\frac{dy}{dx} = b e^{bx} = bc^x$$

LOGARITHMS

In this case b is defined as the logarithm of c. If e^x is a number, x is its logarithm to base e. Logarithm are commonly used either to base, e, or, for ease of numerical calculation, to base 10. If the logarithm of 10 to base e is $\log_e (10) = b$, then the relation between $\log_e (x)$ and $\log_{10} (x)$ is given by taking

$$Z = 10^c \quad \text{so} \quad \log_{10}(Z) = c$$
$$10 = e^b \quad \text{so} \quad \log_e(10) = b$$
$$Z = (e^b)^c = e^{bc}$$

so

$$\log_e(Z) = bc = \log_e(10).\log_{10}(Z)$$

and

$$\log_{10}(Z) = \log_e(Z)/\log_e(10)$$

Throughout this book, unless otherwise specified, logarithms are taken to the base e. These are called 'natural logarithms' and are written $\ln(x)$.

Whatever base is used, all positive numbers have logarithms; from the section on the Properties of the Exponential Function on pp. 233–4, as $x \to \infty$, $\ln x \to \infty$ and as $x \to 0$, $\ln x \to -\infty$. Thus all possible logarithms, positive and negative, are assigned to positive numbers, and negative numbers do not have logarithms.

The Derivative of the Logarithmic Function

If $y = e^x$ and $dy/dx = y$ then

$$x = \ln y \quad \text{and} \quad dx/dy = \frac{1}{dy/dx} = \frac{1}{y}$$

This implies that the integral of $1/y$ as y changes is given by

$$\int \frac{1}{y} dy = (\ln y) + k$$

This enables us to fill the gap left at $n = -1$ in the standard integral formula

$$\int x^n \, dx = \frac{x^{n+1}}{n+1} + k$$

The integral approach also enables us to calculate the logarithm of any number. For any $|x| < 1$, the sum of the series

$$S = 1 - x + x^2 - x^3 + \ldots = \frac{1}{1+x}$$

This is so because

$$xS = x - x^2 + x^3 - x^4 + \ldots$$

so

$$(1 + x)S = 1 \quad \text{and} \quad S = \frac{1}{1+x}$$

Since $|x| < 1$ the series is clearly convergent, thus we can find its integral by integrating each term in turn and summing. Thus

$$\int (1 - x + x^2 - x^3 + \ldots) dx = \int \frac{1}{1+x} dx$$

so that

$$\left(x - \frac{x^2}{2} + \frac{x^3}{3} - \ldots \right) + k_0 = \ln(1 + x) + k_1$$

where the k's are constants of integration. As $e^0 = 1$, we want $\ln(1) = 0$, so we set $k_0 = k_1 = 0$ and get

$$\ln(1 + x) = x - \frac{x^2}{2} + \frac{x^3}{3} - \ldots + (-1)^{n-1}\frac{x^n}{n} + \ldots$$

which is convergent for $|x| < 1$.

ECONOMIC APPLICATIONS OF THE EXPONENTIAL FUNCTION

The Domar Growth Model

The Domar growth model is concerned with finding the rate of change in investment which will equate the rate of change in potential output with the rate of change in aggregate demand and keep capacity fully utilized.

In our discussion of the model it will be assumed that we are dealing with a closed economy without government activity, which has a constant propensity to save s and a constant ratio between output and capital β, and that we are starting from an equilibrium situation where aggregate demand is exactly equal to potential output.

$$\dot{K}_t = \text{rate of change in capital stock}$$
$$\beta = \text{output/capital ratio}$$

Thus

$$\beta \dot{K}_t = \text{rate of change in potential output}$$

But investment is the rate of change in capital stock, i.e. $\dot{K}_t = I_t$, so

$$\beta I_t = \text{rate of change in potential output}$$

The rate of change in aggregate demand is equal to the rate of change in investment $\dot{I}_t$ times the multiplier $1/s$, i.e.

$$\dot{I}_t . \frac{1}{s} = \text{rate of change in aggregate demand}$$

The economy is therefore in equilibrium when

$$\beta I_t = \dot{I}_t . \frac{1}{s}$$

or

$$\dot{I}_t = \beta s I_t \tag{1}$$

In equation (1) time is the independent variable and investment the dependent variable. The equation is of the first order and linear. It is of the first order because it contains only the first derivative $\dot{I}_t$. It is of the first degree or linear as the derivative $\dot{I}_t$ and the dependent variable I_t are both raised to the power 1. Equation (1) is also homogeneous because the coefficient of the dependent variable is a constant, βs, and the equation has no constant term. A non-homogeneous equation has both a constant coefficient and a constant term. The latter type of first-order linear differential equation will be dealt with later in this chapter.

We will have solved equation (1) if we can find the time path of investment I_t. As with first-order difference equations it is necessary to know the value of investment at some initial time, say I_0.

$$\dot{I}_t = \beta s I_t$$

thus

$$\frac{\dot{I}_t}{I_t} = \beta s = \text{constant}$$

This tells us that the economy will be in equilibrium if the proportional rate of change of I_t, $\dot{I}_t / I_t$, is βs. From the section on constant proportional growth rates we know that the investment function which will keep the economy in equilibrium will take the form

$$I_t = A e^{\beta s t}$$

When $t = 0$ then $I_0 = A e^0 = A$, thus

$$I_t = I_0 e^{\beta s t}$$

where I_0 is the value of investment at some initial time. This is the solution to the differential equation. From this it is possible to find the required investment for any value of t, e.g. $I_{20} = I_0 e^{\beta s(20)}$. Notice that the solution does not contain a derivative.

CHANGES IN CAPITAL STOCK

When the investment function takes the form

$$I_t = I_0 e^{\beta s t}$$

the change in capital stock from period 0 to period t^* is given by

$$\int_0^{t^*} I_t \, \mathrm{d}t = \int_0^{t^*} I_0 \, \mathrm{e}^{\beta st} \, \mathrm{d}t = I_0 \int_0^{t^*} \mathrm{e}^{\beta st} \, \mathrm{d}t$$

$$= I_0 \left[\frac{\mathrm{e}^{\beta st}}{\beta s} \right]_0^{t^*} = \frac{I_0}{\beta s} [\mathrm{e}^{\beta st^*} - \mathrm{e}^0] = \frac{I_0}{\beta s} [\mathrm{e}^{\beta st^*} - 1]$$

Notice that the integral of $\mathrm{e}^{\beta st}$ is $(\mathrm{e}^{\beta st}/\beta s) + k$, where k is the constant of integration. $\mathrm{e}^{\beta st}$ must be the derivative of $\mathrm{e}^{\beta st}/(\beta s)$ since $\mathrm{d}\,(\mathrm{e}^{\beta st})/\mathrm{d}t = (\beta s)\,\mathrm{e}^{\beta st}$.

NON-HOMOGENEOUS DIFFERENTIAL EQUATIONS

Suppose we have a model in which we assume consumers adjust their actual consumption expenditure towards a desired level C^*. It is likely that C^* will be proportional to current income, i.e. $C_t^* = \alpha Y_t$, where $\alpha > 0$. However, in this example we will assume C^* is constant so that Y_t is assumed constant for all t. It would seem plausible to assume that the speed at which actual consumption C_t adjusts is proportional to the gap between C^* and C_t, i.e.

$$\dot{C}_t = \lambda(C^* - C_t)$$

where $\lambda > 0$ or

$$\dot{C}_t = \lambda C^* - \lambda C_t \tag{2}$$

$\dot{C}_t$ is the rate of change in actual consumption. It is positive when $C^* > C_t$ and negative when $C^* < C_t$. The larger the gap between C^* and C_t the greater the rate of change, given λ.

Equation (2) is a first-order linear differential equation. The coefficient of the dependent term C_t is a constant, i.e. λ, but the equation also contains a constant term λC^*. Consequently the equation is non-homogeneous.

To solve equation (2), i.e. to find how C_t behaves over time, let us assume $C_t \neq C^*$ and $Z_t = C_t - C^*$, where Z_t is the deviation.

$$C_t = C^* + Z_t \tag{3}$$

Consequently to find C_t it is necessary to solve for Z_t. From equation (3),

$$\dot{C}_t = \frac{\mathrm{d}(C^* + Z_t)}{\mathrm{d}t} = \dot{C}_t^* + \dot{Z}_t = \dot{Z}_t$$

since C^* is a constant and $\dot{C}_t^* = 0$. Inserting these values for C_t and $\dot{C}_t$ into equation (2) gives

$$\dot{Z}_t = \lambda C^* - \lambda C_t = \lambda C^* - \lambda(C^* + Z_t) = - \lambda Z_t$$

i.e.

$$\frac{\dot{Z}_t}{Z_t} = - \lambda$$

thus

$$Z_t = K \mathrm{e}^{-\lambda t}$$

When $t = 0$, $Z_0 = K \mathrm{e}^{-0} = K$

thus

$$Z_t = Z_0 e^{-\lambda t}$$

where $Z_0 = C_0 - C^*$, assuming we are given C_0, and

$$C_t = C^* + Z_0 e^{-\lambda t}$$

This is the solution to equation (2)

$$e^{-\lambda t} \to 0 \quad \text{as} \quad t \to \infty$$

since $\lambda > 0$, thus

$$Z_0 e^{-\lambda t} \to 0 \quad \text{as} \quad t \to \infty$$

i.e.

$$C_t \to C^* \quad \text{as} \quad t \to \infty$$

The model $\dot{C}_t = \lambda(C^* - C_t)$ with $\lambda > 0$ is therefore stable.

STABILITY OF THE INCOME ADJUSTMENT MODEL

Suppose we want to examine the stability of a aystem which can be described by the equation

$$\dot{Y}_t = A - \alpha Y_t \tag{4}$$

where A and α are positive constants. To solve equation (4) it is necessary to know the value of Y at some initial time, say Y_0.

If there is an equilibrium income level Y^*, then $\dot{Y}_t$ will equal zero at that income level. Equilibrium is given by

$$0 = A - \alpha Y^* \quad \text{or} \quad Y^* = \frac{A}{\alpha}$$

Let Z_t be the deviation of Y_t from Y^*, i.e.

$$Y_t = Y^* + Z_t$$

and

$$\dot{Y}_t = \dot{Z}_t$$

because $\dot{Y}^* = 0$. Thus

$$\dot{Z}_t = A - \alpha(Y^* + Z_t) = -\alpha Z_t$$

from equation (4), i.e.

$$\frac{\dot{Z}_t}{Z_t} = -\alpha$$

so

$$Z_t = Z_0 e^{-\alpha t}$$

and

$$Y_t = Y^* + Z_0 e^{-\alpha t}$$
$$Z_0 e^{-\alpha t} \to 0 \quad \text{as} \quad t \to \infty$$

since $\alpha > 0$, thus

$$Z_t \to 0 \quad \text{as} \quad t \to \infty$$

i.e.

$$Y_t \to Y^* \quad \text{as} \quad t \to \infty$$

Consequently the system is stable.

DIFFERENTIAL EQUATIONS APPLIED TO MARKET EQUILIBRIUM

Suppose the demand and supply functions for a particular commodity take the form

$$Q_{Dt} = a - bP_t$$
$$Q_{St} = -c + gP_t$$

where a, b, c, and g are positive constants and the rate of change of price with respect to time is directly proportional to excess demand, i.e.

$$\dot{P}_t = \lambda(Q_{Dt} - Q_{St}) \tag{5}$$

where $\lambda > 0$. It is possible to find the time path of P_t, i.e. solve equation (5), given a value for price for some initial time, say P_0.

If there is an equilibrium price P^* and quantity Q^*, then in equilibrium

$$P_t = P^*$$

and

$$Q_{Dt} = Q_{St} = Q^*$$

Equilibrium is given by

$$Q^* = a - bP^*$$

and

$$Q^* = -c + gP^*$$

i.e.

$$a - bP^* = -c + gP^*$$

thus

$$P^* = \frac{a+c}{b+g}$$

If $P_t \neq P^*$ and we let v_t be the deviation then

$$P_t = P^* + v_t$$

and

$$\dot{P}_t = \dot{v}_t$$

since $\dot{P}_t^* = 0$.

$$\dot{v}_t = \lambda(Q_{Dt} - Q_{St}) = \lambda[a + c - (b+g)P_t]$$
$$= \lambda[a + c - (b+g)(P^* + v_t)]$$

but

$$a + c = (b+g)P^*$$

so

$$\dot{v} = -\lambda(b+g)v_t$$

i.e.

$$\frac{\dot{v}_t}{v_t} = -\lambda(b+g)$$

so

$$v_t = v_0 e^{-\lambda(b+g)t}$$

v_0 can be calculated from the initial information on P_0

$$e^{-\lambda(b+g)t} \to 0 \quad \text{as} \quad t \to \infty$$

since $\lambda > 0$ and $(b + g) > 0$, thus

$$v_t \rightarrow 0 \quad \text{as} \quad t \rightarrow \infty$$

and

$$P_t \rightarrow P^* \quad \text{as} \quad t \rightarrow \infty$$

The model is therefore stable.

COMPOUND INTEREST

If interest on a loan is payable at the rate of r, then after one year £1 is worth $£(1 + r)$. If interest at the rate r per annum is added twice yearly, then after a year £1 is worth $£(1 + \frac{1}{2}r)^2$, which is more than $£(1 + r)$. If interest is added quarterly at the rate $\frac{1}{4}r$, after a year £1 is worth $£(1 + \frac{1}{4}r)^4$, which is more than $£(1 + \frac{1}{2}r)^2$. The more frequently interest is added, the larger the sum at the end of the year. There is a limit to this process, however, which we will examine. Let ϕ be this limit, i.e.

$$\phi = \underset{n \rightarrow \infty}{\text{Lim}} \left[\left(1 + \frac{r}{n} \right)^n \right]$$

Let $z = 1/n$; as $n \rightarrow \infty$, $z \rightarrow 0$, so we will find

$$\phi = \underset{z \rightarrow 0}{\text{Lim}} \left[(1 + rz)^{1/z} \right]$$

If $\alpha = e^\beta$, then $\beta = \ln \alpha$ and $\alpha = e^\beta = e^{\ln \alpha}$. Thus

$$1 + rz = e^{\ln(1 + rz)}$$

so

$$(1 + rz)^{1/z} = (e^{\ln(1 + rz)})^{1/z} = e^{(1/z)\ln(1 + rz)}$$

and

$$\phi = \underset{z \rightarrow 0}{\text{Lim}} \left[e^{(1/z)\ln(1 + rz)} \right]$$

From the series expansion of $\ln(1 + x)$ given in the section on the Derivative of the Logarithmic Function, on pp. 235–6, we know that

$$\ln(1 + rz) = rz - \frac{(rz)^2}{2} + \frac{(rz)^3}{3} - \cdots$$

so

$$\frac{\ln(1 + rz)}{z} = r \left(1 - \frac{rz}{2} + \frac{(rz)^2}{3} - \cdots \right)$$

and

$$\underset{z \rightarrow 0}{\text{Lim}} \left[\frac{\ln(1 + rz)}{z} \right] = r$$

Thus

$$\phi = \underset{z \rightarrow 0}{\text{Lim}} \left[e^{(1/z)\ln(1 + rz)} \right] = e^r$$

Consequently £1 compounded continuously amounts to e^r after one year. During the second year every £1 of the new principal e^r will amount to e^r.

Hence after one year e^r will amount to $(e^r)^2$, which means that £1 will amount to e^{2r} at the end of two years. After t years, £1 will amount to e^{rt}.

If £1 amounts to e^{rt} after t years then P, where P is the initial principal, will amount to Pe^{rt} after t years, i.e.

$$F = Pe^{rt}$$

where F is future value.

THE PRESENT VALUE OF A FUTURE RECEIPT

If we are given the future value and asked to find the present value, i.e. the initial principal, when the future value is discounted at a continuous rate r.

$$\text{Present value} \quad P = \frac{F}{e^{rt}} = Fe^{-rt}$$

The Present Value of an Income Stream

Suppose we are given an income stream of £A per annum, spread evenly over n years, and asked to find the present value

$$\text{Present value} = \int_0^n Ae^{-rt}dt$$

where r is the continuous discount rate,

$$= A\left[\frac{e^{-rt}}{-r}\right]_0^n = A\left[\frac{e^{-rn}-1}{-r}\right] = \frac{A}{r}\left[1 - e^{-rn}\right]$$

This present value tends to A/r as $n \to \infty$, i.e.

$$\frac{A}{r}\left[1 - e^{-rn}\right] \to \frac{A}{r} \quad \text{as} \quad n \to \infty$$

since $e^{-rn} \to 0$.

SEMI-LOGARITHMIC GRAPHS

Where any variable grows over time it is often helpful to graph it with time on the horizontal axis and the logarithm of the variable on the vertical axis. The use of a logarithmic vertical scale means that time paths with equal proportional growth rates are represented by lines with equal slope. In the case of ordinary graphs this is not so. In Figure 42 we plot three functions, with an ordinary vertical scale on the left and a logarithmic scale on the right.

The three functions are as follows:

$$y_1 = a + bt$$

thus

$$\dot{y}_1 = b \quad \text{and} \quad \ddot{y}_1 = 0$$

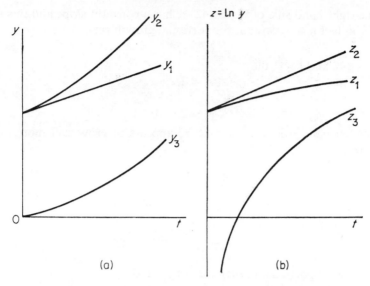

Figure 42   (a) Natural scales.   (b) Semi-logarithmic scales

y_1 as plotted on the left-hand side of Figure 42 appears to have a constant growth rate. However,

$$z_1 = \ln y_1$$

thus

$$\dot{z}_t = \frac{dz_1}{dt} = \frac{d \ln y_1}{dt} = \frac{d \ln y_1}{dy_1} \cdot \frac{dy_1}{dt} = \frac{1}{y_1} \cdot \frac{dy_1}{dt}$$

which is the proportional rate of growth of y_1.

$$\dot{z}_1 = \frac{b}{a + bt} \quad \text{and} \quad \ddot{z}_1 = \frac{-b^2}{(a + bt)^2} < 0$$

The graph of z_1 on the right-hand side of Figure 42 makes it clear that the proportional growth rate of y_1 is steadily decreasing towards zero.

$$y_2 = a e^{bt}$$

thus

$$\dot{y}_2 = ab\, e^{bt} \quad \text{and} \quad \ddot{y}_2 = ab^2 e^{bt} > 0$$

On the left-hand side of Figure 42, y_2 appears to grow ever more rapidly. However $\ln(e^{bt}) = bt$, so

$$z_2 = \ln y_2 = (\ln a) + bt$$

thus

$$\dot{z}_2 = b$$

This is the proportional growth rate of y_2, and

$$\ddot{z}_2 = 0$$

On the right-hand side of Figure 42, z_2 has a constant slope and this shows that y_2 in fact has a constant proportional growth rate.

$$y_3 = ct^2$$

where $c > 0$ and $t > 0$, thus

$$\dot{y}_3 = 2ct$$

and

$$\ddot{y}_3 = 2c > 0$$

On the left-hand side of Figure 42, y_3 appears to grow ever more rapidly. However,

$$z_3 = \ln y_3$$

i.e.

$$z_3 = (\ln c) + 2 \ln t$$

thus

$$\dot{z}_3 = \frac{2}{t}$$

This is the proportional growth rate of y_3, and

$$\ddot{z}_3 = \frac{-2}{t^2} < 0$$

The right-hand side of Figure 42 makes it clear that in fact the proportional growth rate of y_3 is steadily decreasing.

For similar reasons, semi-logarithmic graphs are also useful in comparing the severity of oscillations in different variables over time.

EXERCISES

For convenience in answering these questions, approximate values of e^x are as follows:

$$e^{0 \cdot 2} = 1 \cdot 2214 \quad e^{-0 \cdot 2} = 0 \cdot 8187$$
$$e^{0 \cdot 4} = 1 \cdot 4918 \quad e^{-0 \cdot 4} = 0 \cdot 6703$$
$$e^{0 \cdot 6} = 1 \cdot 8221 \quad e^{-0 \cdot 6} = 0 \cdot 5488$$
$$e^{0 \cdot 8} = 2 \cdot 2255 \quad e^{-0 \cdot 8} = 0 \cdot 4493$$
$$e^{1 \cdot 0} = 2 \cdot 7183 \quad e^{-1 \cdot 0} = 0 \cdot 3679$$

1. Find the future value of

   (i) £500 compounded continuously for 10 years at an interest rate of 4 per cent per annum.

   (ii) £300 compounded continuously for 6 years at 10 per cent per annum.

2. Find the present value of

   (i) £5000 in 10 year's time if it is discounted continuously at 8 per cent per annum.

   (ii) £400 in 5 year's time if it is discounted continuously at 4 per cent per annum.

3. Find the present value of a steady income stream of £500 per annum for 8 years, and discounted continuously at

(i) 5 per cent per annum,

(ii) 10 per cent per annum.

4. The value of an investment project X at time t is

$$X = 500.e^{t^{0.5}}$$

i.e. the value now, when $t = 0$, is 500 and this increases over time. If the present value is found by discounting continuously at a rate of 10 per cent per annum, find the t which will maximize the present value of X.

5. If

$$\dot{Y_t} = 100 - 0.5\,Y_t$$

where Y_t is actual income at time t and $Y_0 = 300$, find Y_t and Y_2 and comment on the stability of the model.

6. If

$$\dot{P_t} = 3 - 0.2P_t$$

where P_t is the market price at time t and $P_0 = 20$, find P_t and P_3 and comment on the stability of the system.

7. The quantity of a good demand Q_{Dt} and the quantity supplied Q_{St} are respectively given by

$$Q_{Dt} = 16 - 4P_t$$

and

$$Q_{St} = -2 + 2P_t$$

The rate of change of price with respect to time is directly proportional to excess demand, i.e.

$$\dot{P_t} = \tfrac{1}{3}(Q_{Dt} - Q_{St})$$

and

$$P_0 = 4$$

Show that P_t tends towards equilibrium.

SOLUTIONS TO EXERCISES

1. Future value $F = Pe^{rt}$ where P is the initial principal, r is the rate of interest, and t is time.

(i) $F = 500\,e^{(0.04)t} = 500\,e^{(0.04)10} = 500\,e^{0.4}$

$$= 500(1.4918) \quad \text{approx.}$$

$$= 745.9 \qquad\qquad *$$

(ii) $F = 300\,e^{(0.1)6} = 300\,e^{0.6} = 300(1.8221) \quad \text{approx.}$

$$= 546.6 \qquad\qquad *$$

2. $F = Pe^{rt}$

$$\text{Present value } P = \frac{F}{e^{rt}} = Fe^{-rt}$$

(i) $P = 5000\,e^{-(0.08)10} = 5000\,e^{-0.8} = 5000(0.4493) \quad \text{approx.}$

$$= 2246.6 \qquad\qquad *$$

(ii) $P = 400 \, e^{-(0.04)5} = 400 \, e^{-0.2} = 400(0.8187)$ approx.
$$= 327.5 \qquad *$$

3. $P = Fe^{-rt}$

The total present value of an income stream is

$$\int_0^{t^*} Fe^{-rt}dt = F\int_0^{t^*} e^{-rt}dt = F\left[\frac{e^{-rt}}{-r}\right]_0^{t^*}$$

(i) When $F = 500$

$$t^* = 8$$

and

$$r = 0.05$$

Present value of income stream $= 500\left[\dfrac{e^{-(0.05)t}}{-0.05}\right]_0^8$

$$= 500\left[\frac{e^{-0.4} - e^0}{-0.05}\right] = 10{,}000[1 - e^{-0.4}]$$

$$= 10{,}000[1 - 0.6703] \text{ approx.}$$

$$= 3297 \qquad *$$

(ii) $F = 500$

$$t^* = 8$$

and

$$r = 0.1$$

Present value of income stream $= 500\left[\dfrac{e^{-(0.1)t}}{-0.1}\right]_0^8$

$$= 500\left[\frac{e^{-0.8} - e^0}{-0.1}\right] = 5000[1 - e^{-0.8}]$$

$$= 5000[1 - 0.4493] \text{ approx.}$$

$$= 2753.4 \qquad *$$

4. Let the present value of X be P. P is then a function of time t, i.e.

$$P = \frac{X}{e^{rt}} = \frac{500.e^{t^{0.5}}}{e^{rt}}$$

$r = 0.1$, thus

$$P = 500 \, e^{t^{0.5}} \left(e^{-0.1t}\right) = 500 \, e^{(t^{0.5} - 0.1t)}$$

This is now in the form $500 \, e^x$, where $x = t^{0.5} - 0.1t$. If x is maximized then $500 \, e^x$ is automatically maximized. x is maximized when $dx/dt = 0$ and $d^2x/dt^2 < 0$.

$$x = t^{0.5} - 0.1t$$

thus

$$dx/dt = 0.5t^{-0.5} - 0.1$$

$dx/dt = 0$ if $t^{-0.5} = 0.2$, i.e.

$$\frac{1}{\sqrt{t}} = \frac{1}{5} \text{ or } \quad t = 25$$

$$d^2x/dt^2 = -0.5(0.5)t^{-1.5} < 0$$

Thus P is maximized when $t = 25$. $\qquad *$

5. $\dot{Y}_t = 100 - 0.5\,Y_t$

If there is an equilibrium level of income Y^* then $\dot{Y}_t$ will equal zero at that income level. Equilibrium is given by

$$0 = 100 - 0.5 Y_t$$

i.e. equilibrium income $Y^* = 200$.

Let Z_t be the deviation of Y_t from Y^*, i.e.

$$\dot{Y}_t = \dot{Z}_t + \dot{Y}^*$$

or

$$\dot{Y}_t = \dot{Z}_t$$

since $\dot{Y}^* = 0$, thus

$$\dot{Z}_t = 100 - 0.5 Y_t$$
$$= 100 - 0.5(200 + Z_t)$$

i.e.

$$\dot{Z}_t = -0.5 Z_t$$

or

$$\frac{\dot{Z}_t}{Z_t} = -0.5$$

The only form of Z_t, as a function of t, which has this property is

$$Z_t = A e^{-0.5t}$$

thus

$$Z_0 = A e^0 = A$$

But $Z_0 = Y_0 - Y^* = 300 - 200 = 100$, thus

$$Z_t = 100 e^{-0.5t}$$
$$Y_t = Y^* + Z_t$$

thus

$$Y_t = 200 + 100 e^{-0.5t} \qquad\qquad *$$

and

$$Y_2 = 200 + 100 e^{-1.0}$$
$$= 200 + 100(0.3679) \quad \text{approx.}$$
$$= 236.8 \qquad\qquad *$$
$$e^{-0.5t} \to 0 \quad \text{as} \quad t \to \infty$$

thus

$$Z_t \to 0 \quad \text{as} \quad t \to \infty$$

i.e.

$$Y_t \to Y^* \quad \text{as} \quad t \to \infty$$

so the system is stable.

6. $\dot{P}_t = 3 - 0.2 P_t$

If there is an equilibrium price level P^* then $\dot{P}_t$ will equal zero when $P_t = P^*$. Equilibrium is given by

$$0 = 3 - 0.2 P_t$$

or

$$P^* = 15$$

If v_t is the deviation of P_t from P^* then

$$\dot{v}_t = \dot{P}_t$$

and

$$\dot{v}_t = 3 - 0.2 P_t = -0.2 v_t$$

i.e.

$$\frac{\dot{v}_t}{v_t} = -0.2$$

thus

$$v_t = v_0 e^{-0.2t} \quad \text{and} \quad v_0 = 5$$

so

$$v_t = 5e^{-0.2t}$$

and

$$P_t = 15 + 5e^{-0.2t} \qquad *$$
$$P_3 = 15 + 5(0.5488) \quad \text{approx.}$$
$$= 17.744 \qquad *$$
$$e^{-0.2t} \to 0 \quad \text{as} \quad t \to \infty$$

thus

$$v_t \to 0 \quad \text{as} \quad t \to \infty$$

and

$$P_t \to P^* \quad \text{as} \quad t \to \infty$$

7. If there is an equilibrium then $P_t = P^*$ and $Q_{Dt} = Q_{St} = Q^*$.
   Equilibrium is given by

$$Q^* = 16 - 4P^*$$

and

$$Q^* = -2 + 2P^*$$

i.e.

$$16 - 4P^* = -2 + 2P^*$$

or

$$P^* = 3$$

If v_t is the deviation of P_t from P^*, then

$$\dot{P}_t = \dot{v}_t$$
$$\dot{P}_t = \tfrac{1}{3}(Q_{Dt} - Q_{St}) = \tfrac{1}{3}(16 - 4P_t + 2 - 2P_t) = 6 - 2P_t$$

thus

$$\dot{v}_t = -2v_t$$

and

$$\frac{\dot{v}_t}{v_t} = -2$$

so

$$v_t = v_0 e^{-2t}$$
$$= e^{-2t}$$

since $v_0 = 1$ and

$$P_t = 3 + e^{-2t}$$

P_t tends towards equilibrium as $t \to \infty$ since $e^{-2t} \to 0$ as $t \to \infty$. $\qquad *$

CHAPTER 15

Second-order Difference Equations

THE ACCELERATOR-MULTIPLIER MODEL

When discussing first-order linear difference equations it was assumed either that investment did not vary with time or that it was related to the previous period's income. The accelerator model of investment states that investment is a lagged function of the rate of change in income

$$I_t = v(Y_{t-1} - Y_{t-2})$$

where v is the accelerator. It would be desirable to have

$$I_t = v(Y_t - Y_{t-1})$$

but since firms cannot know Y_t at the beginning of this period they use the most recent rate of change in Y, i.e. $Y_{t-1} - Y_{t-2}$, when planning investment for period t.

A closed economic system without government activity gives

$$Y_t = C_t + I_t$$

If

$$C_t = C^* + cY_{t-1}$$

and

$$I_t = v(Y_{t-1} - Y_{t-2})$$

then

$$Y_t = C^* + cY_{t-1} + v(Y_{t-1} - Y_{t-2})$$

or

$$Y_t = C^* + (c + v)Y_{t-1} - vY_{t-2} \tag{1}$$

Equation (1) is a second-order linear difference equation since income in any period t is related to the income of the preceding two periods. To discover how Y changes with respect to t it is necessary to know the value of income in two initial periods, say Y_0 and Y_1.

SOLUTION OF A SECOND-ORDER LINEAR DIFFERENCE EQUATION

If there is an equilibrium level of income Y^*, this income level will continue once attained. If $Y_{t-2} = Y^*$ and $Y_{t-1} = Y^*$, then $Y_t = Y^*$. Taking

249

equation (1), this implies

$$Y^* = C^* + (c + v)Y^* - vY^* \tag{2}$$

i.e.

$$Y^* = \frac{C^*}{1 - c}$$

A finite equilibrium level of income exists if $c \neq 1$. The special case where $c = 1$ will be dealt with later.

We know that a finite equilibrium level of income exists, but will actual income Y_t approach Y^* and if so how fast? If $Y_t \neq Y^*$ and we let Z_t equal the deviation of actual income in period t from the equilibrium level Y^*, then

$$Z_t = Y_t - Y^* \quad \text{or} \quad Y_t = Y^* + Z_t$$

As with first-order difference equations it is necessary to solve for Z_t since Y^* is constant with respect to time. Substituting the value for Y_t in terms of Z_t into equation (1) gives

$$Y^* + Z_t = C^* + (c + v)Y^* + (c + v)Z_{t-1} - vY^* - vZ_{t-2} \tag{3}$$

Subtracting equation (2) from equation (3) gives

$$Z_t = (c + v)Z_{t-1} - vZ_{t-2} \tag{4}$$

We now have to solve equation (4), which refers only to the deviation Z_t. If we let $Z_t = Z_0 r^t$, where $Z_0 = Y_0 - Y^*$ and information on this initial income Y_0 is available, it is necessary to find a value or values for r so that $Z_t = Z_0 r^t$ will satisfy equation (4).

If $Z_t = Z_0 r^t$, then

$$Z_{t-1} = Z_0 r^{t-1}, \quad Z_{t-2} = Z_0 r^{t-2}, \quad \text{etc.}$$

$Z_t = Z_0 r^t$ will satisfy equation (4) if

$$Z_0 r^t = (c + v)Z_0 r^{t-1} - vZ_0 r^{t-2}$$

Dividing through by $Z_0 r^{t-2}$ gives

$$r^2 = (c + v)r - v$$

or

$$r^2 - (c + v)r + v = 0 \tag{5}$$

This is called the 'auxiliary equation' of the original difference equation (1). Thus

$$r = \frac{c + v \pm \sqrt{[(c + v)^2 - 4v]}}{2}$$

If $(c + v)^2 > 4v$, equation (5) will have two distinct real roots, which will be labelled r_1 and r_2. If $(c + v)^2 = 4v$, equation (5) will have two equal roots. If $(c + v)^2 < 4v$, equation (5) will have no real roots. This latter case will be considered in Chapter 16.

AUXILIARY EQUATION WITH TWO DISTINCT REAL ROOTS

If we take the case where $(c + v)^2 > 4v$, then $Z_t = Z_0 r^t$ will satisfy equation (4) when $r = r_1$ or r_2. In our simple model these values depend upon the value of the marginal propensity to consume c and the accelerator v.

Since $Z_t = Z_0 r_1^t$ and $Z_t = Z_0 r_2^t$ both satisfy equation (4) it will also be satisfied by

$$Z_t = ar_1^t + br_2^t$$

where a and b are constants to be determined. With the available information on Y_0 and Y_1 and hence on Z_0 and Z_1, it will be possible to find values for a and b. We will then have solved for the deviation Z_t.

Checking the Solution

Before solving for a and b let us check that

$$Z_t = ar_1^t + br_2^t$$

will satisfy equation (4) if $Z_t = Z_0 r_1^t$ and $Z_t = Z_0 r_2^t$ both satisfy this equation. Equation (4) is satisfied if the right-hand side is equal to the left-hand side when the value for Z_t is substituted into the equation. $Z_t = Z_0 r_1^t$ satisfies equation (4), thus

$$Z_0 r_1^t = (c + v)Z_0 r_1^{t-1} - vZ_0 r_1^{t-2}$$

i.e.

$$r_1^t = (c + v)r_1^{t-1} - vr_1^{t-2} \tag{6}$$

$Z_t = Z_0 r_2^t$ satisfies equation (4), thus

$$r_2^t = (c + v)r_2^{t-1} - vr_2^{t-2} \tag{7}$$

If $Z_t = ar_1^t + br_2^t$ satisfies equation (4), $(c + v)Z_{t-1} - vZ_{t-2}$ must equal $ar_1^t + br_2^t$, i.e. Z_t, when this value for Z_t is inserted. Taking $(c + v)Z_{t-1} - vZ_{t-2}$ and substituting for Z_t gives

$$(c + v)[ar_1^{t-1} + br_2^{t-1}] - v[ar_1^{t-2} + br_2^{t-2}]$$
$$= a[(c + v)r_1^{t-1} - vr_1^{t-2}] + b[(c + v)r_2^{t-1} - vr_2^{t-2}]$$
$$= ar_1^t + br_2^t$$

from equation (6) and equation (7). Consequently $Z_t = ar_1^t + br_2^t$ does satisfy equation (4).

Fitting Weights to the Solution

To solve for a and b we use the initial income levels Y_0 and Y_1, i.e.

$$Z_0 = Y_0 - Y^* \quad \text{and} \quad Z_1 = Y_1 - Y^*$$
$$Z_t = ar_1^t + br_2^t$$

thus

$$Z_0 = ar_1^0 + br_2^0 = a + b \tag{8}$$

and

$$Z_1 = ar_1^1 + br_2^1 = ar_1 + br_2 \tag{9}$$

thus $a = Z_0 - b$ from equation (8). Substituting this value for a into equation (9) gives

$$(Z_0 - b)r_1 + br_2 = Z_1$$

thus

$$b(r_2 - r_1) = Z_1 - Z_0 r_1$$

i.e.

$$b = \frac{Z_1 - Z_0 r_1}{r_2 - r_1}$$

and

$$a = \frac{Z_0 r_2 - Z_1}{r_2 - r_1}$$

from equation (8), thus

$$Z_t = \left(\frac{Z_0 r_2 - Z_1}{r_2 - r_1}\right) r_1^t + \left(\frac{Z_1 - Z_0 r_1}{r_2 - r_1}\right) r_2^t$$

This is the solution to equation (4), which satisfies the initial conditions Z_0 and Z_1, since all terms have the same time period

$$Y_t = Y^* + Z_t$$

thus

$$Y_t = \frac{C^*}{1 - c} + \left(\frac{Z_0 r_2 - Z_1}{r_2 - r_1}\right) r_1^t + \left(\frac{Z_1 - Z_0 r_1}{r_2 - r_1}\right) r_2^t$$

This is solution to equation (1). From this it is possible to find a value for actual income in any time period, e.g.

$$Y_5 = \frac{C^*}{1 - c} + \left(\frac{Z_0 r_2 - Z_1}{r_2 - r_1}\right) r_1^5 + \left(\frac{Z_1 - Z_0 r_1}{r_2 - r_1}\right) r_2^5$$

Since $Y_t = Y^* + Z_t$ then actual income $Y_t \to Y^*$ if $Z_t \to 0$ as $t \to \infty$, and the system will be stable. The deviation Z_t will tend to zero as t tends to infinity only if both roots are less than unity in absolute magnitude, i.e. if $|r_1| < 1$ and $|r_2| < 1$.

If both r_1 and r_2 are $\geqslant 0$ and < 1, the system is stable and non-oscillatory, while if either root is negative, i.e. $-1 < r_1 < 0$, $|r_2| < 1$, the system is stable but with damped oscillations.

If $r_1 > 1, 0 \leqslant r_2 < 1$ the system is unstable but non-oscillatory.

If $r_1 > 1, -1 < r_2 < 0$ the system is unstable, with damped oscillations.

If $r_1 < -1$ the system is unstable with increasing oscillations.

AUXILIARY EQUATION WITH TWO EQUAL ROOTS

If $(c + v)^2 = 4v$, then equation (5) will have two equal roots, which will be labelled r.

$$r = \frac{c + v}{2} \tag{10}$$

$Z_t = Z_0 r^t$ will therefore satisfy equation (4); and equation (4) is also satisfied by $Z_t = ar^t + br^t$. However, this latter solution does not enable us to fit a and

b to satisfy the initial conditions, e.g.

$$Z_0 = ar^0 + br^0 = a + b$$
$$Z_1 = ar^1 + br^1 = (a + b)r$$

It is impossible to solve for a and b. In this situation a function of the form

$$Z_t = ar^t + btr^t$$

will work. It can be shown that this solution will satisfy

$$Z_t = (c + v)Z_{t-1} - vZ_{t-2}$$

i.e. equation (4).

Checking the Solution

If

$$Z_t = ar^t + btr^t = (ar^2 + btr^2)r^{t-2}$$
$$Z_{t-1} = ar^{t-1} + b(t-1)r^{t-1} = (ar + btr - br)r^{t-2}$$
$$Z_{t-2} = ar^{t-2} + b(t-2)r^{t-2} = (a + bt - 2b)r^{t-2}$$

Substitute these into $0 = Z_t - (c + v)Z_{t-1} + vZ_{t-2}$ to obtain

$$0 = \{(ar^2 + btr^2) - (c + v)(ar + btr - br) + v(a + bt - 2b)\}r^{t-2}$$
$$= \{a[r^2 - (c + v)r + v] + bt[r^2 - (c + v)r + v] + b[(c + v)r - 2v]\}r^{t-2}$$

But by equations (5) and (10) the square brackets in the last expression are all zero, so $0 = 0$ and equation (4) is satisfied.

Fitting Weights to the Solution

By using $Z_t = ar^t + btr^t$ it is possible to find values for a and b which satisfy the initial conditions Z_0 and Z_1, i.e.

$$Z_0 = ar^0 + b(0)r^0 = a$$
$$Z_1 = ar^1 + b(1)r^1 = ar + br \tag{11}$$

Substituting the value for a into equation (11) gives

$$b = \frac{Z_1 - Z_0 r}{r} = \frac{Z_1}{r} - Z_0$$

thus

$$Z_t = Z_0 r^t + \left(\frac{Z_1}{r} - Z_0\right)tr^t$$

Actual income will approach the equilibrium level if $Z_t \to 0$ as $t \to \infty$, i.e. if $|r| < 1$. It is obvious that $Z_0 r^t \to 0$ as $t \to \infty$ if $|r| < 1$. To get from tr^t to $(t+1)r^{t+1}$ and $(t+2)r^{t+2}$ it is necessary to multiply tr^t by $[(t+1)/t]$. r and $[(t+2)/t]$. r^2 respectively. Since $(t+1)/t$ and $(t+2)/t$ tend to one as $t \to \infty$ and since $|r| < 1$ then in the limit $|[(t+1)/t].r| < 1$ and the absolute value of $[(t+2)/t].r^2$ is even smaller. Consequently $tr^t \to 0$ as $t \to \infty$ if $|r| < 1$.

THE CASE WHERE NO FINITE EQUILIBRIUM EXISTS

If the marginal propensity to consume is unity our second-order difference equation takes the form

$$Y_t = C^* + (1 + v)Y_{t-1} - vY_{t-2} \tag{12}$$

If a finite Y^* exists, we will have $Y_{t-2} = Y_{t-1} = Y^*$; substituting into equation (12) gives

$$Y^* = C^* + (1 + v)Y^* - vY^*$$

i.e.

$$C^* = 0$$

Clearly there is no Y^* which once attained will continue. Y^* will vary with respect to time. If the relation between Y^* and t takes the form $Y_t^* = \alpha t$ and we assume

$$Y_t = Y_t^* = \alpha t$$
$$Y_{t-1} = Y_{t-1}^* = \alpha(t - 1) \quad \text{etc.}$$

Substituting these values into equation (12) will give a value for α, i.e.

$$\alpha t = C^* + (1 + v)\alpha(t - 1) - v\alpha(t - 2)$$

thus

$$\alpha(1 - v) = C^*$$

i.e.

$$\alpha = \frac{C^*}{1 - v}$$

and

$$Y_t^* = \left(\frac{C^*}{1 - v}\right)t$$

The next problem is to find whether actual income Y_t approaches Y_t^*. If $Y_t \neq Y_t^*$ and we let Z_t equal the deviation, then

$$Z_t = Y_t - Y_t^* \quad \text{or} \quad Y_t = Y_t^* + Z_t$$

To solve for the deviation, the value of Y_t in terms of the deviation is inserted into equation (12). The procedure is exactly the same as that described earlier in this chapter.

EXERCISES

1. In a closed economy without government activity

$$C_t = 100 + 0 \cdot 9\, Y_{t-1}$$
$$I_t = 0 \cdot 4(Y_{t-1} - Y_{t-2})$$
$$Y_0 = Y_1 = 1300$$

Find Y_t and Y_4 and comment on the stability of the system.

2. Given a closed dynamic system where

$$C_t = 200 + 0.75(Y_d)_{t-1}$$
$$I_t = 0.5[(Y_d)_{t-1} - (Y_d)_{t-2}]$$
$$G_t = 0.35 Y_{t-1}$$
$$T_t = 0.4 Y_t$$
$$Y_0 = 500 \quad \text{and} \quad Y_1 = 550$$

Find Y_t and Y_2. Is this model stable?

3. In a closed dynamic system without government activity

$$C_t = 100 + 0.75 Y_{t-1}$$
$$I_t = 2.25(Y_{t-1} - Y_{t-2})$$
$$Y_0 = 200 \quad \text{and} \quad Y_1 = 250$$

Find Y_t and comment on the stability of the system.

SOLUTIONS TO EXERCISES

1. $Y_t = C_t + I_t$
$$= 100 + 0.9 Y_{t-1} + 0.4(Y_{t-1} - Y_{t-2})$$
i.e.

$$Y_t = 100 + 1.3 Y_{t-1} - 0.4 Y_{t-2} \qquad (13)$$

If there is an equilibrium level of income Y^*, then

$$Y^* = Y_t = Y_{t-1} = Y_{t-2}$$

thus

$$Y^* = 100 + 1.3 Y^* - 0.4 Y^*$$

i.e.

$$Y^* = 1000$$

Let Z_t be the deviation of Y_t from Y^*, then

$$Y_t = Y^* + Z_t = 1000 + Z_t$$

e.g.

$$Y_{t-2} = 1000 + Z_{t-2}$$

Substitution into equation (13) gives

$$Z_t = 1.3 Z_{t-1} - 0.4 Z_{t-2} \qquad (14)$$

If we let $Z_t = Z_0 r^t$, then

$$Z_0 r^t = 1.3 Z_0 r^{t-1} - 0.4 Z_0 r^{t-2}$$

Dividing through by $Z_0 r^{t-2}$ gives

$$r^2 - 1.3r + 0.4 = 0$$

thus

$$r = \frac{1.3 \pm \sqrt{[(1.3)^2 - 4(0.4)]}}{2} = \frac{1.3 \pm \sqrt{0.09}}{2}$$

$$= 0.8 \quad \text{or} \quad 0.5$$

i.e.

$$Z_t = Z_0(0.8)^t \quad \text{or} \quad Z_0(0.5)^t$$

Since $Z_t = Z_0(0.8)^t$ and $Z_t = Z_0(0.5)^t$ both satisfy equation (14), it will also be satisfied by $Z_t = a(0.8)^t + b(0.5)^t$, where a and b are constants. We have two initial values of

Z_t to determine the values of a and b.

$$Z_0 = Y_0 - Y^* = 1300 - 1000 = 300$$
$$Z_1 = Y_1 - Y^* = 1300 - 1000 = 300$$
$$Z_0 = a(0\cdot8)^0 + b(0\cdot5)^0 = 300 \quad \text{i.e.} \quad a + b = 300 \qquad (15)$$
$$Z_1 = a(0\cdot8)^1 + b(0\cdot5)^1 = 300 \quad \text{i.e.} \quad 0\cdot8a + 0\cdot5b = 300 \qquad (16)$$

Multiplying equation (15) by 0·8 and subtracting equation (16) from equation (15) gives

$$0\cdot3b = -60 \quad \text{i.e.} \quad b = -200$$

thus

$$a = 500$$

from equation (15) and

$$Z_t = 500(0\cdot8)^t - 200(0\cdot5)^t$$

so

$$Y_t = Y^* + 500(0\cdot8)^t - 200(0\cdot5)^t$$
$$= 1000 + 500(0\cdot8)^t - 200(0\cdot5)^t \qquad *$$
$$Y_4 = 1000 + 500(0\cdot8)^4 - 200(0\cdot5)^4$$
$$= 1000 + 204\cdot8 - 12\cdot5 = 1192\cdot3 \qquad *$$

Both roots are < 1, thus

$$500(0\cdot8)^t - 200(0\cdot5)^t \to 0 \quad \text{as} \quad t \to \infty$$

i.e.

$$Y_t \to Y^*$$

The system is therefore stable.

2. $Y_t = C_t + I_t + G_t$
$$= 200 + 0\cdot75(Y_d)_{t-1} + 0\cdot5[(Y_d)_{t-1} - (Y_d)_{t-2}] + 0\cdot35\,Y_{t-1}$$
$$= 200 + 0\cdot75(Y_{t-1} - 0\cdot4\,Y_{t-1}) + 0\cdot5[(Y_{t-1} - 0\cdot4\,Y_{t-1}) - (Y_{t-2} - 0\cdot4\,Y_{t-2})]$$
$$+ 0\cdot35\,Y_{t-1}$$
$$= 200 + 0\cdot45\,Y_{t-1} + 0\cdot3\,Y_{t-1} - 0\cdot3\,Y_{t-2} + 0\cdot35\,Y_{t-1}$$

i.e.

$$Y_t = 200 + 1\cdot1\,Y_{t-1} - 0\cdot3\,Y_{t-2} \qquad (17)$$

Equilibrium is given by

$$Y^* = 200 + 1\cdot1\,Y^* - 0\cdot3\,Y^*$$

thus

$$Y^* = 1000$$

Let Z_t be the deviation of Y_t from Y^*; then

$$Y_t = Y^* + Z_t = 1000 + Z_t$$

Substitution into equation (17) gives

$$Z_t = 1\cdot1Z_{t-1} - 0\cdot3Z_{t-2}$$

If we let $Z_t = Z_0 r^t$, then

$$Z_0 r^t - 1\cdot1Z_0 r^{t-1} + 0\cdot3Z_0 r^{t-2} = 0$$

Dividing through by $Z_0 r^{t-2}$ gives

$$r^2 - 1\cdot1r + 0\cdot3 = 0$$

thus

$$r = \frac{1\cdot1 \pm \sqrt{[(1\cdot1)^2 - 4(0\cdot3)]}}{2} = \frac{1\cdot1 \pm \sqrt{0\cdot01}}{2}$$
$$= 0\cdot6 \quad \text{or} \quad 0\cdot5$$

i.e.

$$Z_t = Z_0(0.6)^t \quad \text{or} \quad Z_0(0.5)^t$$

Equation (18) will also be satisfied by $Z_t = a(0.6)^t + b(0.5)^t$

$$Z_0 = Y_0 - Y^* = 500 - 1000 = -500$$
$$Z_1 = Y_1 - Y^* = 550 - 1000 = -450$$
$$Z_0 = a + b$$
$$Z_1 = 0.6a + 0.5b$$

Thus

$$a + b = -500 \tag{19}$$
$$0.6a + 0.5b = -450 \tag{20}$$

Multiplying equation (19) by 0.6 and subtracting equation (20) from it gives

$$0.1b = 150 \quad \text{or} \quad b = 1500$$

thus

$$a = -2000$$

from equation (19), and

$$Z_t = -2000(0.6)^t + 1500(0.5)^t$$

so

$$Y_t = Y^* - 2000(0.6)^t + 1500(0.5)^t \qquad *$$
$$= 1000 - 2000(0.6)^t + 1500(0.5)^t \qquad *$$
$$Y_2 = 1000 - 720 + 375 = 655 \qquad *$$

The system is stable since both roots are < 1, i.e

$$-2000(0.6)^t + 1500(0.5)^t \to 0 \quad \text{as} \quad t \to \infty$$

3. $Y_t = C_t + I_t$
$$= 100 + 0.75\,Y_{t-1} + 2.25\,Y_{t-1} - 2.25\,Y_{t-2}$$

i.e.

$$Y_t = 100 + 3\,Y_{t-1} - 2.25\,Y_{t-2} \tag{21}$$

Equilibrium is given by

$$Y^* = 100 + 3\,Y^* - 2.25\,Y^*$$

thus

$$Y^* = 400$$

Let Z_t be the deviation of Y_t from Y^*, then

$$Y_t = Y^* + Z_t = 400 + Z_t$$

and from equation (21)

$$Z_t = 3Z_{t-1} - 2.25Z_{t-2} \tag{22}$$

Let $Z_t = Z_0 r^t$; then

$$Z_0 r^t = 3Z_0 r^{t-1} - 2.25Z_0 r^{t-2}$$

and

$$r^2 - 3r + 2.25 = 0$$

thus

$$(r - 1.5)(r - 1.5) = 0$$

There are thus two identical real roots. In this case equation (22) will be satisfied by

$$Z_t = (1.5)^t + bt(1.5)^t$$
$$Z_0 = Y_0 - Y^* = 200 - 400 = -200$$
$$Z_1 = Y_1 - Y^* = 250 - 400 = -150$$

$$Z_0 = a(1 \cdot 5)^0 + b(0)(1 \cdot 5)^0 = a = -200$$
$$Z_1 = a(1 \cdot 5)^1 + b(1)(1 \cdot 5)^1 = -150$$

thus

$$(1.5)(-200) + 1 \cdot 5b = -150$$

so

$$b = 100$$

Thus

$$Z_t = -200(1 \cdot 5)^t + 100t(1 \cdot 5)^t$$

and

$$Y_t = Y^* - 200(1 \cdot 5)^t + 100t(1 \cdot 5)^t$$
$$= 400 - 200(1 \cdot 5)^t + 100t(1 \cdot 5)^t$$

$1 \cdot 5 > 1$ thus $Z_t \to \infty$ as $t \to \infty$. The system is therefore unstable.

CHAPTER 16

Complex Numbers

ECONOMIC FLUCTUATIONS

In Chapter 15 we dealt with a closed economic system without government activity. This gave us the second-order difference equation

$$Y_t = C^* + (c + v)Y_{t-1} - vY_{t-2} \tag{1}$$

where c stood for the propensity to consume and v was the accelerator. When solving this second-order difference equation it was assumed that the roots of the auxiliary equation

$$r^2 - (c + v)r + v = 0 \tag{2}$$

were real. We will now examine this same economic system under the assumption that equation (2) has no real roots.

$$r = \frac{(c + v) \pm \sqrt{[(c + v)^2 - 4v]}}{2}$$

thus there will be no real roots if $4v > (c + v)^2$, i.e. the square root of a negative number is not real. The two roots can be expressed in the form

$$r_1 = \frac{(c + v)}{2} + \frac{i\sqrt{[4v - (c + v)^2]}}{2}$$

$$r_2 = \frac{(c + v)}{2} - \frac{i\sqrt{[4v - (c + v)^2]}}{2}$$

where $i = \sqrt{-1}$ by definition i.e.

$$\sqrt{[(c + v)^2 - 4v]} = \sqrt{-1}\sqrt{[4v - (c + v)^2]} = i\sqrt{[4v - (c + v)^2]}$$
$$i = \sqrt{-1}$$

so

$$i^2 = -1$$
$$i^3 = (i^2)i = (-1)i = -i$$

and

$$i^4 = (i^3)i = -i(i) = 1$$

Since the $\sqrt{-1}$ is not real, i is called an imaginary number. i times any constant is also imaginary number. If we examine the roots of equation (2) we will see that each root is made up of a real part, $(c + v)/2$, and an imaginary part, $\pm i\sqrt{[4v - (c + v)^2]}/2$.

COMPLEX NUMBERS

These roots are therefore complex since a complex number is one with both a real part and an imaginary part. The general form of a complex number is

$$d + if$$

where d is the real part and if is the imaginary part. Both d and f stand for any real number.

Specific Example

Equation (2) becomes

$$r^2 - 2 \cdot 5r + 2 = 0$$

if we let $c = 0 \cdot 5$ and $v = 2$. This gives

$$r = \frac{2 \cdot 5 \pm \sqrt{[(2 \cdot 5)^2 - 8]}}{2}$$

$$= 1 \cdot 25 \pm \frac{\sqrt{-1 \cdot 75}}{2}$$

$\sqrt{-1 \cdot 75}$ is not real but it can be reduced to the form $\sqrt{-1}\sqrt{1 \cdot 75} = i\sqrt{1 \cdot 75}$ thus

$$r = 1 \cdot 25 \pm \frac{i\sqrt{1 \cdot 75}}{2}$$

i.e.

$$\frac{c + v}{2} = 1 \cdot 25 \quad \text{and} \quad \frac{\sqrt{[4v - (c + v)^2]}}{2} = \frac{\sqrt{1 \cdot 75}}{2}$$

Consequently

$$r_1 = 1 \cdot 25 + \frac{i\sqrt{1 \cdot 75}}{2}$$

and

$$r_2 = 1 \cdot 25 - \frac{i\sqrt{1 \cdot 75}}{2}$$

r_1 and r_2 are now in the form $d \pm if$.

In each case the real part $d = 1 \cdot 25$ and the imaginary part $if = \pm i[(\sqrt{1 \cdot 75})/2]$, so that $f = (\sqrt{1 \cdot 75})/2$. Notice that d and f are real.

Addition, Subtraction, and Multiplication of Complex Numbers

If we take any two complex numbers m and n, where $m = d + if$ and $n = g + ih$, their *sum* is a complex number, e.g.

$$m + n = d + if + g + ih = (d + g) + i(f + h)$$

$(d + g)$ is the real part and $i(f + h)$ the imaginary part. $(d + g)$ and $(f + h)$ are both real numbers, e.g. if

$$m = 2 + i\sqrt{3} \quad \text{and} \quad n = 6 + i\sqrt{7}$$

then

$$m + n = 8 + i(\sqrt{3} + \sqrt{7})$$

which is a complex number.

Subtracting two complex numbers leaves us with a complex number, e.g.

$$m - n = d + if - g - ih = (d - g) + i(f - h)$$

Multiplying two complex numbers gives a complex number, e.g.

$$mn = (d + if)(g + ih) = dg + idh + ifg + i^2fh$$
$$= (dg - fh) + i(dh + fg)$$

since $i^2 = -1$.

COMPLEX CONJUGATES

Given the complex number $m = d + if$, its complex conjugate is $d - if$ and is written $\bar{m}$. If $s = x + iy$, then $\bar{s} = x - iy$ is its complex conjugate.

The *sum* of a complex number and its complex conjugate is always real, e.g.

$$m + \bar{m} = d + if + d - if = 2d$$

The *product* of a complex number and its complex conjugate is also real, e.g.

$$m\bar{m} = (d + if)(d - if) = d^2 - idf + idf - i^2f^2$$
$$= d^2 + f^2$$

since $-i^2 = +1$.

If the product of m and n is v, i.e.

$$v = mn = (d + if)(g + ih)$$
$$= (dg - fh) + i(dh + fg)$$

then $\bar{v}$ is the product of $\bar{m}$ and $\bar{n}$, because

$$\bar{m}\bar{n} = (d - if)(g - ih)$$
$$= (dg - fh) - i(dh + fg) = \bar{v}$$

Since $v + \bar{v}$ is real, $mn + \bar{m}\bar{n}$ is always real.

Powers of Complex Conjugates

If r and $\bar{r}$ are complex conjugates, then $r.r = r^2$ and $\bar{r}.\bar{r} = (\bar{r})^2$ are complex conjugates also. This can be proved using the results of the previous section, letting $m = r$ and $n = r$. Similarly, since r^2 and $(\bar{r})^2$ are complex conjugates, $r(r^2) = r^3$ and $\bar{r}(\bar{r})^2 = (\bar{r})^3$ are complex conjugates, using the same type of proof, i.e. letting $m = r$ and $n = r^2$. By the same method it can be shown that for any value of t, r^t and $(\bar{r})^t$ are complex conjugates provided we know that r^{t-1} and $(\bar{r})^{t-1}$ are complex conjugates, since $r(r^{t-1}) = r^t$ and $\bar{r}(\bar{r})^{t-1} = (\bar{r})^t$

are complex conjugates by the same proof. This proof shows that if the property holds for $t - 1$, it holds for t; we have already shown that it holds for $t = 2$, so it holds also for all $t \geqslant 2$. This type of proof is known as mathematical induction.

Division of Complex Numbers

Given any two complex numbers their quotient is also a complex number, assuming the denominator is $\neq 0$. If we take m and n and assume $n \neq 0$, then

$$\frac{m}{n} = \frac{d + if}{g + ih}$$

Multiplying above and below by $\bar{n}$ gives

$$\frac{m}{n} = \frac{m\bar{n}}{n\bar{n}} = \frac{d + if}{g + ih} \cdot \frac{g - ih}{g - ih} = \frac{dg - idh + ifg - i^2 fh}{g^2 + h^2}$$

$$= \frac{dg + fh}{g^2 + h^2} + i\frac{(fg - dh)}{g^2 + h^2}$$

which is a complex number.

THE COMPLEX ROOT PROBLEM

When solving equation (1), in Chapter 15, we assumed that actual income Y_t was not equal to the equilibrium level of income Y^* and we let the deviation in period t equal Z_t. To study the time path of income it was then necessary to solve for Z_t since

$$Y_t = Y^* + Z_t$$

By substituting this value for Y_t into equation (1) we obtained an equation entirely in the deviation, i.e.

$$Z_t = (c + v)Z_{t-1} - vZ_{t-2} \tag{3}$$

To solve for Z_t we let it equal $Z_0 r^t$. We now find that $Z_t = Z_0 r^t$ will satisfy equation (3) when the two roots are complex. If we let $(c + v)/2 = d$ and $\sqrt{[4v - (c + v)^2]}/2 = f$, then

$$r_1 = d + if \quad \text{and} \quad r_2 = d - if$$

Each root is the complex conjugate of the other. If we let $r_1 = r$, then $r_2 = \bar{r}$. Equation (3) is therefore satisfied by

$$Z_t = Z_0 r^t \quad \text{and} \quad Z_t = Z_0 (\bar{r})^t$$

or

$$Z_t = Z_0 (d + if)^t \quad \text{and} \quad Z_t = Z_0 (d - if)^t$$

Since $Z_t = Z_0 r^t$ and $Z_t = Z_0 (\bar{r})^t$ both satisfy equation (3) it will also be satisfied by

$$Z_t = ar^t + b(\bar{r})^t$$

Using the method described in Chapter 15, the reader should check that this is so. As in Chapter 15, a and b are to determined from the information on Y_0 and Y_1, which is necessary for the solution of a second-order difference equation.

Z_t is the deviation of Y_t from Y^* and must therefore be real. As shown in the section on complex conjugates on p. 261, if a and b are complex conjugates and r^t and $(\bar{r})^t$ are complex conjugates, then ar^t and $b(\bar{r})^t$ are complex conjugates, and Z_t is always real. $b = \bar{a}$ so

$$Z_t = ar^t + \bar{a}(\bar{r})^t$$

This is real for all $t \geqslant 1$.

Solving for a and b

If we let $a = g + ih$, then b must equal $g - ih$ if Z_t is to be real, i.e.

$$Z_t = (g + ih)(d + if)^t + (g - ih)(d - if)^t$$
$$Z_0 = (g + ih)(d + if)^0 + (g - ih)(d - if)^0$$
$$= g + ih + g - ih = 2g$$

thus

$$g = \frac{Z_0}{2}$$

Information on Z_0 is available since $Z_0 = Y_0 - Y^*$

$$Z_1 = (g + ih)(d + if) + (g - ih)(d - if)$$
$$= dg + ifg + idh + i^2fh + dg - ifg - idh + i^2fh$$
$$= 2dg - 2fh$$

since $i^2fh = -fh$

$$= 2(dg - fh)$$

thus

$$Z_1 = dZ_0 - 2fh$$

since $2g = Z_0$ and

$$h = \frac{dZ_0 - Z_1}{2f}$$

Information on Z_1 is available and both d and f depend upon c and v. We have now solved for a and b, i.e.

$$a = \frac{Z_0}{2} + i\left(\frac{dZ_0 - Z_1}{2f}\right)$$

and

$$b = \frac{Z_0}{2} - i\left(\frac{dZ_0 - Z_1}{2f}\right)$$

SPECIFIC EXAMPLE

If we let $C^* = 10$, $c = 0.95$, $v = 1.05$, $Y_0 = 250$, and $Y_1 = 300$, then equation (1) will take the form

$$Y_t = 10 + 2Y_{t-1} - 1 \cdot 05 Y_{t-2}$$

Equilibrium is given by

$$Y^* = \frac{10}{0 \cdot 05} = 200$$

Equation (3) will take the form

$$Z_t = 2Z_{t-1} - 1 \cdot 05 Z_{t-2}$$

and the auxiliary equation becomes

$$r^2 - 2r + 1 \cdot 05 = 0$$

The auxiliary equation is satisfied by $r = [2 \pm \sqrt{(4 - 4 \cdot 2)}]/2$ thus

$$r = d + if = 1 + i\sqrt{0 \cdot 05}$$

and $\bar{r} = 1 - i\sqrt{0 \cdot 05}$, i.e. $d = 1$ and $f = (\sqrt{0 \cdot 2})/2 = \sqrt{0 \cdot 05}$.

Solving for the Deviation

Each root is the complex conjugate of the other. Since

$$Z_t = Z_0(1 + i\sqrt{0 \cdot 05})^t \quad \text{and} \quad Z_t = Z_0(1 - i\sqrt{0 \cdot 05})^t$$

both satisfy the above equation in Z_t it will also be satisfied by

$$Z_t = a(1 + i\sqrt{0 \cdot 05})^t + b(1 - i\sqrt{0 \cdot 05})^t$$

So that Z_t shall be real for all t, a and b must be complex conjugates. Let $a = g + ih$ and $b = \bar{a} = g - ih$, i.e.

$$Z_t = (g + ih)(1 + i\sqrt{0 \cdot 05})^t + (g - ih)(1 - i\sqrt{0 \cdot 05})^t$$

We must now find values for g and h which enable us to determine the weights a and b. For this we use Y_0 and Y_1.

$$Z_0 = Y_0 - Y^* = 50$$

and

$$Z_1 = Y_1 - Y^* = 100$$
$$Z_0 = (g + ih) + (g - ih) = 2g = 50$$

so

$$g = 25$$
$$
\begin{aligned}
Z_1 &= (g + ih)(1 + i\sqrt{0 \cdot 05}) + (g - ih)(1 - i\sqrt{0 \cdot 05}) \\
&= g + ig\sqrt{0 \cdot 05} + ih - h\sqrt{0 \cdot 05} + g - ig\sqrt{0 \cdot 05} - ih - h\sqrt{0 \cdot 05} \\
&= 2g - 2h\sqrt{0 \cdot 05} \\
&= 2g - h\sqrt{0 \cdot 2}
\end{aligned}
$$

i.e.

$$100 = 50 - h\sqrt{0 \cdot 2}$$

thus

$$h = -50/\sqrt{0 \cdot 2} = -25\sqrt{20}$$

i.e.

$$a = 25(1 - i\sqrt{20}) \quad \text{and} \quad b = \bar{a} = 25(1 + i\sqrt{20})$$

so
$$Z_t = 25[(1 - i\sqrt{20})(1 + i\sqrt{0\cdot05})^t + (1 + i\sqrt{20})(1 - i\sqrt{0\cdot05})^t]$$

The problem is to find how Z_t behaves with respect to time. As it stands there is little one can say. Consequently, to solve for the deviation we need to make use of some elementary trigonometry.

TRIGONOMETRIC FUNCTIONS

Consider a right-angle triangle ABC, see Figure 43. If the length of AB is x and the length of BC is y, then $\sqrt{(x^2 + y^2)}$ gives the length of the hypotenuse AC.

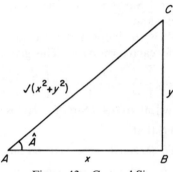

Figure 43 Cos and Sin
of angle A

The Sine of the angle A, written Sin A, is the ratio of the opposite side to the hypotenuse,
$$\text{Sin } A = \frac{BC}{AC} = \frac{y}{\sqrt{(x^2 + y^2)}}$$

The Cosine of the angle A, written Cos A, is the ratio of the adjacent side to the hypotenuse,
$$\text{Cos } A = \frac{AB}{AC} = \frac{x}{\sqrt{(x^2 + y^2)}}$$

Clearly
$$(\text{Cos } A)^2 + (\text{Sin } A)^2 = \frac{x^2}{x^2 + y^2} + \frac{y^2}{x^2 + y^2} = 1$$

If we have any two numbers α and β such that $\alpha^2 + \beta^2 = 1$, we can find an angle θ such that Cos $\theta = \alpha$ and Sin $\theta = \beta$. The numbers $m/\sqrt{(m^2 + n^2)}$ and $n/\sqrt{(m^2 + n^2)}$ are such that their squares sum to one, so we can always find an angle θ such that Cos $\theta = m/\sqrt{(m^2 + n^2)}$ and Sin $\theta = n/\sqrt{(m^2 + n^2)}$.

Cos and Sin of any Sum of Two Angles

If we take any two angles A and B then
$$\text{Sin }(A + B) = \text{Sin } A \text{ Cos } B + \text{Cos } A \text{ Sin } B$$

and
$$\text{Cos}\,(A+B) = \text{Cos}\,A\,\text{Cos}\,B - \text{Sin}\,A\,\text{Sin}\,B$$

For a proof of this see Appendix 3.

APPLICATION OF TRIGONOMETRIC FUNCTIONS

Since the general expressions for a and b derived above are rather awkward to handle, we will continue to use a and b when expressing the deviation, i.e.
$$Z_t = a(d+if)^t + b(d-if)^t$$

$w = \sqrt{(d^2+f^2)}$ gives the length of the hypotenuse if d and f are the lengths of the other two sides of a right-angled triangle. We can find an angle θ such that $\text{Cos}\,\theta = d/w$ and $\text{Sin}\,\theta = f/w$. The value of d/w and thus of θ depends on the values of c and v.

The expression for Z_t is unchanged if we divide the contents of each bracket by w and multiply each term by w^t. This give us

$$Z_t = aw^t\left[\frac{d}{w}+i\frac{f}{w}\right]^t + bw^t\left[\frac{d}{w}-i\frac{f}{w}\right]^t$$
$$= w^t[a(\text{Cos}\,\theta + i\,\text{Sin}\,\theta)^t + b(\text{Cos}\,\theta - i\,\text{Sin}\,\theta)^t]$$

It can be shown, however, that
$$(\text{Cos}\,\theta + i\,\text{Sin}\,\theta)^t = (\text{Cos}\,t\theta + i\,\text{Sin}\,t\theta)$$

and
$$(\text{Cos}\,\theta - i\,\text{Sin}\,\theta)^t = (\text{Cos}\,t\theta - i\,\text{Sin}\,t\theta)$$

This is known as de Moivre's Theorem; this is proved in Appendix 4. It enables us to express the deviation Z_t in the form
$$Z_t = w^t[a(\text{Cos}\,t\theta + i\,\text{Sin}\,t\theta) + b(\text{Cos}\,t\theta - i\,\text{Sin}\,t\theta)]$$

If c and v are constant, θ will be constant, but the angle $t\theta$ will increase with t. $a = g+ih$ and $b = g-ih$, thus
$$Z_t = w^t[(g+ih)(\text{Cos}\,t\theta + i\,\text{Sin}\,t\theta) + (g-ih)(\text{Cos}\,t\theta - i\,\text{Sin}\,t\theta)]$$
$$= w^t[g\,\text{Cos}\,t\theta + ig\,\text{Sin}\,t\theta + ih\,\text{Cos}\,t\theta + i^2h\,\text{Sin}\,t\theta$$
$$+ g\,\text{Cos}\,t\theta - ig\,\text{Sin}\,t\theta - ih\,\text{Cos}\,t\theta + i^2h\,\text{Sin}\,t\theta]$$
$$= w^t[2g\,\text{Cos}\,t\theta - 2h\,\text{Sin}\,t\theta]$$
$$= 2w^t[g\,\text{Cos}\,t\theta - h\,\text{Sin}\,t\theta]$$

Let $\sqrt{(g^2+h^2)} = k$. Let g/k be the Cosine and h/k the Sine of an angle ϕ. As both g and h are real and take definite values, see the section on 'solving for a and b', it is possible to find the angle ϕ.

Divide through the expression for Z_t by k and take it outside the brackets. Thus

$$Z_t = 2w^t k\left[\frac{g}{k}\,\text{Cos}\,t\theta - \frac{h}{k}\,\text{Sin}\,t\theta\right]$$
$$= 2w^t k[\text{Cos}\,\phi\,\text{Cos}\,t\theta - \text{Sin}\,\phi\,\text{Sin}\,t\theta]$$
$$= 2w^t k[\text{Cos}\,(\phi+t\theta)]$$

see rule for the Cosine of a sum of two angles, above.

CYCLICAL BEHAVIOUR OF THE COSINE

The angle $(\phi + t\theta)$ will vary with t since ϕ and θ are fixed. θ depends on c and v, and ϕ depends on Z_0, Z_1, c, and v, all of which are constant. If we assume that $\phi = 30°$ and $\theta = 20°$, then $(\phi + t\theta)$ is given by

$$30° + 1(20°) = 50° \quad \text{when } t = 1$$
$$30° + 2(20°) = 70° \quad \text{when } t = 2$$

$$\cdot \qquad \cdot \qquad \cdot$$
$$\cdot \qquad \cdot \qquad \cdot$$
$$\cdot \qquad \cdot \qquad \cdot$$

$$30° + 10(20°) = 230° \quad \text{when } t = 10$$

$$\cdot \qquad \cdot \qquad \cdot$$
$$\cdot \qquad \cdot \qquad \cdot$$
$$\cdot \qquad \cdot \qquad \cdot$$

$$30° + 20(20°) = 430° \quad \text{when } t = 20 \text{ etc.}$$

Now the Cosine of any angle varies between $+1$ and -1 every $360°$, see Figure 44.

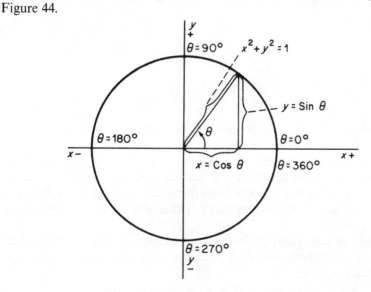

Figure 44 Cyclical behaviour of the Cosine

Cos $0° = +1$	Cos $360° = +1$	Cos $720° = +1$
Cos $90° = 0$	Cos $450° = 0$	etc.
Cos $180° = -1$	Cos $540° = -1$	
Cos $270° = 0$	Cos $630° = 0$	

Consequently Cos $(\phi + t\theta)$ will vary from $+1$ to -1 every $360°$. This oscillation between constant limits will be due to changes in t, i.e. the angle will increase as t increases. However, this is but part of the deviation.

STABILITY OF THE SYSTEM

$$Z_t = 2w^t\sqrt{(g^2 + h^2)}[\text{Cos}\,(\phi + t\theta)]$$

Clearly the value of $2\sqrt{(g^2 + h^2)}$ will not vary with t so that this will not affect the stability of the system. The amplitude of the oscillations will depend upon w^t. Since w is raised to the power t it will determine whether the system will have increasing, decreasing, or constant oscillations.

If $w < 1$ then $w^t \to 0$ as $t \to \infty$ and $Z_t \to 0$ even though $\text{Cos}\,(\phi + t\theta)$ oscillates between constant limits, the length of the cycle depending on the value of θ. In this case the oscillations fade away due to w^t and the whole system is stable, i.e. $Y_t \to Y^*$.

If $w > 1$ then $w^t \to \infty$ as $t \to \infty$ and $Z_t \to \pm \infty$. The system will have ever increasing oscillations as w^t increases with t.

If $w = 1$ then $w^t = 1$, for all t, and the $\text{Cos}\,(\phi + t\theta)$ part of the deviation takes over, giving perpetual oscillations of constant amplitude.

Let us now examine w. If $w < 1$ then $w^2 < 1$, if $w > 1$ then $w^2 > 1$; and if $w = 1$ then $w^2 = 1$.

$$w = \sqrt{(d^2 + f^2)}$$

thus

$$w^2 = d^2 + f^2$$

$$= \left(\frac{c+v}{2}\right)^2 + \left(\frac{\sqrt{[4v - (c+v)^2]}}{2}\right)^2$$

$$= \frac{(c+v)^2}{4} + \frac{4v - (c+v)^2}{4} = v$$

where v is the accelerator and the constant term of the auxiliary equation (2). Thus if $v < 1$ the system has oscillations which will fade away; if $v > 1$ the system has increasing oscillations; and if $v = 1$ the system has perpetual oscillations of constant amplitude. In summary we can say that complex roots always give rise to oscillations but these may be damped, explosive, or of constant amplitude depending on the value of the constant term of the auxiliary equation.

For a further discussion of the relation between the parameters of the auxiliary equation of a second-order difference equation and the stability of the system, see Appendix 5.

EXERCISES

1. Given a closed dynamic system in which

$$C_t = 200 + 0{\cdot}8(Y_d)_{t-1}$$
$$I_t = 0{\cdot}8(Y_{t-1} - Y_{t-2})$$
$$T_t = 0{\cdot}25\,Y_t$$
$$G_t = 0{\cdot}2\,Y_{t-1}$$
$$Y_0 = 880 \quad \text{and} \quad Y_1 = 968$$

Find Y_t and Y_4 and comment on the time path of income.

2. Given a closed economy without government activity in which
$$C_t = 100 + 0.8\, Y_{t-1}$$
$$I_t = 2(Y_{t-1} - Y_{t-2})$$
$$Y_0 = 600 \quad \text{and} \quad Y_1 = 592$$

Find Y_t and Y_3 and comment on the time path of income.

SOLUTIONS TO EXERCISES

1. $Y_t = C_t + I_t + G_t$
$$= 200 + 0.8(Y_{t-1} - 0.25\, Y_{t-1}) + 0.8(Y_{t-1} - Y_{t-2}) + 0.2\, Y_{t-1}$$
$$= 200 + 1.6\, Y_{t-1} - 0.8\, Y_{t-2} \tag{4}$$

Equilibrium is given by
$$Y^* = \frac{200}{1 - 0.8} = 1000$$

Let Z_t be the deviation of Y_t from Y^*, then
$$Y_t = Y^* + Z_t = 1000 + Z_t$$

thus equation (4) gives
$$Z_t = 1.6\, Z_{t-1} - 0.8\, Z_{t-2} \tag{5}$$

If we let $Z_t = Z_0 r^t$ then
$$r^2 - 1.6r + 0.8 = 0$$

thus
$$r = \frac{1.6 \pm \sqrt{(2.56 - 3.2)}}{2} = \frac{1.6 \pm \sqrt{-0.64}}{2}$$

i.e.
$$r = 0.8 + i(0.4) \quad \text{or} \quad 0.8 - i(0.4)$$

We now have two complex roots, each the complex conjugate of the other. Since
$$Z_t = Z_0[0.8 + i(0.4)]^t \quad \text{and} \quad Z_t = Z_0[0.8 - i(0.4)]^t$$
both satisfy equation (5) thus it will also be satisfied by
$$Z_t = a[0.8 + i(0.4)]^t + b[0.8 - i(0.4)]^t$$

$\sqrt{[(0.8)^2 + (0.4)^2]} = \sqrt{0.8}$ gives the length of the hypotenuse if the lengths of the other sides of a right-angled triangle are 0.8 and 0.4. Consequently we let θ be an angle such that
$$\text{Cos}\,\theta = \frac{0.8}{\sqrt{0.8}} \quad \text{and} \quad \text{Sin}\,\theta = \frac{0.4}{\sqrt{0.8}}$$

The angle θ whose Sine and Cosine take the above values is $26.5651°$ approx.
$$Z_t = a(\sqrt{0.8})^t\left[\frac{0.8}{\sqrt{0.8}} + i\frac{0.4}{\sqrt{0.8}}\right]^t + b(\sqrt{0.8})^t\left[\frac{0.8}{\sqrt{0.8}} - i\frac{0.4}{\sqrt{0.8}}\right]^t$$

thus
$$Z_t = (\sqrt{0.8})^t[a(\text{Cos}\,\theta + i\,\text{Sin}\,\theta)^t + b(\text{Cos}\,\theta - i\,\text{Sin}\,\theta)^t]$$

thus
$$Z_t = (\sqrt{0.8})^t[a(\text{Cos}\,t\theta + i\,\text{Sin}\,t\theta) + b(\text{Cos}\,t\theta - i\,\text{Sin}\,t\theta)] \tag{6}$$

To find values for a and b let us return to
$$Z_t = a[0.8 + i(0.4)]^t + b[0.8 - i(0.4)]^t$$

Since Z_t is the deviation from an economic variable income then it will be real.

Z_t will be real if a and b are complex conjugates, i.e.
$$a = d + if \quad \text{and} \quad b = d - if$$
where d and f are real numbers.

This gives
$$Z_t = (d + if)[0\cdot8 + i(0\cdot4)]^t + (d - if)[0\cdot8 - i(0\cdot4)]^t$$

To find d and f we will use the information on Y_0 and Y_1.
$$Z_0 = Y_0 - Y^* = 880 - 1000 = -120$$
$$Z_1 = Y_1 - Y^* = 968 - 1000 = -32$$
$$Z_0 = d + if + d - if = 2d$$

i.e.
$$2d = -120$$

thus
$$d = -60$$
$$Z_1 = (0\cdot8)d + i(0\cdot4)d + i(0\cdot8)f - 0\cdot4f + (0\cdot8)d - i(0\cdot4)d$$
$$\quad - i(0\cdot8)f - 0\cdot4f$$

i.e.
$$Z_1 = 1\cdot6d - 0\cdot8f$$

or
$$-32 = 1\cdot6(-60) - 0\cdot8f$$

thus
$$f = -80$$

so
$$a = -60 + i(-80) \quad \text{and} \quad b = -60 - i(-80)$$

Inserting these values into (6) gives
$$Z_t = (\sqrt{0\cdot8})^t[(-60 + i(-80))\,(\text{Cos } t\theta + i\,\text{Sin } t\,\theta) + (-60 - i(-80))(\text{Cos } t\theta - i\,\text{Sin } t\,\theta)]$$
$$\quad = (\sqrt{0\cdot8})^t[-120\,\text{Cos } t\theta + 160\,\text{Sin } t\theta]$$

Thus
$$Y_t = Y^* + Z_t$$
$$\quad = 1000 + 0\cdot8^{t/2}[-120\,\text{Cos } t\theta + 160\,\text{Sin } t\theta] \qquad\qquad *$$

These oscillations will fade away as $t \to \infty$ since Cos $t\theta$ and Sin $t\theta$ will oscillate between constant limits while $0\cdot8^{t/2} \to 0$ as $t \to \infty$, so that $Z_t \to 0$ and $Y_t \to 1000$.

When $t = 4$, $t\theta = 4(26\cdot5651°) = 106\cdot2602°$
$$\text{Cos } 4\theta = -0\cdot28 \quad \text{and} \quad \text{Sin } 4\theta = 0\cdot96$$

thus
$$Y_4 = 1000 + 0\cdot64[-120(-0\cdot28) + 160(0\cdot96)]$$
$$\quad = 1000 + 0\cdot64(187\cdot2)$$
$$\quad = 1119\cdot81 \qquad\qquad *$$

We can check that this gives the same answer as can be obtained be obtained by repeated application of equation (4).
$$Y_t = 200 + 1\cdot6Y_{t-1} - 0\cdot8Y_{t-2}$$
$$Y_0 = 880, \quad Y_1 = 968$$

thus
$$Y_2 = 200 + 1\cdot6(968) - 0\cdot8(880) = 1044\cdot8$$
$$Y_3 = 200 + 1\cdot6(1044\cdot8) - 0\cdot8(968) = 1097\cdot28$$
$$Y_4 = 200 + 1\cdot6(1097\cdot28) - 0\cdot8(1044\cdot8) = 1119\cdot81$$

2.　$Y_t = 100 + 0\cdot8\,Y_{t-1} + 2(Y_{t-1} - Y_{t-2})$
$$\quad = 100 + 2\cdot8\,Y_{t-1} - 2Y_{t-2} \qquad\qquad (7)$$

Equilibrium is given by $Y^* = 500$. Let Z_t be the deviation of Y_t from Y^*, then

$$Y_t = 500 + Z_t$$

thus equation (7) gives

$$Z_t = 2·8Z_{t-1} - 2Z_{t-2}$$

If we let $Z_t = Z_0 r^t$, then

$$r^2 - 2·8r + 2 = 0$$

i.e.

$$r = \frac{2·8 \pm \sqrt{(7·84 - 8)}}{2} = 1·4 \pm i(0·2)$$

We have two roots, each the complex conjugate of the other. Thus

$$Z_t = a[1·4 + i(0·2)]^t + b[1·4 - i(0·2)]^t \qquad (8)$$

Let $1·4/\sqrt{[(1·4)^2 + (0·2)^2]}$ or $1·4/\sqrt{2} = \text{Cos } A$ and $0·2/\sqrt{2} = \text{Sin } A$. The angle A whose Sine and Cosine take these values is $8·1301°$ approx.

$$Z_t = (\sqrt{2})^t \left[a\left(\frac{1·4}{\sqrt{2}} + i\frac{0·2}{\sqrt{2}}\right)^t + b\left(\frac{1·4}{\sqrt{2}} - i\frac{0·2}{\sqrt{2}}\right)^t \right]$$
$$= (\sqrt{2})^t [a(\text{Cos } A + i \text{ Sin } A)^t + b(\text{Cos } A - i \text{ Sin } A)^t]$$
$$= (\sqrt{2})^t [a(\text{Cos } tA + i \text{ Sin } tA) + b(\text{Cos } tA - i \text{ Sin } tA)] \qquad (9)$$

To find values for a and b we return to equation (8). Z_t will be real if $a = d + if$ and $b = d - if$. This gives

$$Z_t = (d + if)[1·4 + i(0·2)]^t + (d - if)[1·4 - i(0·2)]^t$$
$$Z_0 = 600 - 500 = 100 \quad \text{and} \quad Z_1 = 592 - 500 = 92$$
$$Z_0 = 2d = 100 \quad \text{i.e.} \quad d = 50$$

and

$$Z_1 = 2·8d - 0·4f = 92$$

so

$$140 - 0·4f = 92 \quad \text{or} \quad f = 120$$

Thus

$$a = 50 + i(120) \quad \text{and} \quad b = 50 - i(120)$$

Inserting these values into equation (9) gives

$$Z_t = (\sqrt{2})^t [(50 + i(120))(\text{Cos } tA + i \text{ Sin } tA) + (50 - i(120))(\text{Cos } tA - i \text{ Sin } tA)]$$
$$= (\sqrt{2})^t [100 \text{ Cos } tA - 240 \text{ Sin } tA]$$

Thus

$$Y_t = 500 + Z_t$$
$$= 500 + 2^{t/2} [100 \text{ Cos } tA - 240 \text{ Sin } tA] \qquad \text{*}$$

As $2^{t/2} \to \infty$ as $t \to \infty$ while Cos tA and Sin tA oscillate between fixed limits, the system will have ever increasing oscillations and must soon break down.

When $t = 3$, $3A = 24·3903°$, and Cos $3A = 0·9108$ and Sin $3A = 0·4130$.

$$Y_3 = 500 + 2^{1·5} [100(0·9108) - 240(0·4130)]$$
$$= 477·28$$

This can be checked by repeated application of equation (7). $\qquad$ *

$$Y_t = 100 + 2·8Y_{t-1} - 2Y_{t-2}$$
$$Y_0 = 600 \quad \text{and} \quad Y_1 = 592$$
$$Y_2 = 100 + 2·8(592) - 2(600) = 557·6$$
$$Y_3 = 100 + 2·8(557·6) - 2(592) = 477·28$$

CHAPTER 17

Difference Equations in Two Variables

This chapter will show how the technique of second-order difference equations can be used to analyse models in which two variables are each single lagged functions of themselves and of each other.

A FOREIGN TRADE MULTIPLIER MODEL

Suppose we take two countries and assume that Y is income in country 1 and R is income in country 2. Taking country 1 we have

$$Y_t = C_t + I_t + G_t + X_t - M_t$$

Assume that

$$C_t = C^* + cY_{t-1}$$
$$I_t = I^*$$
$$G_t = G^*$$

and

$$M_t = M^* + \mu Y_{t-1}$$

where $\mu > 0$ is the propensity to import and C^*, I^*, G^*, and M^* are the autonomous parts of the relevant functions.

If country 1's exports X_t are a function of country 2's income lagged one period then

$$X_t = aR_{t-1}$$

where $a > 0$. This gives

$$Y_t = C^* + I^* + G^* - M^* + (c - \mu)Y_{t-1} + aR_{t-1}$$

If we let

$$k = C^* + I^* + G^* - M^*$$

and

$$b = c - \mu$$

then

$$Y_t = k + bY_{t-1} + aR_{t-1} \tag{1}$$

This is a first-order linear difference equation in two variables, Y and R, where Y is determined by a single lag as a function of R and Y.

Taking country 2 we have

$$R_t = C'_t + I'_t + G'_t + X'_t - M'_t$$

Since country 2's exports are the same as country 1's imports we have

$$X'_t = M_t = M^* + \mu Y_{t-1}$$

and the functions are as above, then the identity can be reduced to the form

$$R_t = l + gR_{t-1} + \mu Y_{t-1} \tag{2}$$

where l stands for the autonomous components of the functions, seek k above, and $g = c' - a$, similar to b above.

We now have two first-order difference equations, i.e. equations (1) and (2), in two variables where each is a single lagged function of itself and of the other. It is possible to examine the time paths of Y and R using the same methods as those employed to deal with difference equations in a single variable.

EQUILIBRIUM INCOME LEVELS

If there is an equilibrium income level in each country, these levels are such that if both are attained simultaneously they will both be repeated, i.e. if

$$Y_{t-1} = Y^* \quad \text{and} \quad R_{t-1} = R^*$$

then

$$Y_t = Y^* \quad \text{and} \quad R_t = R^*$$

Taking equation (1) this implies

$$Y^* = k + bY^* + aR^* \tag{3}$$

Taking equation (2) it implies

$$R^* = l + gR^* + \mu Y^* \tag{4}$$

Equations (3) and (4) can now be solved using any of the methods described in Chapter 3.

If we let $k = 400$, $l = 200$, $a = 0.1$, $b = 0.8$, $\mu = 0.1$ and $g = 0.7$, equation (3) becomes

$$Y^* = 400 + 0.8 Y^* + 0.1 R^*$$

or

$$0.2 Y^* - 0.1 R^* = 400 \tag{5}$$

Equation(4) becomes

$$R^* = 200 + 0.7 R^* + 0.1 Y^*$$

or

$$-0.1 Y^* + 0.3 R^* = 200 \tag{6}$$

Multiplying equation (6) by 2 and adding to equation (5) gives

$$0.5 R^* = 800 \quad \text{or} \quad R^* = 1600$$

Inserting this value into equation (5) gives

$$0.2 Y^* = 560 \quad \text{or} \quad Y^* = 2800$$

SOLVING FOR THE DEVIATIONS

Once the equilibrium is found the problem is to discover whether actual incomes, i.e. Y_t and R_t, approach it, and if so how fast. If $Y_t \neq Y^*$ and we let m_t equal the deviation then

$$m_t = Y_t - Y^* \quad \text{or} \quad Y_t = Y^* + m_t$$

If $R_t \neq R^*$ and we let n_t equal the deviation, then

$$n_t = R_t - R^* \quad \text{or} \quad R_t = R^* + n_t$$

To find the time path of Y_t and R_t it is necessary to solve for m_t and n_t since Y^* and R^* are constant with respect to time. Substituting the value for Y_t in terms of m_t and R_t in terms of n_t into equation (1) gives

$$Y^* + m_t = k + b(Y^* + m_{t-1}) + a(R^* + n_{t-1}) \tag{7}$$

Subtracting equation (3) from equation (7) gives

$$m_t = bm_{t-1} + an_{t-1} \tag{8}$$

Substituting the same values into equation (2) gives

$$R^* + n_t = l + g(R^* + n_{t-1}) + \mu(Y^* + m_{t-1}) \tag{9}$$

Subtracting equation (4) from equation (9) gives

$$n_t = gn_{t-1} + \mu m_{t-1} \tag{10}$$

From equations (8) and (10) we can obtain a second-order difference equation in one variable, i.e. either in n or m.

Second-order Difference Equation in m

$$n_{t-1} = \frac{1}{a}m_t - \frac{b}{a}m_{t-1}$$

from equation (8). This holds for all t, e.g.

$$n_4 = \frac{1}{a}m_5 - \frac{b}{a}m_4$$

or

$$n_t = \frac{1}{a}m_{t+1} - \frac{b}{a}m_t$$

Inserting these values for n_t and n_{t-1} into equation (10) gives an equation in one variable, m, in three time periods,

$$\frac{1}{a}m_{t+1} - \frac{b}{a}m_t = g\left[\frac{1}{a}m_t - \frac{b}{a}m_{t-1}\right] + \mu m_{t-1}$$

$$= \frac{g}{a}m_t - \frac{bg}{a}m_{t-1} + \mu m_{t-1}$$

thus

$$\frac{1}{a}m_{t+1} - \left(\frac{b+g}{a}\right)m_t + \left(\frac{bg - a\mu}{a}\right)m_{t-1} = 0$$

Multiplying through by a gives

$$m_{t+1} - (b+g)m_t + (bg - a\mu)m_{t-1} = 0$$

This is true for all t, e.g.

$$m_6 - (b+g)m_5 + (bg - a\mu)m_4 = 0$$

thus

$$m_t - (b+g)m_{t-1} + (bg - a\mu)m_{t-2} = 0$$

This is a second-order difference equation in one variable which can be solved by letting $m_t = m_0 r^t$. If $m_t = m_0 r^t$, then

$$m_0 r^t - (b+g)m_0 r^{t-1} + (bg - a\mu)m_0 r^{t-2} = 0$$

Dividing through by $m_0 r^{t-2}$ gives the auxiliary equation

$$r^2 - (b+g)r + bg - a\mu = 0 \qquad (11)$$

If $(b+g)^2 > 4(bg - a\mu)$ this equation will have two distinct real roots. As a and μ are both positive this must be the case in the present example. The deviation will take the form

$$m_t = \alpha r_1^t + \beta r_2^t$$

where α and β are weights to be determined. To find α and β it is necessary to know the value of income in an initial period in both countries, say Y_0 and R_0. From this it is possible to find m_0 and n_0. With this information on m_0 and n_0 it is possible to find m_1 and n_1.

from equation (8) and

$$m_1 = bm_0 + an_0$$

$$n_1 = gn_0 + \mu m_0$$

from equation (10). In this case information on m_0 and m_1 is all that is required to solve for α and β.

The deviation $m_t \to 0$ as $t \to \infty$ only if both roots are less than unity in absolute magnitude. When this is so $Y_t \to Y^*$ or the time path of income is stable in country 1. One negative root with absolute value > 1 will give rise to increasing oscillations and render the system unstable, see Chapter 15, section on auxiliary equations with two distinct real roots.

If $(b+g)^2 = 4(bg - a\mu)$ equation (11) will have two equal roots which will be labelled r, see relevant section Chapter 15. Actual income in country 1 will approach the equilibrium level Y^* if $|r| < 1$.

If $(b+g)^2 < 4(bg - a\mu)$ equation (11) will have complex roots, see Chapter 16. This will give rise to oscillations in Y_t. If $bg - a\mu = 1$ the system will have perpetual oscillations of constant amplitude. If $bg - a\mu < 1$ the oscillations will fade away and m_t will tend to zero as $t \to \infty$, i.e. $Y_t \to Y^*$. If $bg - a\mu > 1$ the system will have increasing oscillations so that actual income in country 1 will never approach the equilibrium level Y^*.

From equation (8) we can see that if $m_t \to 0$ as $t \to \infty$, $bm_{t-1} \to 0$ and $an_{t-1} \to 0$. As a is a constant, $n_{t-1} \to 0$, i.e. $n_t \to 0$ as $t \to \infty$ and $R_t \to R^*$.

Alternatively we can examine the time path of income in country 2 by solving for the deviation n_t. From equations (8) and (10) it is possible to obtain a second-order difference equation in n. This will give the exact same auxiliary equation as the one considered above, i.e. equation (11), and hence the same roots.

Second-order Difference Equation in n

$$m_{t-1} = \frac{1}{\mu}n_t - \frac{g}{\mu}n_{t-1}$$

from equation (10). Since this holds for all t then

$$m_t = \frac{1}{\mu}n_{t+1} - \frac{g}{\mu}n_t$$

Inserting these values for m_t and m_{t-1} into equation (8) gives

$$\frac{1}{\mu}n_{t+1} - \frac{g}{\mu}n_t = b\left(\frac{1}{\mu}n_t - \frac{g}{\mu}n_{t-1}\right) + an_{t-1}$$

so

$$\frac{1}{\mu}n_{t+1} - \left(\frac{b+g}{\mu}\right)n_t + \left(\frac{bg - a\mu}{\mu}\right)n_{t-1} = 0$$

Multiplying through by μ gives

$$n_{t+1} - (b+g)n_t + (bg - a\mu)n_{t-1} = 0$$

or

$$n_t - (b+g)n_{t-1} + (bg - a\mu)n_{t-2} = 0$$

This will give us equation (11). However, the value of the weights α and β, which do not affect the stability of the system, will depend upon n_0 and n_1 not m_0 and m_1. Hence the value of the deviation n_t may differ from m_t for any given t, i.e. n_3 need not equal m_3. As mentioned earlier, it is possible to find n_1 from equation (10), given m_0 and n_0, i.e. given Y_0 and R_0.

CAPITAL STOCK ADJUSTMENT THEORY OF INVESTMENT

In Chapters 15 and 16 it was assumed that investment is determined by a simple accelerator model. In such a model investment is a function of one variable, income. This was convenient for difference equations in one variable. The capital stock adjustment model assumes investment is a function of two variables, income and capital stock. Investment is assumed to be proportional to the difference between desired capital stock K^* this period and actual capital stock of the previous period. K_t^* is a function of Y_t as it gives the most appropriate level of capital needed to produce an output of Y_t, i.e.

$$K_t^* = \alpha Y_t$$

where $\alpha > 0$. It would be desirable to have

$$I_t = \lambda(K_t^* - K_{t-1})$$

i.e.

$$I_t = \lambda(\alpha Y_t - K_{t-1})$$

where K_{t-1} is total capital stock at the end of period $t-1$. However, since firms do not know Y_t when investment in period t has to be decided upon, the investment function is assumed to take the form

$$I_t = \lambda(K_{t-1}^* - K_{t-1})$$

i.e.

$$I_t = \lambda(\alpha Y_{t-1} - K_{t-1})$$

Owing to costs of adjustment and to uncertainty about the optimal capital stock, firms are unlikely to attempt to eliminate the difference between K_{t-1}^* and K_{t-1} completely in one period. Thus λ will be < 1 but > 0.

Investment in period t is a net addition to capital stock during that period, i.e.

$$\begin{aligned} K_t &= K_{t-1} + I_t \\ &= K_{t-1} + \lambda(\alpha Y_{t-1} - K_{t-1}) \end{aligned}$$

thus

$$K_t = (1 - \lambda)K_{t-1} + \alpha\lambda Y_{t-1} \tag{12}$$

Equation (12) is a first-order difference equation in two variables. Capital stock is a function of Y and K lagged one period.

The national income identity

$$Y_t = C_t + I_t$$

where

$$C_t = C^* + cY_{t-1}$$

and

$$I_t = \lambda(\alpha Y_{t-1} - K_{t-1})$$

gives another difference equation in the same two variables with income a function of Y and K lagged one period

$$Y_t = C^* + cY_{t-1} + \lambda(\alpha Y_{t-1} - K_{t-1})$$

i.e.

$$Y_t = C^* + (c + \alpha\lambda)Y_{t-1} - \lambda K_{t-1} \tag{13}$$

Using the method given above we can solve for Y_t and K_t. If there is an equilibrium pair of income Y^* and capital stock K^*, these levels will continue if both are simultaneously achieved, i.e. if $Y_{t-1} = Y^*$ and $K_{t-1} = K^*$, then $Y_t = Y^*$ and $K_t = K^*$. We can see from equations (12) and (13) that achievement of its equilibrium level by one of the variables, if the other does not achieve equilibrium, will not result in the system staying in equilibrium. Taking equation (12) this gives

$$K^* = (1 - \lambda)K^* + \alpha\lambda Y^*$$

i.e.

$$\lambda K^* = \alpha\lambda Y^* \quad \text{or} \quad K^* = \alpha Y^*$$

Taking equation (13) we have

$$Y^* = C^* + (c + \alpha\lambda)Y^* - \lambda K^*$$
$$= C^* + (c + \alpha\lambda)Y^* - \alpha\lambda Y^*$$

thus

$$Y^* = \frac{C^*}{1 - c}$$

$$K^* = \alpha Y^*$$

thus

$$K^* = \frac{\alpha C^*}{1 - c}$$

A finite equilibrium level of income and capital stock will exist if $c \neq 1$. To determine the time path of income and capital stock let z_t equal the deviation of Y_t from Y^* and v_t equal the deviation of K_t from K^*, i.e.

$$Y_t = Y^* + z_t = \frac{C^*}{1 - c} + z_t$$

and

$$K_t = K^* + v_t = \frac{\alpha C^*}{1 - c} + v_t$$

Equation (12) gives

$$v_t = (1 - \lambda)v_{t-1} + \alpha\lambda z_{t-1} \tag{14}$$

Equation (13) gives

$$z_t = (c + \alpha\lambda)z_{t-1} - \lambda v_{t-1} \tag{15}$$

From equations (14) and (15) it is possible to obtain a second-order difference equation in one variable.

Second-order Difference Equation in z

$$v_{t-1} = -\frac{1}{\lambda}z_t + \frac{(c + \alpha\lambda)}{\lambda}z_{t-1}$$

from equation (15), thus

$$v_t = -\frac{1}{\lambda}z_{t+1} + \frac{(c + \alpha\lambda)}{\lambda}z_t$$

Substituting these values for v_t and v_{t-1} into equation (14) gives

$$-\frac{1}{\lambda}z_{t+1} + \frac{(c + \alpha\lambda)}{\lambda}z_t = -\frac{(1 - \lambda)}{\lambda}z_t + \frac{(1 - \lambda)(c + \alpha\lambda)}{\lambda}z_{t-1} + \alpha\lambda z_{t-1}$$

Multiplying both sides by $-\lambda$ gives

$$z_{t+1} - (c + \alpha\lambda)z_t = (1 - \lambda)z_t - (c + \alpha\lambda - c\lambda - \alpha\lambda^2)z_{t-1} - \alpha\lambda^2 z_{t-1}$$

thus

$$z_{t+1} - (c + \alpha\lambda + 1 - \lambda)z_t + (c + \alpha\lambda - c\lambda)z_{t-1} = 0$$

or

$$z_t - (c + a\lambda + 1 - \lambda)z_{t-1} + (c + \alpha\lambda - c\lambda)z_{t-2} = 0$$

The auxiliary equation will take the form

$$r^2 - (c + \alpha\lambda + 1 - \lambda)r + (c + \alpha\lambda - c\lambda) = 0 \qquad (16)$$

The roots of the system may be real or complex depending upon the values of α, c and λ. If $(c + \alpha\lambda + 1 - \lambda)^2 \geqslant 4(c + \alpha\lambda - c\lambda)$, equation (16) will have real roots and actual income will approach the equilibrium level if both roots are less than unity in absolute magnitude. If $(c + \alpha\lambda + 1 - \lambda)^2 < 4(c + \alpha\lambda - c\lambda)$, equation (16) will have no real roots. This will give rise to oscillations which will be damped if $c + \alpha\lambda - c\lambda < 1$, explosive if $c + \alpha\lambda - c\lambda > 1$ or of constant amplitude if $c + \alpha\lambda - c\lambda = 1$.

We could have worked with the deviation v_t instead of z_t. As with the previous example this would have yielded the same auxiliary equation and the same roots.

To solve completely for z_t and v_t we need to know Y_0 and K_0. This will give us values for z_0 and v_0. Substituting these values into equations (14) and (15) will provide us with values for v_1 and z_1 respectively. We can then proceed to find the weights, i.e. α and β of the previous example, which fit these initial conditions. The weights for z_t which will satisfy z_0 and z_1 need not take the same values as those for v_t which satisfy v_0 and v_1.

MARKET MODEL WITH STOCKS

Suppose we have a model where the price of a good is set by dealers who hold stocks of the good. We will assume that these dealers adjust their price upwards when they feel stocks are running low and downwards when stocks are getting high. If the desired or normal level of stocks is S^*, where S^* takes a certain value, and P is price, then

$$P_t = P_{t-1} - \lambda(S_{t-1} - S^*) \qquad (17)$$

where $\lambda > 0$. If stocks at the end of period $t - 1$, S_{t-1}, are greater than the desired level, i.e. $S_{t-1} > S^*$, dealers will lower their price in the following period, so that $P_t < P_{t-1}$. If $S_{t-1} < S^*$ then P_t will be greater than P_{t-1}. If $S_{t-1} = S^*$ then P_t will equal P_{t-1} and price will be stable. λ is the coefficient of adjustment. Equation (17) is a difference equation in two variables with price a function of P and S lagged one period.

We know that stocks in period t will equal stocks at the end of period $t - 1$ plus excess supply in period t. If X_t, excess supply in period t, is an increasing function of P_{t-1} and takes the form

$$X_t = -A + aP_{t-1}$$

where a and A are positive constants,

i.e.
$$S_t = S_{t-1} + X_t$$
$$S_t = S_{t-1} - A + aP_{t-1} \qquad (18)$$

Equation (18) is another difference equation in two variables with stocks a function of P and S lagged one period.

If $S_t \neq S^*$ and $P_t \neq P^*$ and we let u_t be the deviation of S_t from S^* and w_t be the deviation of P_t from P^*, equation (17) becomes

$$w_t = w_{t-1} - \lambda u_{t-1} \tag{19}$$

and equation (18) becomes

$$u_t = u_{t-1} + a w_{t-1} \tag{20}$$

From equations (19) and (20) we can obtain a second-order difference equation in u

$$w_{t-1} = \frac{1}{a} u_t - \frac{1}{a} u_{t-1}$$

from equation (20), thus

$$w_t = \frac{1}{a} u_{t+1} - \frac{1}{a} u_t$$

Inserting these values for w_t and w_{t-1} into equation (19) gives

$$\frac{1}{a} u_{t+1} - \frac{1}{a} u_t = \frac{1}{a} u_t - \frac{1}{a} u_{t-1} - \lambda u_{t-1}$$

Multiplying across by a and grouping gives

$$u_{t+1} - 2u_t + (1 + a\lambda)u_{t-1} = 0$$

or

$$u_t - 2u_{t-1} + (1 + a\lambda)u_{t-2} = 0$$

Letting $u_t = u_0 r^t$ and dividing through by $u_0 r^{t-2}$ we get the auxiliary equation

$$r^2 - 2r + 1 + a\lambda = 0$$

This equation will have no real roots since $(-2)^2 < 4(1 + a\lambda)$. Consequently the system will have oscillations. These oscillations will be ever increasing since $1 + a\lambda > 1$ so that $u_t \to \pm \infty$ as $t \to \infty$ and S_t will never approach S^*. In the same way P_t will never approach P^*.

As this model must eventually imply negative prices, outputs, or both, it could clearly only describe reality over a limited range, beyond which the linear equations cannot hold.

EXERCISES

1. Suppose we have a two-country model where Y is income in country 1 and R is income in country 2. In country 1

$$C_t = 10 + 0.9\, Y_{t-1}$$
$$I_t = 40$$
$$M_t = 150 + 0.1\, Y_{t-1}$$
$$X_t = 200 + 0.3\, R_{t-1}$$

In country 2

$$C_t' = 200 + 0.9\, R_{t-1}$$
$$I_t' = 50$$
$$M_t' = 200 + 0.3\, R_{t-1}$$
$$X_t' = 150 + 0.1\, Y_{t-1}$$

Comment on the stability of this system. If $Y_0 = 2030$ and $R_0 = 1090$, find Y_t, R_t, Y_4 and R_4.

2. Given that the relation between the price of a good and stocks of that good held by dealers is

$$P_t = P_{t-1} - 0.8(S_{t-1} - 120)$$

where 120 is the desired level of stocks, and that excess supply, X_t, of this good in period t is a function of P_{t-1}, i.e.

$$X_t = -52 + 0.2P_{t-1}$$

Excess supply is added to stocks so that

$$S_t = S_{t-1} + X_t$$

Comment on the stability of the system and find S_t and S_2 when $S_0 = 180$ and $P_0 = 100$.

3. Given a closed system without government activity where

$$C_t = 100 + 0.8\, Y_{t-1}$$
$$I_t = 0.5(K^*_{t-1} - K_{t-1})$$

where K^*_{t-1} is desired capital stock in period $t - 1$ and

$$K^* = 2Y_t$$

Is this system stable? Find Y_t and Y_6 when $Y_0 = 520$ and $K_0 = 1008$.

SOLUTION TO EXERCISES

1. Country 1

$$Y_t = 10 + 0.9\, Y_{t-1} + 40 + 200 + 0.3R_{t-1} - 150 - 0.1\, Y_{t-1}$$

i.e.

$$Y_t = 100 + 0.8\, Y_{t-1} + 0.3R_{t-1} \tag{21}$$

Country 2

$$R_t = 200 + 0.1\, Y_{t-1} + 0.6R_{t-1} \tag{22}$$

Equilibrium is given by

$$Y^* = 100 + 0.8\, Y^* + 0.3R^*$$

from equation (21), i.e.

$$0.2\, Y^* - 0.3R^* = 100 \tag{23}$$

and

$$R^* = 200 + 0.1\, Y^* + 0.6R^*$$

from equation (22), i.e.

$$-0.1\, Y^* + 0.4R^* = 200 \tag{24}$$

Multiplying equation (24) by 2 and adding to equation (23) gives

$$R^* = 1000$$

If $R^* = 1000$ then $Y^* = 2000$ from equation (23).

Let the deviation of Y_t from Y^* equal m_t and $R_t - R^* = n_t$. Inserting Y_t and R_t in terms of m_t and n_t into equation (21) gives

$$m_t = 0.8m_{t-1} + 0.3n_{t-1} \tag{25}$$

and applying the same process to equation (22) gives

$$n_t = 0.1m_{t-1} + 0.6n_{t-1} \tag{26}$$

From equations (25) and (26) we get a second-order difference equation

$$m_{t-1} = 10n_t - 6n_{t-1}$$

from equation (26), thus

$$m_t = 10n_{t+1} - 6n_t$$

Substituting into equation (25) gives

$$10n_{t+1} - 6n_t = 8n_t - 4\cdot8n_{t-1} + 0\cdot3n_{t-1}$$

thus

$$10n_{t+1} - 14n_t + 4\cdot5n_{t-1} = 0$$

or

$$n_t - 1\cdot4n_{t-1} + 0\cdot45n_{t-2} = 0$$

This gives the auxiliary equation

$$r^2 - 1\cdot4r + 0\cdot45 = 0$$

thus

$$r = \frac{1\cdot4 \pm \sqrt{(1\cdot96 - 1\cdot8)}}{2} = 0\cdot9 \quad \text{or} \quad 0\cdot5$$

Both roots are real and < 1. Consequently $n_t \to 0$ as $t \to \infty$ and $R_t \to R^*$. Y_t also approaches Y^* as $t \to \infty$ as a second-order difference equation in m will yield the same roots.

With roots of $0\cdot9$ and $0\cdot5$ the deviation n_t will take the form

$$n_t = \alpha(0\cdot9)^t + \beta(0\cdot5)^t$$

where α and β are weights to be determined.

$$n_0 = R_0 - R^* = 1090 - 1000 = 90$$
$$m_0 = Y_0 - Y^* = 2030 - 2000 = 30$$

But

$$n_1 = 0\cdot1m_0 + 0\cdot6n_0$$

from equation (26), i.e.

$$n_1 = 3 + 54 = 57$$

and

$$m_1 = 0\cdot8m_0 + 0\cdot3n_0$$

from equation (25) thus

$$m_1 = 24 + 27 = 51$$

With values for n_0 and n_1 we can find α and β.

$$n_0 = \alpha(0\cdot9)^0 + \beta(0\cdot5)^0 = \alpha + \beta = 90$$
$$n_1 = 0\cdot9\alpha + 0\cdot5\beta = 57$$

Multiplying n_1 by 2 and subtracting n_0 gives

$$0\cdot8\alpha = 24 \quad \text{or} \quad \alpha = 30$$

then

$$\beta = 60$$

thus

$$n_t = 30(0\cdot9)^t + 60(0\cdot5)^t$$

and

$$R_t = 1000 + 30(0\cdot9)^t + 60(0\cdot5)^t \qquad *$$
$$R_4 = 1000 + 30(0\cdot9)^4 + 60(0\cdot5)^4$$
$$= 1000 + 19\cdot68 + 3\cdot75 = 1023\cdot43 \qquad *$$

The deviation m_t also takes the form

$$m_t = \gamma(0\cdot9)^t + \delta(0\cdot5)^t$$
$$m_0 = \gamma + \delta = 30$$
$$m_1 = 0\cdot9\gamma + 0\cdot5\delta = 51$$

Multiplying m_1 by 2 and subtracting m_0 gives

$$0\cdot8\gamma = 72 \quad \text{or} \quad \gamma = 90$$

When $\gamma = 90$ then $\delta = -60$, thus

$$m_t = 90(0\cdot9)^t - 60(0\cdot5)^t$$

and

$$Y_t = 2000 + 90(0\cdot9)^t - 60(0\cdot5)^t \qquad *$$
$$Y_4 = 2000 + 59\cdot05 - 3\cdot75 = 2055\cdot30 \qquad *$$

2. $P_t = P_{t-1} - 0\cdot8S_{t-1} + 96$ (27)

Another equation with stocks as a function of P and S lagged one period will enable us to solve for S_t.

$$S_t = S_{t-1} + X_t$$

thus

$$S_t = S_{t-1} - 52 + 0\cdot2P_{t-1} \qquad (28)$$

Equilibrium is given by

$$S^* = 120$$

from equation (27), and

$$P^* = \frac{52}{0\cdot2} = 260$$

from equation (28). Let u_t be the deviation of S_t from S^* and w_t be the deviation of P_t from P^*. Equation (27) becomes

$$w_t = w_{t-1} - 0\cdot8u_{t-1} \qquad (29)$$

Equation (28) becomes

$$u_t = u_{t-1} + 0\cdot2w_{t-1} \qquad (30)$$

Equation (30) will give w_t and w_{t-1} as a function of the deviation u. Inserting these values into equation (29) gives a second-order difference equation in u, i.e.

$$\frac{1}{0\cdot2}u_{t+1} - \frac{1}{0\cdot2}u_t = \frac{1}{0\cdot2}u_t - \frac{1}{0\cdot2}u_{t-1} - 0\cdot8u_{t-1}$$

Multiplying across by 0·2 and grouping gives

$$u_{t+1} - 2u_t + 1\cdot16u_{t-1} = 0$$

or

$$u_t - 2u_{t-1} + 1\cdot16u_{t-2} = 0$$

This gives the auxiliary equation

$$r^2 - 2r + 1\cdot16 = 0$$

thus

$$r = \frac{2 \pm \sqrt{(4 - 4\cdot64)}}{2} = 1 \pm i(0\cdot4)$$

This auxiliary equation has complex roots which will give rise to ever increasing oscillations since the constant term is > 1. Consequently S_t does not approach the equilibrium level $S^* = 120$. The deviation u_t will take the form

$$u_t = \alpha[1 + i(0\cdot4)]^t + \tilde{\alpha}[1 - i(0\cdot4)]^t$$

So that u is real for all t the weights must be complex conjugates, see Chapter 16. $1^2 + 0.4^2 = 1.16$, so we divide the square brackets through by $\sqrt{1.16}$. The angle θ whose Cosine equals $1/\sqrt{1.16}$ is $21.8014°$ approx.

$$u_t = (\sqrt{1.16})^t \left[\alpha\left(\frac{1}{\sqrt{1.16}} + i\frac{(0.4)}{\sqrt{1.16}}\right)^t + \bar\alpha\left(\frac{1}{\sqrt{1.16}} - i\frac{(0.4)}{\sqrt{1.16}}\right)^t \right]$$

$$= (\sqrt{1.16})^t [\alpha(\text{Cos }\theta + i\text{ Sin }\theta)^t + \bar\alpha(\text{Cos }\theta - i\text{ Sin }\theta)^t]$$

$$= (\sqrt{1.16})^t [\alpha(\text{Cos }t\theta + i\text{ Sin }t\theta) + \bar\alpha(\text{Cos }t\theta - i\text{ Sin }t\theta)]$$

If $\alpha = d + if$ and $\bar\alpha = d - if$, then

$$u_t = 2(\sqrt{1.16})^t [d\text{ Cos }t\theta - f\text{ Sin }t\theta]$$

The values of d and f can be found from information on S_0 and P_0.

$$u_t = (d + if)[1 + i(0.4)]^t + (d - if)[1 - i(0.4)]^t$$

thus

$$u_0 = 2d$$

and

$$u_1 = (d + if)[1 + i(0.4)] + (d - if)[1 - i(0.4)] = 2d - 0.8f$$
$$u_0 = S_0 - S^* = 60$$

thus

$$d = 30$$
$$u_1 = u_0 + 0.2w_0$$

from equation (30) and

$$w_0 = P_0 - P^* = 100 - 260 = -160$$

so

$$u_1 = 60 + 0.2(-160) = 28$$

and

$$28 = 60 - 0.8f$$

thus $f = 40$.

Thus

$$u_t = 2(1.16)^{t/2}[30\text{ Cos }t\theta - 40\text{ Sin }t\theta]$$

and

$$S_t = 120 + 2(1.16)^{t/2}[30\text{ Cos }t\theta - 40\text{ Sin }t\theta] \qquad *$$

When $t = 2$, $2\theta = 43.6028°$, $\text{Cos }2\theta = 0.7241$ and $\text{Sin }2\theta = 0.6897$

$$S_2 = 120 + 2(1.16)[30(0.7241) - 40(0.6897)]$$
$$= 120 + 2.32(-5.8632) = 106.40 \qquad *$$

This can be compared with the result obtained by application of equations (27) and (28).

$$P_t = P_{t-1} - 0.8S_{t-1} + 96$$
$$S_t = S_{t-1} - 52 + 0.2P_{t-1}$$
$$S_0 = 180 \quad \text{and} \quad P_0 = 100$$
$$S_1 = S_0 - 52 + 0.2P_0$$
$$= 180 - 52 + 20 = 148$$
$$P_1 = P_0 - 0.8S_0 + 96$$
$$= 100 - 0.8(180) + 96$$
$$= 52$$
$$S_2 = S_1 - 52 + 0.2P_1$$
$$= 148 - 52 + 0.2(52) = 106.40$$

3. $Y_t = C_t + I_t$

$\quad = 100 + 0.8\,Y_{t-1} + 0.5(2\,Y_{t-1} - K_{t-1})$

or

$$Y_t = 100 + 1.8\,Y_{t-1} - 0.5K_{t-1} \tag{31}$$

Another equation with K_t as a function of Y and K lagged one period will enable us to solve for Y_t.

$$K_t = K_{t-1} + I_t$$

thus

$$K_t = K_{t-1} + 0.5(2\,Y_{t-1} - K_{t-1})$$

or

$$K_t = 0.5K_{t-1} + Y_{t-1} \tag{32}$$

Equations (31) and (32) are specific forms of equations (13) and (12) p. 277, respectively. Consequently equilibrium Y and K will take the values

$$Y^* = \frac{100}{1 - 0.8} = 500 \quad \text{and} \quad K^* = 2(500) = 1000$$

Given equations (31) and (32) the auxiliary equation will take the form

$$r^2 - [0.8 + 2(0.5) + 1 - 0.5]r + [0.8 + 2(0.5) - (0.8)(0.5)] = 0$$

see equation (16), p. 279, i.e.

$$r^2 - 2.3r + 1.4 = 0$$

thus

$$r = \frac{2.3 \pm \sqrt{(5.29 - 5.6)}}{2} = \frac{2.3}{2} \pm \frac{i\sqrt{0.31}}{2}$$

The auxiliary equation has no real roots so that the system will have oscillations which will be explosive since the constant term of the equation is > 1. Hence Y_t will never approach Y^*. The same is true for K_t. If we let z_t equal the deviation of Y_t from Y^* and v_t equal the deviation of K_t from K^*, as in the chapter, then

$$z_t = \alpha\left(\frac{2.3}{2} + \frac{i\sqrt{0.31}}{2}\right)^t + \bar{\alpha}\left(\frac{2.3}{2} - \frac{i\sqrt{0.31}}{2}\right)^t$$

If $\alpha = d + if$, then

$$z_t = 2(\sqrt{1.4})^t[d\,\mathrm{Cos}\,t\theta - f\,\mathrm{Sin}\,t\theta]$$

The angle θ whose Cosine equals

$$\frac{2.3/2}{\sqrt{[(2.3/2)^2 + (\sqrt{0.31}/2)^2]}} = \frac{2.3}{2\sqrt{1.4}}$$

is $13.6082°$ approx.

$$z_t = 2(1.4)^{t/2}[d\,\mathrm{Cos}\,t\theta - f\,\mathrm{Sin}\,t\theta]$$

$$z_0 = 2d \quad \text{but} \quad z_0 = Y_0 - Y^* = 520 - 500 = 20$$

so

$$d = 10$$

By equation (31), $z_1 = 1.8z_0 - 0.5v_0$ and

$$v_0 = K_0 - K^* = 1008 - 1000 = 8$$

thus

$$z_1 = 1.8(20) - 0.5(8) = 36 - 4 = 32$$

$$z_1 = 2\sqrt{(1.4)}\left[10\left(\frac{2.3}{2\sqrt{1.4}}\right) - f\left(\frac{\sqrt{0.31}}{2\sqrt{1.4}}\right)\right] = 32$$

thus

$$10(2{\cdot}3) - f(\sqrt{0{\cdot}31}) = 32$$

$$f = -\frac{9}{\sqrt{0{\cdot}31}} = -16{\cdot}1645 \text{ approx.}$$

Thus

$$Y_t = 500 + 2(1{\cdot}4)^{t/2}[10\,\text{Cos}\,t\theta + 16{\cdot}1645\,\text{Sin}\,t\theta] \qquad *$$

When $t = 6, 6\theta = 81{\cdot}6492°$ approx, and

$$\text{Cos}\,6\theta = 0{\cdot}1452, \quad \text{Sin}\,6\theta = 0{\cdot}9894$$

$$Y_6 = 500 + 2(1{\cdot}4)^3[10(0{\cdot}1452) + 16{\cdot}1645(0{\cdot}9894)]$$

$$= 500 + 2(1{\cdot}4)^3(17{\cdot}4452)$$

$$= 595{\cdot}7390 \qquad *$$

This is the same as the result obtained by repeated application of equations (31) and (32).

CHAPTER 18

Second-order Differential Equations

Any of the models which give rise to second-order difference equations when set up in terms of discrete time periods could alternatively have been set up in differential equation form. This chapter will show second-order differential equations can be used to analyse models consisting of two first-order differential equations in two variables.

MARKET MODEL WITH STOCKS

Suppose we have a model where the rate of change in stocks of a good $\dot{S}_t$ is directly proportional to excess supply X_t and

$$X_t = -A + aP_t$$

where A and a are positive constants, then

$$\dot{S}_t = \lambda(-A + aP_t)$$

where $\lambda > 0$, i.e.

$$\dot{S}_t = -A\lambda + a\lambda P_t \tag{1}$$

If we assume that dealers adjust their actual price P_t towards a target price P'_t at a speed which is proportional to the gap between P'_t and P_t, then

$$\dot{P}_t = \beta(P'_t - P_t) \quad \text{where} \quad \beta > 0$$

$\dot{P}_t$ is positive when $P'_t > P_t$ and negative when $P'_t < P_t$

If the target price is related to stocks so that

$$P'_t = M - bS_t$$

where M and b are positive constants, then

$$\dot{P}_t = \beta(M - bS_t - P_t)$$

i.e.

$$\dot{P}_t = M\beta - b\beta S_t - \beta P_t \tag{2}$$

Equations (1) and (2) are first-order differential equations in two variables, i.e. S and P. In solving for S_t and P_t equations (1) and (2) will give rise to a second-order differential equation.

If an equilibrium S^* and P^* are simultaneously achieved then $\dot{S}_t = \dot{P}_t = 0$ at these levels. At equilibrium equation (1) becomes

$$-A\lambda + a\lambda P^* = 0 \quad \text{or} \quad P^* = \frac{A}{a} \tag{3}$$

and equation (2) becomes

$$M\beta - b\beta S^* - \beta P^* = 0 \tag{4}$$

Inserting the value for P^* from equation (3) into equation (4) gives

$$M\beta - b\beta S^* - \frac{\beta A}{a} = 0$$

or

$$S^* = \frac{aM - A}{ab}$$

If $S_t \neq S^*$ and we let u_t be the deviation, then

$$S_t = S^* + u_t = \frac{aM - A}{ab} + u_t$$

and

$$\dot{S}_t = \dot{u}_t$$

since $\dot{S}^* = 0$. If $P_t \neq P^*$ and we let w_t be the deviation, then

$$P_t = P^* + w_t = \frac{A}{a} + w_t$$

and

$$\dot{P}_t = \dot{w}_t$$

since $\dot{P}^* = 0$. Substituting these values for S_t, $\dot{S}_t$, P_t and $\dot{P}_t$ into equations (1) and (2) gives

$$\dot{u}_t = a\lambda w_t \tag{5}$$

and

$$\dot{w}_t = -b\beta u_t - \beta w_t \tag{6}$$

From equations (5) and (6) it is possible to obtain a second-order differential equation in either u or w.

Second-order Differential Equation in u

$$w_t = \frac{1}{a\lambda}\dot{u}_t$$

from equation (5), thus

$$\dot{w}_t = \frac{d}{dt}\left(\frac{1}{a\lambda}\dot{u}_t\right) = \frac{1}{a\lambda}\ddot{u}_t,$$

where $\ddot{u}_t = d\dot{u}_t/dt$. Substituting these values for w_t and $\dot{w}_t$ into equation (6) gives

$$\frac{1}{a\lambda}\ddot{u}_t = -b\beta u_t - \frac{\beta}{a\lambda}\dot{u}_t$$

Multiplying both sides by $a\lambda$ gives

$$\ddot{u}_t + \beta\dot{u}_t + ab\beta\lambda u_t = 0 \tag{7}$$

Equation (7) is a second-order differential equation in u_t. To find the time path of S_t and P_t we must solve for u_t.

If $u_t \to 0$ as $t \to \infty$ then $S_t \to S^*$. When $u_t \to 0$ then $\dot{u}_t \to 0$, thus from equation (5) $a\lambda w_t$ or $w_t \to 0$ since $a\lambda$ is constant. Consequently $P_t \to P^*$ when $u_t \to 0$.

We must find a value for u_t which satisfies equation (7). If we let $u_t = u_0 e^{rt}$, where $u_0 = S_0 - S^*$ and information on S_0 is available, it is necessary to obtain a value or values for r so that $u_t = u_0 e^{rt}$ will satisfy equation (7). If $u_t = u_0 e^{rt}$, then

$$\dot{u}_t = r u_0 e^{rt}$$

and

$$\ddot{u}_t = r^2 u_0 e^{rt}$$

Inserting these values into equation (7) gives

$$r^2 u_0 e^{rt} + \beta r u_0 e^{rt} + ab\beta\lambda u_0 e^{rt} = 0$$

Dividing through by $u_0 e^{rt}$ gives

$$r^2 + \beta r + ab\beta\lambda = 0 \tag{8}$$

Auxiliary Equation with Two Distinct Real Roots

If $\beta^2 > 4ab\beta\lambda$, equation (8) will have two distinct real roots which will be labelled r_1 and r_2. Consequently $u_t = u_0 e^{r_1 t}$ and $u_t = u_0 e^{r_2 t}$ will satisfy equation (7). As in Chapter 15, equation (7) is also satisfied by

$$u_t = g_1 e^{r_1 t} + g_2 e^{r_2 t} \tag{9}$$

where g_1 and g_2 are weights to be determined. If r_1 and r_2 are both negative then $u_t \to 0$ as $t \to \infty$ since $e^{r_1 t}$ and $e^{r_2 t}$ tend to zero as $t \to \infty$. The system will be unstable if either root is positive.

To determine the weights g_1 and g_2 we must have information on stocks and price at some initial time, say S_0 and P_0. With this information it is possible to find a value for u_0 and $\dot{u}_0$ which will give us two equations in two unknowns, i.e. g_1 and g_2. From equation (9) we have

$$u_0 = g_1 e^0 + g_2 e^0$$

i.e.

$$u_0 = g_1 + g_2 \tag{10}$$

From equation (9) we also have

$$\dot{u}_t = r_1 g_1 e^{r_1 t} + r_2 g_2 e^{r_2 t}$$

thus

$$\dot{u}_0 = r_1 g_1 e^0 + r_2 g_2 e^0$$

i.e.

$$\dot{u}_0 = r_1 g_1 + r_2 g_2 \tag{11}$$

But $u_0 = S_0 - S^*$ and

$$\dot{u}_0 = \dot{S}_0 = -A\lambda + a\lambda P_0$$

from equation (1). u_0 and $\dot{u}_0$ are therefore given if S_0 and P_0 are given. It is then a matter of solving equations (10) and (11) to find g_1 and g_2.

We could have examined the time path of price by solving for the deviation w_t. This will give us the same auxiliary equation, i.e. equation (8) and hence the same roots, but the weights g_1 and g_2 need not take the same values since u_0 need not equal w_0 and $\dot{u}_0$ need not equal $\dot{w}_0$.

Specific Example

Let $\lambda = 1$, $A = 68$, $a = 0.4$, $\beta = 2$, $M = 250$, $b = 0.8$, $P_0 = 30$ and $S_0 = 120$. Equilibrium S^* is given by

$$S^* = \frac{0.4(250) - 68}{0.4(0.8)} = 100$$

Equation (8) becomes

$$r^2 + 2r + 0.64 = 0$$

i.e.

$$r = \frac{-2 \pm \sqrt{(4 - 2.56)}}{2}$$

or

$$r_1 = -1.6 \quad \text{and} \quad r_2 = -0.4$$

thus

$$u_t = g_1 e^{-1.6t} + g_2 e^{-0.4t}$$

i.e.

$$u_0 = g_1 + g_2 \tag{12}$$

and

$$\dot{u}_t = -1.6g_1 e^{-1.6t} - 0.4g_2 e^{-0.4t}$$

i.e.

$$\dot{u}_0 = 1.6g_1 - 0.4g_2 \tag{13}$$
$$u_0 = S_0 - S^* = 120 - 100 = 20$$
$$\dot{u}_0 = \dot{S}_0 = -A\lambda + a\lambda P_0 = -68 + 12 = -56$$

From equation (12)

$$g_1 + g_2 = 20 \tag{14}$$

From equation (13)

$$-1.6g_1 - 0.4g_2 = -56 \tag{15}$$

Multiplying equation (14) by 0.4 and adding to equation (15) gives

$$-1.2g_1 = -48 \quad \text{or} \quad g_1 = 40$$

and

$$g_2 = -20$$

from equation (14), thus

$$u_t = 40e^{-1.6t} - 20e^{-0.4t}$$

and

$$S_t = 100 + 40e^{-1.6t} - 20e^{-0.4t}$$

Since r_1 and r_2 are both negative then $u_t \to 0$ as $t \to \infty$ and $S_t \to S^* = 100$.

Auxiliary Equation with Two Equal Roots

If $\beta^2 = 4ab\beta\lambda$, equation (8) will have two equal roots, i.e. $r_1 = r_2 = r$. As in Chapter 15, equation (7) will therefore be satisfied by

$$u_t = g_1 e^{rt} + g_2 t e^{rt} \tag{16}$$

thus

$$\dot{u}_t = rg_1 e^{rt} + g_2 e^{rt} + rg_2 t e^{rt}$$

i.e.

$$\dot{u}_t = rg_1 e^{rt} + g_2 e^{rt}(1 + rt) \tag{17}$$

To obtain the derivative of $g_2 t e^{rt}$ with respect to t we use the product rule, i.e.

$$\frac{d(g_2 t e^{rt})}{dt} = g_2 e^{rt}\frac{dt}{dt} + t\frac{d(g_2 e^{rt})}{dt}$$

$$= g_2 e^{rt} + trg_2 e^{rt}$$

$$u_0 = g_1 e^0 + g_2(0)e^0 = g_1$$

from equation (16), and

$$\dot{u}_0 = rg_1 e^0 + g_2 e^0(1 + 0) = rg_1 + g_2$$

from equation (17). Given S_0 and P_0 we can find u_0 and $\dot{u}_0$ and solve for g_1 and g_2. For this system to be stable r must be negative.

Auxiliary Equation with Complex Roots

If $\beta^2 < 4ab\beta\lambda$, equation (8) will have complex roots. If we let $r_1 = m + if$ then $r_2 = m - if$, where $m = -\frac{1}{2}\beta$ and $f = \frac{1}{2}\sqrt{(4ab\beta\lambda - \beta^2)}$, and

$$u_t = g_1 e^{mt+ift} + g_2 e^{mt-ift}$$

or

$$u_t = e^{mt}[g_1 e^{ift} + g_2 e^{-ift}]$$

Clearly $e^{mt} \to 0$ as $t \to \infty$ if $m < 0$. Consequently, to find how u_t behaves over time we must examine $e^{\pm ift}$.

We saw in Chapter 14 that

$$e^x = 1 + x + \frac{x^2}{2!} + \frac{x^3}{3!} + \frac{x^4}{4!} + \dots$$

Applying this to e^{ift}, where $ft = x$, gives

$$e^{ix} = 1 + ix + \frac{(ix)^2}{2!} + \frac{(ix)^3}{3!} + \frac{(ix)^4}{4!} + \frac{(ix)^5}{5!} + \frac{(ix)^6}{6!} + \frac{(ix)^7}{7} + \dots$$

$$= 1 + ix - \frac{x^2}{2!} - i\frac{x^3}{3!} + \frac{x^4}{4!} + i\frac{x^5}{5!} - \frac{x^6}{6!} - i\frac{x^7}{7!} + \dots$$

If we let

$$1 - \frac{x^2}{2!} + \frac{x^4}{4!} - \frac{x^6}{6!} + \dots = A$$

and

$$x - \frac{x^3}{3!} + \frac{x^5}{5!} - \frac{x^7}{7!} + \ldots = B$$

then

$$e^{ix} = A + iB$$

Series A and B will converge, i.e. the sum of the first n terms will come very close to a definite number.

$$\frac{dA}{dx} = -x + \frac{x^3}{3!} - \frac{x^5}{5!} + \ldots = -B$$

and

$$\frac{dB}{dx} = 1 - \frac{x^2}{2!} + \frac{x^4}{4!} - \frac{x^6}{6!} + \ldots = A$$

We can now identify A with Cos x and B with Sin x. From Appendix 3 we know that

$$\frac{d(\text{Cos } x)}{dx} = -\text{Sin } x \quad \text{and} \quad \frac{d(\text{Sin } x)}{dx} = \text{Cos } x$$

Also when $x = 0$, $A(0) = 1$ and $B(0) = 0$. Thus $A = \text{Cos } x$ and $B = \text{Sin } x$, so

$$e^{ix} = \text{Cos } x + i \text{ Sin } x$$

i.e.

$$e^{ift} = \text{Cos } ft + i \text{ Sin } ft$$

and

$$u_t = e^{mt}[g_1(\text{Cos } ft + i \text{ Sin } ft) + g_2(\text{Cos } ft - i \text{ Sin } ft)]$$

u_t is the deviation of S_t from S^* and must therefore be real. It will be real if g_1 and g_2 are complex conjugates. If we let $g_1 = v + in$ and $g_2 = v - in$, then

$$u_t = e^{mt}[(v + in)(\text{Cos } ft + i \text{ Sin } ft) + (v - in)(\text{Cos } ft - i \text{ Sin } ft)]$$

i.e.

$$u_t = 2e^{mt}(v \text{ Cos } ft - n \text{ Sin } ft) \tag{18}$$

or

$$u_t = 2e^{mt}\sqrt{(v^2 + n^2)}\left[\frac{v}{\sqrt{(v^2 + n^2)}} \text{Cos } ft - \frac{n}{\sqrt{(v^2 + n^2)}} \text{Sin } ft\right]$$

We can always find an angle θ such that

$$\text{Cos } \theta = \frac{v}{\sqrt{(v^2 + n^2)}} \quad \text{and} \quad \text{Sin } \theta = \frac{n}{\sqrt{(v^2 + n^2)}}$$

thus

$$u_t = 2e^{mt}\sqrt{(v^2 + n^2)}(\text{Cos } \theta \text{ Cos } ft - \text{Sin } \theta \text{ Sin } ft)$$
$$= 2e^{mt}\sqrt{(v^2 + n^2)}[\text{Cos }(\theta + ft)]$$

see Appendix 3. We saw in Chapter 16 that the Cosine of an angle varies from $+1$ to -1 every $360°$. Consequently Cos $(\theta + ft)$ oscillates between constant limits. $2\sqrt{(v^2 + n^2)}$ will not change over time so that this will not affect the stability of the system. Whether the system will have damped, increasing, or constant oscillations will depend on e^{mt}.

If $m < 0$ then $e^{mt} \to 0$ as $t \to \infty$ and $u_t \to 0$. The oscillations due to Cos $(\theta + ft)$ will fade away and $S_t \to S^*$.

If $m > 0$ the system will have increasing oscillations as e^{mt} increases with t.

Finding Values for the Weights g_1 and g_2

To solve for v and n and hence for g_1 and g_2 we use u_0 and $\dot{u}_0$. Cos $0 = 1$ and Sin $0 = 0$, thus

$$u_0 = 2e^0(v \text{ Cos } 0 - n \text{ Sin } 0) = 2v$$

from equation (18), i.e.

$$v = 0 \cdot 5 u_0 \tag{19}$$

We have now solved for v since u_0 is given once we know S_0 and S^*. Using the product rule we can obtain $\dot{u}_t$ from equation (18), i.e.

$$\dot{u}_t = 2e^{mt}[- fv \text{ Sin } ft - fn \text{ Cos } ft] + 2me^{mt}[v \text{ Cos } ft - n \text{ Sin } ft] \tag{20}$$

$d(\text{Cos } ft)/dt = - f \text{ Sin } ft$ because $d(\text{Cos } x)/dx = - \text{Sin } x$, and if we let $x = ft$, then the function of a function rule gives

$$\frac{d(\text{Cos } x)}{dt} = \frac{d(\text{Cos } x)}{dx} \cdot \frac{dx}{dt} = - (\text{Sin } x).f = - f \text{ Sin } ft$$

In the same way it can be shown that $d(\text{Sin } ft)/dt = f \text{ Cos } ft$. From equation (20),

$$\dot{u}_0 = 2e^0(- fv \text{ Sin } 0 - fn \text{ Cos } 0) + 2me^0(v \text{ Cos } 0 - n \text{ Sin } 0)$$

i.e.

$$\dot{u}_0 = - 2fn + 2mv = - 2fn + u_0 m$$

or

$$n = \frac{u_0 m - \dot{u}_0}{2f} \tag{21}$$

We have now solved for n since we know the value of m and f and can find u_0 and $\dot{u}_0$ given S_0 and P_0.

Specific Example

Let $\lambda = 1$, $A = 10$, $a = 1$, $\beta = 1 \cdot 6$, $M = 20$, $b = 0 \cdot 5$, $P_0 = 14$, and $S_0 = 40$. Equilibrium S^* is given by

$$\frac{1(20) - 10}{1(0 \cdot 5)} = 20$$

Equation (8) becomes

$$r^2 + 1 \cdot 6r + 0 \cdot 8 = 0$$

i.e.

$$r = - 0 \cdot 8 \pm i(0 \cdot 4)$$

i.e.

$$r_1 = -0.8 + i(0.4) \quad \text{and} \quad r_2 = -0.8 - i(0.4)$$

or

$$m = -0.8 \quad \text{and} \quad f = 0.4$$

Thus

$$u_t = 2e^{-0.8t}\sqrt{(v^2 + n^2)}[\text{Cos}(\theta + 0.4t)] \tag{22}$$

$e^{-0.8t} \to 0$ as $t \to \infty$ so that $u_t \to 0$ as $t \to \infty$. Consequently the system is stable. To find values for v and n we use u_0 and $\dot{u}_0$.

$$u_0 = S_0 - S^* = 40 - 20 = 20$$
$$\dot{u} = -A\lambda + a\lambda P_0 = -10 + 14 = 4$$

But $v = 0.5u_0 = 10$ from equation (19) and

$$n = \frac{20(-0.8) - 4}{2(0.4)} = -25$$

from equation (21). $\sqrt{(v^2 + n^2)} = \sqrt{725}$ and the angle θ whose Cos is $10/\sqrt{725}$ is $68.20°$ approximately, thus

$$u_t = 10\sqrt{29}\,e^{-0.8t}[\text{Cos}(68.20° + 0.4t)]$$

from equation (22), and

$$S_t = 20 + 10\sqrt{29}\,e^{-0.8t}[\text{Cos}(68.20° + 0.4t)]$$

NATIONAL INCOME MODEL

Suppose we have a closed model without government activity where consumption is a function of 'normal' income W_t, i.e.

$$C_t = A + cW_t$$

and investment is assumed to be proportional to the difference between desired capital stock K_t^* and actual capital stock, i.e.

$$\dot{K}_t = I_t = \lambda(K_t^* - K_t) = \lambda(vW_t - K_t)$$

or

$$\dot{K}_t = v\lambda W_t - \lambda K_t \tag{23}$$

where λ and v are positive constants and $K_t^* = vW_t$.

If 'normal' income W_t adjusts towards actual income Y_t at a speed which is proportional to the gap between W_t and Y_t, i.e.

$$\dot{W}_t = \beta(Y_t - W_t)$$

where $\beta > 0$

$$= \beta(C_t + I_t - W_t)$$

Inserting the above values for C_t and I_t gives

$$\dot{W}_t = \beta[A + cW_t + v\lambda W_t - \lambda K_t - W_t]$$

i.e.

$$\dot{W}_t = A\beta + \beta(c + v\lambda - 1)W_t - \beta\lambda K_t \tag{24}$$

Equations (23) and (24) are first-order differential equations in two variables, i.e. W and K. We can solve for W_t and K_t in the same way as we did for S_t and P_t.

Readers who want to make frequent use of the techniques of differential and difference equations will find that the derivation of their characteristic equations can be made simpler and speedier by the use of 'operators'; these are described in Appendix 6.

EXERCISES

1. Assume a closed system without government activity where consumption is a function of normal income W_t, i.e.

$$C_t = 20 + 0.5 W_t$$

If W_t adjusts towards actual income Y_t so that

$$\dot{W}_t = Y_t - W_t$$

and investment is related to the difference between desired capital stock K_t^* and actual capital stock K_t, i.e.

$$I_t = \dot{K}_t = 0.2(K_t^* - K_t)$$

where $K_t^* = 2W_t$. Find W_t when $W_0 = 80$ and $K_0 = 10$. Comment on the stability of this system.

2. Given a model where the rate of change in stock $\dot{S}_t$ is related to excess supply X_t so that

$$\dot{S}_t = X_t = -10 + 2P_t$$

If dealers adjust their actual price P_t towards a target price P_t' at a speed which depends on the gap between P_t' and P_t so that

$$\dot{P}_t = P_t' - P_t$$

and P_t' is related to stocks so that

$$P_t' = 6 - 0.25 S_t$$

Find P_t and S_t when $P_0 = 6$ and $S_0 = 4$.

SOLUTIONS TO EXERCISES

1. $\dot{K}_t = 0.2(2W_t - K_t)$

or

$$\dot{K}_t = 0.4 W_t - 0.2 K_t \tag{25}$$
$$\dot{W}_t = Y_t - W_t$$
$$= C_t + I_t - W_t$$
$$= 20 + 0.5 W_t + 0.4 W_t - 0.2 K_t - W_t$$

i.e.

$$\dot{W}_t = 20 - 0.1 W_t - 0.2 K_t \tag{26}$$

Equations (25) and (26) are first-order differential equations in two variables.

If an equilibrium W^* and K^* are simultaneously achieved then $\dot{W}_t = \dot{K}_t = 0$ at these levels.

At equilibrium

$$0.4 W^* - 0.2 K^* = 0$$

from equation (25) and

$$-0.1W^* - 0.2K^* = -20$$

from equation (26)

Subtraction gives

$$W^* = 40$$

thus

$$K^* = 80$$

If $W_t \neq W^*$ and we let m_t be the deviation, then

$$W_t = 40 + m_t$$

and

$$\dot{W}_t = \dot{m}_t$$

If $K_t \neq K^*$ and we let n_t be the deviation, then

$$K_t = 80 + n_t$$

and

$$\dot{K}_t = \dot{n}_t$$

Inserting these values for W_t and K_t in terms of m_t and n_t into equations (25) and (26) we get, from equation (25)

$$\dot{n}_t = 0.4m_t - 0.2n_t \tag{27}$$

from equation (26)

$$\dot{m}_t = -0.1m_t - 0.2n_t \tag{28}$$

From equations (27) and (28) we get a second-order differential equation in m. From equation (28)

$$n_t = -5\dot{m}_t - 0.5m_t$$

thus

$$\dot{n}_t = -5\ddot{m}_t - 0.5\dot{m}_t$$

Inserting these values for n_t and $\dot{n}_t$ into equation (27), gives

$$-5\ddot{m}_t - 0.5\dot{m}_t = 0.4m_t + \dot{m}_t + 0.1m_t$$

or

$$\ddot{m}_t + 0.3\dot{m}_t + 0.1m_t = 0 \tag{29}$$

Let $m_t = m_0 \, e^{rt}$, i.e.

$$\dot{m}_t = rm_0 \, e^{rt} \quad \text{and} \quad \ddot{m}_t = r^2 m_0 \, e^{rt}$$

Substituting these values into equation (29) gives

$$r^2 m_0 \, e^{rt} + 0.3rm_0 \, e^{rt} + 0.1m_0 \, e^{rt} = 0$$

Dividing through by $m_0 \, e^{rt}$ gives

$$r^2 + 0.3r + 0.1 = 0$$

i.e.

$$r = \frac{-0.3 \pm \sqrt{(0.09 - 0.4)}}{2} = -0.15 \pm i(0.0775)^{1/2}$$

If we let $(0.0775)^{1/2} = \alpha$, then

$$m_t = g_1 \, e^{-0.15t + i\alpha t} + g_2 \, e^{-0.15t - i\alpha t}$$

where g_1 and g_2 are weights to be determined. m_t will be real if we let $g_1 = v + ih$ and $g_2 = v - ih$, i.e.

$$m_t = 2e^{-0.15t}(v \cos \alpha t - h \sin \alpha t) \tag{30}$$

Equation (30) is similar to equation (18) of this chapter. From equation (30) we can find m_0 and $\dot{m}_0$ and thus solve for v and h, i.e.

$$m_0 = 2\,e^0(v\cos 0 - h\sin 0) = 2v$$

and

$$\dot{m}_0 = 2\,e^0(-av\sin 0 - \alpha h\cos 0) + (-0{\cdot}15)2\,e^0(v\cos 0 - h\sin 0)$$
$$= -2\alpha h - 0{\cdot}3v \tag{31}$$

$$m_0 = W_0 - W^* = 80 - 40 = 40$$

and

$$\dot{m}_0 = \dot{W}_0 = 20 - 0{\cdot}1\,W_0 - 0{\cdot}2K_0$$

from equation (26), i.e.

$$\dot{m}_0 = 20 - 8 - 2 = 10$$

Thus $v = 0{\cdot}5(40) = 20$, and from equation (31)

$$10 = -2(0{\cdot}0775)^{1/2}h - 0{\cdot}3(20)$$

i.e.

$$h = -8(0{\cdot}0775)^{-1/2}$$

Equation (30) can be reduced to the form

$$m_t = 2\,e^{-0{\cdot}15t}(v^2 + h^2)^{1/2}(\cos\phi\cos\alpha t - \sin\phi\sin\alpha t)$$

where $v/(v^2 + h^2)^{1/2} = \cos\phi = 20/35$ approx, and $\sin\phi = 28{\cdot}74/35$ approx. i.e. $\phi = 55{\cdot}16°$ approx. Thus where the symbol $\simeq$ means 'is approximately equal to',

$$m_t \simeq 70\,e^{-0{\cdot}15t}[\cos\{55{\cdot}16° + (0{\cdot}0775)^{1/2}t\}]$$

and

$$W_t \simeq 40 + 70\,e^{-0{\cdot}15t}[\cos\{55{\cdot}16° + (0{\cdot}0775)^{1/2}t\}] \qquad *$$

The system is stable table since $e^{-0{\cdot}15t} \to 0$ as $t \to \infty$ and $m_t \to 0$. When $m_t \to 0$ then $n_t \to 0$, from equation (28). Consequently $W_t \to W^*$ and $K_t \to K^*$.

2. $\dot{S}_t = -10 + 2P_t$ (32)
    $\dot{P}_t = 6 - 0{\cdot}25S_t - P_t$ (33)

Equations (32) and (33) are two first-order differential equations in two variables. At equilibrium $\dot{S}_t = \dot{P}_t = 0$, i.e.

$$P^* = 5$$

from equation (32), and

$$S^* = 4$$

from equation (33). If $S_t \neq S^*$ and we let u_t be the deviation, then

$$S_t = 4 + u_t \quad \text{and} \quad \dot{S}_t = \dot{u}_t$$

If $P_t \neq P^*$ and we let v_t be the deviation, then

$$P_t = 5 + v_t \quad \text{and} \quad \dot{P}_t = \dot{v}_t$$

Inserting these values for S_t and P_t into equations (32) and (33) we get

$$\dot{u}_t = 2v_t \tag{34}$$

and

$$\dot{v}_t = -0{\cdot}25u_t - v_t$$

From equation (34), $v_t = 0{\cdot}5\dot{u}_t$ and $\dot{v}_t = 0{\cdot}5\ddot{u}_t$, thus from equation (35)

$$0{\cdot}5\ddot{u}_t + 0{\cdot}5\dot{u}_t + 0{\cdot}25u_t = 0$$

i.e.

$$\ddot{u}_t + \dot{u}_t + 0{\cdot}5u_t = 0 \tag{36}$$

Let $u_t = u_0\,e^{rt}$, so that $\dot{u}_t = ru_0\,e^{rt}$ and $\ddot{u}_t = r^2u_0\,e^{rt}$. Substituting these values into equation (36) and dividing through by $u_0\,e^{rt}$ gives

$$r^2 + r + 0{\cdot}5 = 0$$

thus

$$r = \frac{-1 \pm \sqrt{(1-2)}}{2} = -0{\cdot}5 \pm i(0{\cdot}5)$$

A second-order differential equation in v_t will give exactly the same roots. Consequently

$$u_t = g_1 e^{-0{\cdot}5t + i(0{\cdot}5)t} + g_2 e^{-0{\cdot}5t - i(0{\cdot}5)t}$$
$$v_t = f_1 e^{-0{\cdot}5t + i(0{\cdot}5)t} + f_2 e^{-0{\cdot}5t - i(0{\cdot}5)t}$$

where the gs and fs are weights to be determined in each case.

Solving for u_t

If we let $g_1 = m + in$ and $g_2 = m - in$, u_t will be real, i.e.

$$u_t = 2 e^{-0{\cdot}5t}[m \cos(0{\cdot}5t) - n \sin(0{\cdot}5t)] \tag{37}$$
$$u_0 = 2 e^0(m \cos 0 - n \sin 0) = 2m$$

and

$$\dot{u}_0 = 2 e^0(-0{\cdot}5m \sin 0 - 0{\cdot}5n \cos 0) - e^0(m \cos 0 - n \sin 0)$$
$$= -n - m$$

But

$$u_0 = S_0 - S^* = 4 - 4 = 0$$

and

$$\dot{u}_0 = \dot{S}_0 = -10 + 2P_0 = 2$$

Consequently $m = 0$ and $n = -2$.

Substituting these values into equation (37) gives

$$u_t = 2 e^{-0{\cdot}5t}[2 \sin(0{\cdot}5t)]$$

thus

$$S_t = 4 + 4 e^{-0{\cdot}5t} \sin(0{\cdot}5t) \qquad\qquad *$$

Solving for v_t

If $f_1 = x + iy$ and $f_2 = x - iy$, v_t will be real, i.e.

$$v_t = 2 e^{-0{\cdot}5t}[x \cos(0{\cdot}5t) - y \sin(0{\cdot}5t)] \tag{38}$$
$$v_0 = 2x$$
$$\dot{v}_0 = -x - y$$

But $v_0 = P_0 - P^* = 6 - 5 = 1$ and

$$\dot{v}_0 = \dot{P}_0 = 6 - 0{\cdot}25 S_0 - P_0 = -1$$

Consequently $x = y = 0{\cdot}5$ and

$$v_t = e^{-0{\cdot}5t}[\cos(0{\cdot}5t) - \sin(0{\cdot}5t)]$$

from equation (38), i.e.

$$P_t = 5 + e^{-0{\cdot}5t}[\cos(0{\cdot}5t) - \sin(0{\cdot}5t)] \qquad\qquad *$$

The Derivative of $y=x^n$

n A POSITIVE INTEGER

We assume first that $y = x^n$, where n is a positive integer. We know that where $y = x^1 = x$,

$$dy/dx = 1$$

and that where $y = x^2$,

$$dy/dx = 2x$$

Let y be the product of two functions of x,

$$y = u.v$$

where $u = f(x)$ and $v = g(x)$. The products rule for derivatives tells us that

$$\frac{dy}{dx} = u\left(\frac{dv}{dx}\right) + v\left(\frac{du}{dx}\right)$$

Thus as $x^3 = x.x^2$

$$d(x^3)/dx = x(2x) + x^2(1) = 2x^2 + x^2 = 3x^2$$

Similarly $x^4 = x.x^3$ so

$$d(x^4)/dx = x(3x^2) + x^3(1) = 3x^3 + x^3 = 4x^3$$
$$x^{n+1} = x.x^n$$

Suppose that, for some value of n, we know that

$$d(x^n)/dx = nx^{n-1}$$
$$d(x^{n+1})/dx = x\left(\frac{d(x^n)}{dx}\right) + x^n\left(\frac{dx}{dx}\right)$$
$$= x.nx^{n-1} + x^n = nx^n + x^n$$
$$= (n+1)x^n$$

Thus if the formula holds for n it also holds for $n + 1$. But we know that it holds for $n = 1$ and $n = 2$, so it also holds for $n = 3$, etc, and for n any positive integer. This method of proof is known as mathematical induction.

n A NEGATIVE INTEGER

We can now prove that the derivative of x^n is nx^{n-1} for any negative integer

n. When $n < 0$, let $m = -n$ and $z = 1/x$. Then

$$y = x^n = z^{-n} = z^m$$

By the function of a function rule, where y is a function of z and z is a function of x,

$$\frac{dy}{dx} = \frac{dy}{dz} \cdot \frac{dz}{dx}$$

$$dy/dz = mz^{m-1} \quad \text{and} \quad dz/dx = \frac{-1}{x^2}$$

as shown on p. 77, thus combining these results we have

$$\frac{dy}{dx} = \frac{dy}{dz} \cdot \frac{dz}{dx} = mz^{m-1}\left(\frac{-1}{x^2}\right) = -mx^{1-m-2} = -mx^{-m-1} = nx^{n-1}$$

Thus the rule applies to negative as well as positive integer powers.

n ANY RATIONAL NUMBER

We can now prove it for any rational index. Let $y = x^{n/m}$, where n and m are integers. Let $z = y^m = x^n$. By the function of a function rule,

$$\frac{dy}{dx} = \frac{dy}{dz} \cdot \frac{dz}{dx} = \frac{dz}{dx} \Big/ \frac{dz}{dy}$$

Thus

$$dy/dx = nx^{n-1}/(my^{m-1}) = \frac{n}{m}x^{n-1}y^{1-m} = \frac{n}{m}x^{n-1+(n/m)-n}$$

$$= \frac{n}{m}x^{(n/m)-1}$$

For example, if $y = x^{2/3}$, let $z = y^3 = x^2$. Then $dz/dy = 3y^2$ and $dz/dx = 2x$, thus

$$\frac{dy}{dx} = \frac{dz}{dx} \Big/ \frac{dz}{dy} = 2x/(3y^2) = \tfrac{2}{3}xy^{-2} = \tfrac{2}{3}x \cdot x^{-4/3} = \tfrac{2}{3}x^{-1/3}$$

Thus the rule that $d(x^n)/dx = nx^{n-1}$ has been shown to hold for all positive and negative rational powers of x. It can in fact also be shown to hold for non-rational powers, but this involves the use of exponentials (see Chapter 14). For economic purposes it is sufficient to note that any irrational number can be approximated arbitrarily closely by a rational number.

Euler's Theorem

We assume that the function $f(x, y)$ is homogeneous of degree n for all x, y and λ. Thus

$$f(\lambda x, \lambda y) = \lambda^n f(x, y) \tag{1}$$

We use f_1 and f_2 to denote the derivatives of f with respect to its first and second arguments. Thus $f_1(\lambda x, \lambda y)$ indicates the partial derivative of $f(x, y)$ with respect to x, its first argument, evaluated at the point $(\lambda x, \lambda y)$. We now differentiate both sides of equation (1) with respect to λ. Using the fact that $\partial(\lambda x)/\partial\lambda = x$ and $\partial(\lambda y)/\partial\lambda = y$,

$$f_1(\lambda x, \lambda y).x + f_2(\lambda x, \lambda y).y = n\lambda^{n-1}f(x, y) \tag{2}$$

Differentiate both sides of equation (1) with respect to x to get

$$f_1(\lambda x, \lambda y).\lambda = \lambda^n f_1(x, y) \tag{3}$$

and differentiate both sides of equation (1) with respect to y to get

$$f_2(\lambda x, \lambda y).\lambda = \lambda^n f_2(x, y) \tag{4}$$

Substitute equations (3) and (4) into equation (2) to get

$$\lambda^{n-1}f_1(x, y).x + \lambda^{n-1}f_2(x, y).y = n\lambda^{n-1}f(x, y)$$

and divide through by λ^{n-1} to get

$$x.f_1(x, y) + y.f_2(x, y) = n.f(x, y) \tag{5}$$

Equation (5) is Euler's Theorem.

Trigonometric Functions

ABC is a right-angled triangle, with a right angle at *B*, see Figure 45. The angle *BAC* is θ.

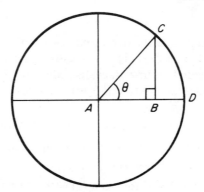

Figure 45   Cos and Sin of angle θ

SINES AND COSINES

Cosine $\theta = AB/AC$ and Sine $\theta = BC/AC$, or if we let $AC = 1$,

$$\text{Cos } \theta = AB \quad \text{and} \quad \text{Sin } \theta = BC$$

COS OF A SUM OF TWO ANGLES

We can see how the Cos and Sin of any sum of two angles is derived from Figure 46. In this figure *ABC* and *ACD* are both right-angled triangles, with right angles at *B* and *C* respectively. The angle *BAC* is θ and the angle *CAD* is ϕ. We let $AD = 1$. Then we have

$$\text{Cos }(\theta + \phi) = AF = AB - BF$$
$$\text{Cos } \theta = AB/AC \quad \text{so} \quad AB = \text{Cos } \theta \, AC$$
$$\text{Cos } \phi = AC \quad \text{so} \quad AB = \text{Cos } \theta \, \text{Cos } \phi$$

The angle CDE = angle $ACE = \theta$ thus $CE/DC = \text{Sin } \theta$
$$BF = CE = \text{Sin } \theta \, DC$$
$$DC = \text{Sin } \phi \quad \text{thus} \quad BF = \text{Sin } \theta \, \text{Sin } \phi$$

Thus
$$\text{Cos}\,(\theta + \phi) = AF = AB - BF = \text{Cos}\,\theta\,\text{Cos}\,\phi - \text{Sin}\,\theta\,\text{Sin}\,\phi$$

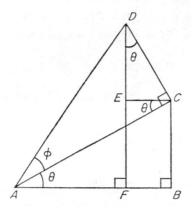

Figure 46   Cos and Sin of sum
of two angles θ and ϕ

DERIVATIVE OF THE COSINE

From this rule we can find the derivative of Cos θ.
$$\text{Cos}\,(\theta + \Delta\theta) = \text{Cos}\,\theta\,\text{Cos}\,\Delta\theta - \text{Sin}\,\theta\,\text{Sin}\,\Delta\theta$$
Thus
$$\Delta\,\text{Cos}\,\theta = \text{Cos}\,\theta\,\text{Cos}\,\Delta\theta - \text{Sin}\,\theta\,\text{Sin}\,\Delta\theta - \text{Cos}\,\theta$$

$$\frac{d\,\text{Cos}\,\theta}{d\theta} = \lim_{\Delta\theta \to 0}\frac{\Delta\,\text{Cos}\,\theta}{\Delta\theta}$$

$$= \lim_{\Delta\theta \to 0}\left(\frac{\text{Cos}\,\theta\,\text{Cos}\,\Delta\theta - \text{Cos}\,\theta}{\Delta\theta} - \text{Sin}\,\theta\,\frac{\text{Sin}\,\Delta\theta}{\Delta\theta}\right)$$

To evaluate these limits we refer to Figure 45. If the angle θ is measured by the distance along the curve CD on a circle of unit radius, its Sin is measured by CB. As C moves closer to D, the lengths of the curve CD and the vertical CB become more nearly equal, and as $\theta \to 0$, Sin $\theta \to 0$. Thus as $\Delta\theta \to 0$, Sin $\Delta\theta \to \Delta\theta$ and

$$\lim_{\Delta\theta \to 0}\left(\frac{\text{Sin}\,\Delta\theta}{\Delta\theta}\right) = 1$$

The units of measurement of angles used here are called radians; they can easily be converted to degrees, since $360° = 2\pi$ radians. As $\theta \to 0$, Cos $\theta = AB \to AD = 1$. Thus as $\Delta\theta \to 0$, Cos $\Delta\theta \to 1$ and

$$\lim_{\Delta\theta \to 0}(\text{Cos}\,\theta\,\text{Cos}\,\Delta\theta - \text{Cos}\,\theta) = 0$$

Thus
$$\frac{d\,\text{Cos}\,\theta}{d\theta} = \lim_{\Delta\theta \to 0}\frac{\Delta\,\text{Cos}\,\theta}{\Delta\theta} = 0 - (\text{Sin}\,\theta)(1) = -\,\text{Sin}\,\theta$$

SIN OF A SUM OF TWO ANGLES

From Figure 46 we can also see that

$$\text{Sin}\,(\theta + \phi) = FD = FE + ED = BC + ED$$
$$BC = AC\,\text{Sin}\,\theta = \text{Cos}\,\phi\,\text{Sin}\,\theta$$

and

$$DE = DC\,\text{Cos}\,\theta = \text{Sin}\,\phi\,\text{Cos}\,\theta$$

thus

$$\text{Sin}\,(\theta + \phi) = \text{Sin}\,\theta\,\text{Cos}\,\phi + \text{Cos}\,\theta\,\text{Sin}\,\phi$$

DERIVATIVE OF THE SINE

Again

$$\text{Sin}\,(\theta + \Delta\theta) = \text{Sin}\,\theta\,\text{Cos}\,\Delta\theta + \text{Cos}\,\theta\,\text{Sin}\,\Delta\theta$$

so

$$\Delta\,\text{Sin}\,\theta = \text{Sin}\,\theta\,\text{Cos}\,\Delta\theta + \text{Cos}\,\theta\,\text{Sin}\,\Delta\theta - \text{Sin}\,\theta$$

$$\frac{d\,(\text{Sin}\,\theta)}{d\theta} = \underset{\Delta\theta \to 0}{\text{Lim}}\left(\frac{\text{Sin}\,\theta\,\text{Cos}\,\Delta\theta - \text{Sin}\,\theta}{\Delta\theta} + \text{Cos}\,\theta\,\frac{\text{Sin}\,\Delta\theta}{\Delta\theta}\right)$$
$$= \text{Cos}\,\theta$$

COS AND SIN OF NEGATIVE ANGLES

In Figure 47 we compare the Cos and Sin of θ with the Cos and Sin of $(-\theta)$. The angles BAC and CAE are equal, but are measured in opposite directions from AD. The Cos of each angle is AC, thus

$$\text{Cos}\,(-\theta) = \text{Cos}\,\theta$$
$$\text{Sin}\,(-\theta) = CE = -BC = -\text{Sin}\,\theta$$

i.e.

$$\text{Sin}\,(-\theta) = -\text{Sin}\,\theta$$

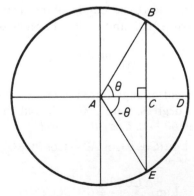

Figure 47   Cos and Sin of $(-\theta)$

COS AND SIN OF A DIFFERENCE OF TWO ANGLES

From this we can derive the formulae for the Cos and Sin of any difference of two angles θ and ϕ.

$$\text{Cos}\,(\theta - \phi) = \text{Cos}\,[\theta + (-\phi)]$$
$$= \text{Cos}\,\theta\,\text{Cos}\,(-\phi) - \text{Sin}\,\theta\,\text{Sin}\,(-\phi)$$
$$= \text{Cos}\,\theta\,\text{Cos}\,\phi + \text{Sin}\,\theta\,\text{Sin}\,\phi$$

and

$$\text{Sin}\,(\theta - \phi) = \text{Sin}\,[\theta + (-\phi)]$$
$$= \text{Sin}\,\theta\,\text{Cos}\,(-\phi) + \text{Cos}\,\theta\,\text{Sin}\,(-\phi)$$
$$= \text{Sin}\,\theta\,\text{Cos}\,\phi - \text{Cos}\,\theta\,\text{Sin}\,\phi$$

COSINE OF $(180° - \theta)$

Cos $180° = -1$ and Sin $180° = 0$, thus

$$\text{Cos}\,(180° - \theta) = \text{Cos}\,180°\,\text{Cos}\,\theta + \text{Sin}\,180°\,\text{Sin}\,\theta$$
$$= -\text{Cos}\,\theta + 0\,(\text{Sin}\,\theta) = -\text{Cos}\,\theta$$

Since Cos $(-\theta) = \text{Cos}\,\theta$

$$\text{Cos}\,(\theta - 180°) = \text{Cos}\,(180° - \theta) = -\text{Cos}\,\theta$$

De Moivre's Theorem

De Moivre's Theorem states that

$$(\text{Cos } \theta + i \text{ Sin } \theta)^t = \text{Cos } t\theta + i \text{ Sin } t\theta$$

This is proved as follows.

Assume that this rule holds for some value of t; as it is true for $t = 1$ the assumption is clearly justified. It can then be shown to hold for $t + 1$.

Suppose we take $\text{Cos } (t + 1)\theta + i \text{ Sin } (t + 1)\theta$ and call it k.

$$k = \text{Cos } (t\theta + \theta) + i \text{ Sin } (t\theta + \theta)$$

Use of the rules for the Cos and Sin of a sum of angles gives

$$k = \text{Cos } t\theta \text{ Cos } \theta - \text{Sin } t\theta \text{ Sin } \theta + i \text{ Sin } t\theta \text{ Cos } \theta + i \text{ Cos } t\theta \text{ Sin } \theta$$

because

$$(i \text{ Sin } t\theta)(i \text{ Sin } \theta) = - \text{Sin } t\theta \text{ Sin } \theta$$

thus

$$k = \text{Cos } t\theta \text{ Cos } \theta + i \text{ Sin } t\theta \, i \text{ Sin } \theta + i \text{ Sin } t\theta \text{ Cos } \theta + i \text{ Cos } t\theta \text{ Sin } \theta$$
$$= \text{Cos } \theta \,(\text{Cos } t\theta + i \text{ Sin } t\theta) + i \text{ Sin } \theta \,(\text{Cos } t\theta + i \text{ Sin } t\theta)$$
$$= (\text{Cos } \theta + i \text{ Sin } \theta)(\text{Cos } t\theta + i \text{ Sin } t\theta)$$

Assume that

$$\text{Cos } t\theta + i \text{ Sin } t\theta = (\text{Cos } \theta + i \text{ Sin } \theta)^t$$

for some value of t, then

$$k = (\text{Cos } \theta + i \text{ Sin } \theta)(\text{Cos } \theta + i \text{ Sin } \theta)^t$$
$$= (\text{Cos } \theta + i \text{ Sin } \theta)^{t+1}$$

i.e.

$$\text{Cos } (t + 1)\theta + i \text{ Sin } (t + 1)\theta = (\text{Cos } \theta + i \text{ Sin } \theta)^{t+1}$$

Consequently if $\text{Cos } t\theta + i \text{ Sin } t\theta = (\text{Cos } \theta + i \text{ Sin } \theta)^t$ for any value of t then it will hold for that value of t plus 1. If it holds for t then it will hold for $t + 1$. If it holds for $t + 1$ then it will hold for $t + 2$, etc. Clearly $(\text{Cos } \theta + i \text{ Sin } \theta)^t = \text{Cos } t\theta + i \text{ Sin } t\theta$ when $t = 1$, i.e.

$$(\text{Cos } \theta + i \text{ Sin } \theta)^1 = \text{Cos } 1\theta + i \text{ Sin } 1\theta = \text{Cos } \theta + i \text{ Sin } \theta$$

Since it holds for $t = 1$ it must therefore hold for $t = 2$. Since it holds for $t = 2$ then it must hold for $t = 3$. Consequently it must hold for all integer values of $t \geqslant 1$.

In the same way, since $\mathrm{Cos}\,(-\,\theta) = \mathrm{Cos}\,\theta$ and $\mathrm{Sin}\,(-\,\theta) = -\,\mathrm{Sin}\,\theta$, it can be shown that

$$(\mathrm{Cos}\,\theta - i\,\mathrm{Sin}\,\theta)^t = \mathrm{Cos}\,t\theta - i\,\mathrm{Sin}\,t\theta$$

The Stability of Second-order Difference Equations

Suppose that the auxiliary or characteristic equation of a second-order difference equation is

$$r^2 + br + c = 0$$

This gives

$$r = \frac{-b \pm \sqrt{(b^2 - 4c)}}{2}$$

These roots will be real if $b^2 \geqslant 4c$ and complex if $4c > b^2$. In Figure 48, combinations of b and c above the parabola yield complex roots. On or below it, the roots are real.

For stability, if the roots are real we require $|r| < 1$. Without loss of generality let $r_1 \geqslant r_2$; stability then requires that

$$1 > r_1 \geqslant r_2 > -1$$

For complex roots, where

$$r = \alpha \pm i\beta$$

where $\alpha = -b/2$ and $\beta = \sqrt{(4c - b^2)}/2$, stability requires that

$$\alpha^2 + \beta^2 < 1$$

With complex roots,

$$\alpha^2 + \beta^2 = \frac{b^2 + 4c - b^2}{4} = c$$

Thus complex roots are stable if $c < 1$ and unstable if $c > 1$. The shaded area in Figure 48 between MN and the parabola shows the combinations of values of b and c which yield complex, stable, roots, i.e. damped oscillations.

In the case of real roots, consider the values of b and c consistent with any real root r. This gives $r^2 + br + c = 0$, or

$$br + c = -r^2$$

which can be regarded as a linear relation between b and c, considering r as a parameter. Along this line, $dc/db = -r$. This line is tangent to the parabola $b^2 = 4c$ at the point $b = -2r$, $c = b^2/4 = r^2$. At this point the slope of the

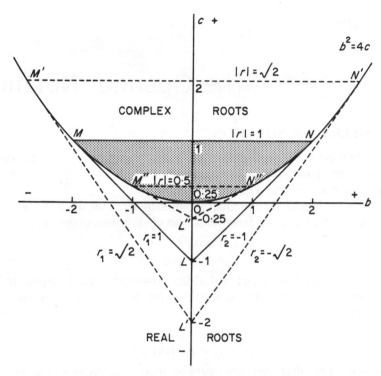

Figure 48   The roots of a second-order
difference equation

parabola is given by $2b\,db = 4\,dc$ so that $dc/db = b/2 = -r$. At all points on this line, one root is equal to r.

At any point b, c, below the parabola $b^2 = 4c$, there will be two lines through the point which are tangents to the parabola; along each tangent one root is the same as at its point of tangency. Thus to have $-1 < r_1 \leqslant r_2 < 1$ we require that b, c lie between the parabola and the lines tangent to the parabola at point M in Figure 48, where $b = -2, c = 1$, and point N, where $b = 2, c = 1$. The line through M is given by $b + c = -1$; this corresponds to $r = 1$, and cuts the c-axis at point L, where $b = 0, c = -1$. The line through N is given by $b - c = 1$; this corresponds to $r = -1$ and also passes through L.

Thus, whether the roots are real or complex, to have both roots with $|r| < 1$, we require to be above both the line LM and the line LN; thus in Figure 48 any point within the triangle LMN will have stable roots, and any point outside will have unstable roots.

In Figure 48, the triangle $L'M'N'$ shows the locus of points with $|r| = \sqrt{2}$, and the triangle $L''M''N''$ shows the locus of points with $|r| = 0.5$.

The Operator Notation

THE OPERATOR NOTATION

This Appendix will explain a simple but powerful notation which can be used to simplify the process of finding the characteristic equations of both differential and difference equation systems.

In the text the differential equation models were set up with the homogeneous part, representing deviations from equilibrium, in the form

$$\dot{x} = ax + by \tag{1}$$
$$\dot{y} = cx + dy \tag{2}$$

This is the form from which solutions were approached above. In real economic models, however, it is quite common to encounter equations in the form

$$a\dot{x} + b\dot{y} + cx + dy = 0 \tag{3}$$
$$f\dot{x} + g\dot{y} + hx + ky = 0 \tag{4}$$

In this case it requires some considerable manipulation to get equations (3) and (4) into the form of equations (1) and (2); this is tedious, and affords ample opportunities for arithmetic blunders. Here use of the operator notation offers a great saving of time. In this method $\dot{x}$ is denoted as Dx, $\ddot{x}$ as D^2x, etc. D is called the differential operator; the equations can be manipulated as though D were a number.

If we start from equation (1) and (2), they can be written in D notation as

$$Dx = ax + by$$
$$Dy = cx + dy$$

so, grouping terms,

$$(D - a)x - by = 0 \tag{5}$$
$$-cx + (D - d)y = 0 \tag{6}$$

For the moment we will treat D as though it were a constant parameter. If equations (5) and (6) are to have any solution other than $x = y = 0$, they must be linearly dependent, i.e.

$$(D - a)(D - d) - bc = 0$$

so that

$$D^2 - (a + d)D + (ad - bc) = 0 \tag{7}$$

This is the same characteristic equation as is obtained by methods of substitution; we can see that it must always work as follows.

To eliminate x from equation (5) and (6), multiply equation (5) by c and equation (6) by $(D - a)$, and add the resulting equations;

$$c(D - a)x \quad - bcy \quad = 0$$
$$- c(D - a)x + (D - a)(D - d)y = 0$$

thus

$$(D - a)(D - d)y - bcy = 0$$

The characteristic equation is thus

$$D^2 - (a + d)D + (ad - bc) = 0 \tag{7}$$

where D takes the place of root r.

We can apply exactly the same operations to equations (1) and (2) without using the D notation; multiply equation (1) by c and equation (2) by D (i.e. differentiate both sides of the equation) and by $- a$, and add the resulting equations;

$$c\dot{x} = acx + bcy$$
$$\ddot{y} = c\dot{x} + d\dot{y}$$
$$- a\dot{y} = - acx - ady$$

so, adding,

$$\ddot{y} - a\dot{y} = bcy + d\dot{y} - ady$$

i.e.

$$\ddot{y} - (a + d)\dot{y} + (ad - bc)y = 0 \tag{8}$$

This gives equation (7) as the characteristic equation. The advantage of the D operator is that it makes it obvious what operations need to be performed on the original equations to produce an equation in y only, rather than relying on trial and error. This advantage becomes really large when we deal with equation (3) and (4). Rewriting them in D notation, and grouping terms,

$$(aD + c)x + (bD + d)y = 0 \tag{9}$$
$$(fD + h)x + (gD + k)y = 0 \tag{10}$$

For equations (9) and (10) to have any solution other than $x = y = 0$, they must be linearly dependent, i.e.

$$(aD + c)(gD + k) - (bD + d)(fD + h) = 0$$

so that the characteristic equation is

$$(ag - bf)D^2 + (ak + cg - bh - df)D + (ck - dh) = 0 \tag{11}$$

Again, we can see that this must work; if we multiply equation (3) by $- fD$ (i.e. differentiate both sides and multiply by $- f$) and by $- h$, and multiply equation (4) by aD and by c, and add the resulting equations, we get

$$- af\ddot{x} - bf\ddot{y} - cf\dot{x} - df\dot{y} = 0$$
$$- ah\dot{x} - bh\dot{y} - chx - dhy = 0$$
$$af\ddot{x} + ag\ddot{y} + ah\dot{x} + ak\dot{y} = 0$$
$$cf\dot{x} + cg\dot{y} + chx + cky = 0$$

Adding,

$$(ag - bf)\ddot{y} + (ak + cg - bh - df)\dot{y} + (ck - dh)y = 0 \qquad (12)$$

equation (12) clearly gives the same characteristic equation as equation (11).
A similar technique works for difference equation systems. Suppose

$$x_{t+1} = ax_t + by_t \qquad (13)$$
$$y_{t+1} = cx_t + dy_t \qquad (14)$$

Let G be the growth operator, so that

$$x_{t+1} = Gx_t, \quad \text{etc.}$$

Rewrite equations (13) and (14) in G operator form, and group terms:

$$(G - a)x_t - by_t = 0 \qquad (15)$$
$$- cx_t + (G - d)y_t = 0 \qquad (16)$$

Thus for equations (15) and (16) to have any solution other than $x_t = y_t = 0$, they must be linearly dependent, so that

$$(G - a)(G - d) - bc = 0$$

This gives the characteristic equation

$$G^2 - (a + d)G + (ad - bc) = 0$$

where G replaces the root r.

Index